T0145355

Terrestrial Environmental Sciences

Series editors

Olaf Kolditz
Hua Shao
Wenqing Wang
Thomas Nagel
Sebastian Bauer

More information about this series at http://www.springer.com/series/13468

Olaf Kolditz · Thomas Nagel
Hua Shao · Wenqing Wang
Sebastian Bauer
Editors

Thermo-Hydro-Mechanical-Chemical Processes in Fractured Porous Media: Modelling and Benchmarking

From Benchmarking to Tutoring

 Springer

Editors
Olaf Kolditz
Environmental Informatics
Helmholtz Centre for Environmental
  Research—UFZ
Leipzig
Germany

and

Technische Universität Dresden
Dresden
Germany

Thomas Nagel
Environmental Informatics
Helmholtz Centre for Environmental
  Research—UFZ
Leipzig
Germany

and

Trinity College Dublin
Dublin
Ireland

Hua Shao
Rock Characterisation for Storage and Final
  Disposal
Federal Institute for Geosciences and Natural
  Resources
Hanover
Germany

Wenqing Wang
Environmental Informatics
Helmholtz Centre for Environmental
  Research—UFZ
Leipzig
Germany

Sebastian Bauer
Institut für Geowissenschaften
Christian-Albrechts-Universität zu Kiel
Kiel
Germany

ISSN 2363-6181   ISSN 2363-619X   (electronic)
Terrestrial Environmental Sciences
ISBN 978-3-030-09826-1   ISBN 978-3-319-68225-9   (eBook)
https://doi.org/10.1007/978-3-319-68225-9

Printed on acid-free paper

This Springer imprint is published by Springer Nature
The registered company is Springer International Publishing AG
The registered company address is: Gewerbestrasse 11, 6330 Cham, Switzerland

Preface

This is the fourth volume of the THMC benchmark book series dealing with benchmarks and examples of thermo-hydro-mechanical-chemical processes in fractured porous media:

1. http://www.springer.com/de/book/9783642271762
2. http://www.springer.com/de/book/9783319118932
3. http://www.springer.com/de/book/9783319292236
4. http://www.springer.com/de/book/9783319682242 (this volume)

Recently, the benchmark books became items of the new book series in "Terrestrial Environmental Sciences" http://www.springer.com/series/13468.

The present book is subtitled "From Benchmarking to Tutoring." The material from benchmark books has also been prepared for tutorials which build the foundation for teaching purposes and training courses as well, highlighting the multipurpose benefits from continuous benchmarking.

The book structure again follows the "classic" scheme, presenting first single processes and then coupled processes with increasing complexity. The list of symbols and an index you will find at the end of the book. With this book, we also want to award the work of merit of distinguished scientists in the field "Modelling and Benchmarking of THMC Processes."

Along with this version, we also provide the input files for self-exercising and enhanced reproducibility. These can be found at the OGS community page http://docs.opengeosys.org/books (Fig. 1). Software engineering has been further improved by integrated Ctests and direct GitHub-Links for OGS code references (Sect. A.2).

The OGS Tutorial series has been extended by additional volumes on Computational Hydrology (I–III) and Geoenergy Modelling (I–III). A new section on Computational Geotechnics was opened starting with a volume on the Storage of Energy Carriers followed by a volume on Models of Thermochemical Heat Storage:

Books & Tutorials

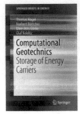

Computational Geotechnics I – Storage of Energy Carriers

`BEGINNER` `INTERMEDIATE` `ADVANCED`

In this book, effective computational methods to facilitate those pivotal simulations using open-source software are introduced and discussed with a special focus on the coupled thermo-mechanical behavior of the rock salt. A cohesive coverage of applying geotechnical modeling to the subsurface storage of hydrogen produced from renewable energy sources is accompanied by specific, reproducible example simulations to provide the reader with direct access to this fascinating and important field. Energy carriers such as natural gas, hydrogen, oil, and even compressed air can be stored in subsurface geological formations such as depleted oil or gas reservoirs, aquifers, and caverns in salt rock. Many challenges have arisen in the design, safety and environmental impact assessment of such systems, not the least of ...

Show details ➡

Computational Hydrology II: Groundwater Quality Modeling

`BEGINNER` `INTERMEDIATE`

This book explores the application of the open-source software OpenGeoSys (OGS) for hydrological numerical simulations concerning conservative and reactive transport modeling. It provides general information on the hydrological and groundwater flow modeling of a real case study and step-by-step model set-up with OGS, while also highlighting related components such as the OGS Data Explorer. The material is based on unpublished manuals and the results of a collaborative project between China and Germany (SUSTAIN H2O). Though the book is primarily intended for graduate students and applied scientists who deal with hydrological modeling, it also offers a valuable source of information for professional geoscientists wishing to expand their knowledge of the numerical modeling of hydrological processes including nitrate reactive transport modeling. This book is ...

Show details ➡

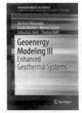

Geoenergy Modeling III: Enhanced Geothermal Systems

`INTERMEDIATE` `ADVANCED`

This tutorial presents the introduction of the open-source software OpenGeoSys for enhanced geothermal reservoir modeling. There are various commercial software tools available to solve complex scientific questions in geothermics. This book will introduce the user to an open source numerical software code for geothermal modeling which can even be adapted and extended based on the needs of the researcher. The book explains basic mathematical equations and numerical methods to modeling flow and heat transport in fractured porous rock formations. In order to help readers gain a system-level understanding of the necessary analysis, the authors include two benchmark examples and two case studies of real deep geothermal test-sites located in Germany and France.

Show details ➡

Fig. 1 Benchmark books and related tutorials

- "Computational Hydrology I: Groundwater Flow Modeling" (Sachse et al. 2015)
  http://www.springer.com/de/book/9783319133348,
- "Computational Hydrology II: Groundwater Quality Modeling" (Sachse et al. 2017)
  http://www.springer.com/us/book/9783319528083,
- "Computational Hydrology III: OGS#IPhreeqc Coupled Reactive Transport Modeling" (Jang et al. 2018)
  http://www.springer.com/us/book/9783319671529,
- "Geoenergy Modeling I: Geothermal Processes in Fractured Porous Media" (Boettcher et al. 2016)
  http://www.springer.com/de/book/9783319313337,

- "Geoenergy Modeling II: Shallow Geothermal Systems" (Shao et al. 2017)
  http://www.springer.com/us/book/9783319450551,
- "Geoenergy Modeling III: Enhanced Geothermal Systems" (Watanabe et al. 2017)
  http://www.springer.com/us/book/9783319465791,
- "Computational Geotechnics—Storage of Energy Carriers" (Nagel et al. 2017a)
  http://www.springer.com/de/book/9783319313337,
- "Models of Thermochemical Heat Storage" (Lehmann et al. 2018)
  http://www.springer.com/de/book/9783319715216,

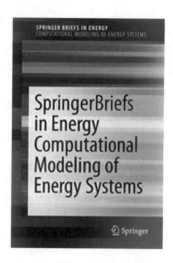

For geo- and energy-related tutorials, a new subseries on "Computational Modeling of Energy Systems" has been launched edited by Thomas Nagel and Haibing Shao. "This subseries puts a spotlight on advanced computational and theoretical methods, tools, and frameworks for the design, analysis, optimization, and assessment of a diverse range of energy technologies and systems. The intention is to make the methods transparent and to allow engineers and scientists from different disciplines to enter the field of energy research enabling them to perform meaningful simulations for the advancement of clean and secure energy systems."

(http://www.springer.com/series/15395)

Enjoy reading, exercising and benchmarking.

Leipzig and Dresden, Germany	Olaf Kolditz
Leipzig and Dublin, Germany	Thomas Nagel
Leipzig, Germany	Wenqing Wang
Hanover, Germany	Hua Shao
Kiel, Germany	Sebastian Bauer
November 2017	

Contents

Contributors

We appreciate the contributions to the 4th benchmark book by:

- Sabine Attinger (Helmholtz Centre for Environmental Research - UFZ, Leipzig, DE)
- Lars Bilke (Helmholtz Centre for Environmental Research - UFZ, Leipzig, DE)
- Mauro Cacace (GFZ, Helmholtz Centre Potsdam, German Research Centre for Geosciences, DE)
- Tao Chen (RWTH, E.ON Energy Research Center, RWTH Aachen University, DE)
- Holger Class (Department of Hydromechanics and Modelling of Hydrosystems, University of Stuttgart, DE)
- Thomas Fischer (Helmholtz Centre for Environmental Research - UFZ, Leipzig, DE)
- Lothar Fuchs (Institute for Technical and Scientific Hydrology (itwh GmbH), Hanover, DE)
- Thomas Graf (Institute of Fluid Mechanics and Environmental Physics in Civil Engineering, Leibniz Universität Hanover, DE)
- Eunseon Jang (Helmholtz Centre for Environmental Research - UFZ, Leipzig, DE)
- Carolin Helbig (Helmholtz Centre for Environmental Research - UFZ, Leipzig, DE)
- Rainer Helmig (Department of Hydromechanics and Modelling of Hydrosystems, University of Stuttgart, DE)
- Falk Heße (Helmholtz Centre for Environmental Research - UFZ, Leipzig, DE)
- Yonghui Huang (Helmholtz Centre for Environmental Research - UFZ, Leipzig, DE)
- Markus Huber (geo:tools, Mittenwald, DE)
- Miao Jing (Helmholtz Centre for Environmental Research - UFZ, Leipzig, DE)
- Vinay Kumar (BGR, Federal Institute for Geosciences and Natural Resources, Hanover, DE)
- Herbert Kunz (BGR, Federal Institute for Geosciences and Natural Resources, Hanover, DE)
- Christoph Lehmann (Helmholtz Centre for Environmental Research - UFZ, Leipzig, DE)
- Falk Lindenmaier (BGR, Federal Institute for Geosciences and Natural Resources, Hanover, DE)
- Renchao Lu (Helmholtz Centre for Environmental Research - UFZ, Leipzig, DE)
- Fabien Magri (Helmholtz Centre for Environmental Research - UFZ, Leipzig, DE)
- Jobst Maßmann (BGR, Federal Institute for Geosciences and Natural Resources, Hanover, DE)
- Xing-Yuan Miao (Helmholtz Centre for Environmental Research - UFZ, Leipzig, DE)

- Thomas Nagel (Helmholtz Centre for Environmental Research - UFZ, Leipzig, DE)
- Dmitri Naumov (Helmholtz Centre for Environmental Research - UFZ, Leipzig, DE)
- Christoph Neukum (BGR, Federal Institute for Geosciences and Natural Resources, Hanover, DE)
- Insa Neuweiler (Institute of Fluid Mechanics and Environmental Physics in Civil Engineering, Leibniz Universität Hanover, DE)
- Peter Ostermeier (Institute for Energy Systems, Technical University of Munich, DE)
- Francesco Parisio (Helmholtz Centre for Environmental Research - UFZ, Leipzig, DE)
- Aaron Peche (Institute of Fluid Mechanics and Environmental Physics in Civil Engineering, Leibniz Universität Hanover, DE)
- Karsten Rink (Helmholtz Centre for Environmental Research - UFZ, Leipzig, DE)
- Gabriele Seitz (Department of Hydromechanics and Modelling of Hydrosystems, University of Stuttgart, DE)
- Haibing Shao (Helmholtz Centre for Environmental Research - UFZ, Leipzig, DE)
- Leonard Stoeckl (BGR, Federal Institute for Geosciences and Natural Resources, Hanover, DE)
- Sara Vassolo (BGR, Federal Institute for Geosciences and Natural Resources, Hanover, DE)
- Peter Vogel (BGR, Federal Institute for Geosciences and Natural Resources, Hanover, DE)
- Wenqing Wang (Helmholtz Centre for Environmental Research - UFZ, Leipzig, DE)
- Marc Walther (TU Dresden, DE; UFZ, Helmholtz Centre for Environmental Research)
- Norihiro Watanabe (AIST, National Institute of Advanced Industrial Science and Technology, Renewable Energy Research Center, Koriyama, Fukushima, Japan)
- Tianyuan Zheng (Helmholtz Centre for Environmental Research - UFZ, Leipzig, DE)
- Gesa Ziefle (BGR, Federal Institute for Geosciences and Natural Resources, Hanover, DE)

Dr. Uwe-Jens Görke[1]

We would like to honor Dr. Uwe-Jens Görke for his scientific contribution to the field of computational mechanics and in particular for his tremendous engagement supporting young scientists and community efforts.

Uwe studied Mathematics and Continuum Mechanics (Theoretical Mechanics) at the Kharkov State University in the former Soviet Union (now Ukraine). At that time our paths crossed for the first time. Dr. Uwe-Jens Görke received his Ph.D. in 1987 from the College of Engineering (Zwickau) in the field of computational mechanics focusing on viscoelastic material models at large strains. Uwe then worked as a Postdoc at the Institute of Mechanics of the Academy of Sciences of the GDR (Karl-Marx-Stadt), where our ways crossed again, when Olaf Kolditz did his my Ph.D. (Uwe was always ahead of Olaf Kolditz …). He worked at several research institutes (Fraunhofer Institute for Structural Durability in Darmstadt) and universities (Chemnitz University of Technology) with main research topics in material modelling, parameter identification, and computational mechanics for solid and porous media with applications in industry and bioengineering. Uwe was also a visiting scientist at the AO Research Institute in Davos (Switzerland) which is dealing with applied Preclinical Research and Development within trauma and disorders of the musculoskeletal system and translation of this knowledge to achieve more effective patient care worldwide. Uwe was working there in the field of biomechanics.

Since 2008, Uwe has been a researcher at the Helmholtz Centre for Environmental Research - UFZ. His research portfolio includes continuum mechanics applied to coupled processes in porous media, thermodynamically consistent material models for inelastic solid structures at large strains, finite element method for thermo-hydro-mechanical coupled processes in porous media,

---

[1]Photo by Andre Künzelmann, UFZ.

porous media applications in geo- and biomechanics. His current research focuses on coupled processes in the subsurface, with a particular interest in geomechanics and material modelling. Since 2012, Uwe is senior scientist and deputy head of the Department of Environmental Informatics and now mainly involved in project management issues.

Uwe belongs to the editor team of the book series "Thermo-Hydro-mechanical-Chemical Processes in Fractured-Porous Media: Modelling and Benchmarking." From this volume, Uwe passes his editorship to Thomas Nagel. We are very grateful to Uwe for his editorial work. And with this opportunity, we would like to thank Uwe deeply for his contribution to porous media science and his support and engagement for the community and promotion of young scientists.

We very much appreciate the contributions to Uwe's Laudatio by two of his academic teachers and longtime companions.

Prof. Dr. sc. techn. Dr. E. h. H. Günther (Chemnitz, Former Director of the Institute of Mechanics with the Academy of Sciences of the GDR)

"Nach dem Studium in der Fachrichtung Mathematik und Mechanik der Universität Charkow und dem Diplom-Abschluss auf dem Gebiet der Kontinuumsmechanik begann Uwe Görke seine Tätigkeit am Lehrstuhl für Technische Mechanik der Ingenieurhochschule Zwickau (heute Westsächsische Hochschule Zwickau). Im Mittelpunkt standen Untersuchungen zu großen Deformationen in der Festkörpermechanik und 1982 promovierte er mit einer Arbeit zur Viskoelastizität kompressibler Werkstoffe. Die Ergebnisse wurden in ein hauseigenes FEMProgramm implementiert und erfolgreich auf Praxisprobleme wie die Berechnung von Hochdrucklippendichtungen und des Spannungs- und Dehnungsverhalten von Patellaknorpel angewendet. Hinzukamen numerische und experimentelle Untersuchungen zur Materialparameteridentifikation aus Relaxationsversuchen. Biomechanische Probleme weicher Gewebe standen immer wieder in seinem Interessenfeld, auch nach seinem Wechsel an das Institut für Mechanik der damaligen Akademie der Wissenschaften. So entstand letzlich unter seiner wesentlichen Mitwirkung, ein nichtlineares, modulares Materialmodell für das komplexe mechanische Verhalten biologischer Gewebe bei großen Deformationen, einschließlich Viskosität, Kompressibilität, Porosität, Isotropie und Anisotropie sowie von biologischen Vorgängen wie Osmose und Gewebe-Remodellierung. Dieses Modell wurde erfolgreich auf biomedizinische Problemstellungen des AO Research Institute Davos und des Rush University Medical-Center Chicago angewendet. Sein außerordentliches kreatives und kooperatives Managment, auch bei der FEM Implementierung, hat wesentlich zum Erfolg der Projekte beigetragen."

Prof. Dr.-Ing. habil. Reiner Kreißig (Chemnitz, Former Chair of Solid Mechanics at TU Chemnitz)

Scientific contributions from Dr.-Ing. Uwe-Jens Görke during his work at the professorship of Solid Mechanics at Chemnitz University of Technology in the years 1995–2007.

Dr. Görke has worked at the professorship Solid Mechanics mainly in research. Particularly noteworthy are his excellent research results, the partial supervision of Ph.D. students, and his contribution to several research proposals to the German Science Foundation DFG (e.g., Collaborative Research Centre (SFB) 393, Package Proposals 47 and 273).

As a scientist in SFB 393, Dr. Görke was substantially involved in the development of the in-house finite element code SPC-PM2AdNl for solving of nonlinear problems of continuum mechanics. His priorities touched many advanced topics in the field of computational mechanics, especially

- the theoretical development and numerical realization of appropriate material laws for elastoplasticity,
- the numerical solution of the nonlinear initial-boundary value problem,
- the hierarchical adaptive strategy (error estimation with respect to the equilibrium, error indicator with respect to the yield condition, mesh refinement and/or coarsening, transfer of the field variables to newly generated nodes and integration points),
- identification of the material parameters based on the analysis of inhomogeneous displacement fields (application of a nonlinear deterministic optimization approach (Levenberg-Marquardt method) for a least squares type objective function, semianalytical sensitivity analysis for the determination of the gradient of the objective function).

The development of a generalized substructure approach based on the ideas of Mandel and Dafalias was supported by Dr. Görke. In this approach, a so-called substructure configuration, which is established in addition to the usual configurations, representing the kinematics of the continuum, is defined. The substantial advantage of this modelling concept lies in its thermodynamic consistency. Aside from the theoretical work itself, the numerical realization of related problems was realized.

In the DFG priority program SPP 1146, Dr. Görke was involved in the simulation of incremental forming processes based on the material model with a substructure approach. The theoretical foundation for the modelling of coupled thermomechanical processes was developed and numerically realized in the in-house finite element code PM2AdNl. The material model is characterized by a comparatively low number of material parameters. These were determined within a gradient-based, nonlinear optimization method as described above.

Coupled multiphysics problems (hyperelastic, nearly incompressible materials, and biphasic saturated porous media at large strains) formed an additional main focus of the work of Dr. Görke. Special linearization techniques have been used to solve these problems. After spatial discretization, a global system for the incremental form of the initial-boundary problem within the framework of a stable mixed U/p-c finite element approach was defined. The global system is solved using an iterative solver with hierarchical preconditioning. Adaptive mesh evolution is controlled by a residual a posteriori error estimator.

The above results have been published in five Preprints of the SFB 393 and five Scientific Computing Preprints (www.tu-chemnitz.de/mathematik/).

Symbols

Greek Symbols

α	Biot constant (–)
α	Thermal expansion coefficient (K^{-1})
α	Intergranular radius (–)
α	Van Genuchten parameter (m^{-1})
α_k	Kinetic isotope fractionation factor (–)
α_L	Longitudinal dispersion length (m)
α_T	Transversal dispersion length (m)
β	Cubic thermal expansion coefficient (K^{-1})
β_c	Burial constant (–)
χ	Bishop coefficient (–)
2δ	Fault width (m)
Δ	Half of aspect ratio (–)
γ	Activity coefficient for dissolved species (–)
γ	Dimensionless temperature (–)
Γ	Domain boundary (–)
ϵ	Strain tensor (–)
$\dot{\epsilon}$	Strain rate (s^{-1})
ϵ	Length scale (m)
ϵ	Isotope enrichment factor (–)
ϵ	Strain (–)
ϵ_v	Volume plastic strain (–)
η	Porosity (–)
η_M	Maxwell viscosity (Pa d)
η_K	Kelvin viscosity (Pa d)
γ	Activity coefficient for dissolved species (–)
γ_l	First-order degradation rate (day^{-1})
κ	Thermal conductivity ($W\ m^{-1}\ K^{-1}$)
λ_c	Virgin compression index (–)
λ	Lamé coefficients (GPa)

λ_p	Hardening parameter (–)
λ	Thermal conductivity (W m^{-1} K^{-1})
λ_{arith}	Arithmetic effective thermal conductivity (W m^{-1} K^{-1})
λ_b	Bulk thermal conductivity (W m^{-1} K^{-1})
λ_{eff}	Effective thermal conductivity (W m^{-1} K^{-1})
λ_f	Fluid thermal conductivity (W m^{-1} K^{-1})
λ_{geom}	Geometric effective thermal conductivity (W m^{-1} K^{-1})
λ_{harm}	Harmonic effective thermal conductivity (W m^{-1} K^{-1})
λ_{pm}	Thermal conductivity of porous medium (W m^{-1} K^{-1})
λ_s	Solid thermal conductivity (W m^{-1} K^{-1})
μ	Lamé coefficients (GPa)
μ	Dynamic viscosity (Pa s)
μ_0	Base dynamic viscosity (Pa s)
ν	Poisson number (–)
Q_ω	Source/sink term (kg m^{-3} s^{-1})
ω	Intergranular thickness (m)
ω	Saturation index (–)
ϕ	Porosity (–)
ϕ	Friction angle (deg)
ψ	Dilatancy angle (deg)
ρ	Density (kg m^{-3})
ϱ_{SR}	Real density of solid (kg m^{-3})
ϱ_{LR}	Real density of liquid (kg m^{-3})
ϱ_{IR}	Real density of ice (kg m^{-3})
ρ_s	Density of solid (kg m^{-3})
ρ_w	Density of water (kg m^{-3})
ρ_d^s	Density of bentonite bulk (kg m^{-3})
ρ_0	Base fluid density (kg m^{-3})
σ	Cauchy stress tensor (Pa)
σ_V	Von Mises equivalent stress (Pa)
σ_{con}	Confining stress (Pa)
σ_{eff}	Effective stress (Pa)
σ	Effective stress (Pa)
σ_{max}^{sw}	Tested maximum swelling stress (Pa)
σ_a	Contact stress (Pa)
σ_c	Critical stress (Pa)
ν_{M}	Maxwell viscosity tensor (Pa d)
ν_{K}	Kelvin viscosity tensor (Pa d)

Roman Symbols

a	Specific surface area (m^2 m^{-3})
a^{σ}	Activity of stressed solid (–)
$\dot{a}$	Effective diameter of ion (m)
A	Surface area (m^2)
$\mathbf{b}$	Body force vector (N)
b_h	Fracture hydraulic aperture (m)
b_m	Fracture mechanical aperture (m)
c	Normalized concentration (–)
c	Concentration (kg m^{-3})
c	Cohesion (–)
C_{eq}^h	Solubility under hydrostatic pressure (mol m^{-3})
C_i	Intergranular concentration (mol m^{-3})
C_p	Pore-space concentration (mol m^{-3})
Cr	Courant number (–)
c_f	Specific fluid heat capacity (J kg^{-1} k^{-1})
c_p	Heat capacity (J kg^{-1} k^{-1})
Cp	Heat capacity (J kg^{-1} k^{-1})
d	Order parameter (–)
D	Diffusivity coefficient (m^2 s^{-1})
D_f	Intergranular diffusion coefficient (m^2 s^{-1})
E	Young's modulus (Pa)
e	Void ratio (–)
g	Gravitational coefficient (m s^{-2})
g	Plastic potential (J)
$\mathbf{g}$	Gravity vector (m s^{-2})
$\mathbf{g}$	Gravitational acceleration (m s^{-2})
g_c	Fracture toughness (N m^{-1})
G	Gibbs energy (J)
G	Gibbs energy (J mol^{-1})
G	Shear modulus (Pa)
G_F	Plastic potential (J)
G_M	Maxwell shear modulus (–)
G_K	Kelvin shear modulus (–)
H	Fault height (m)
H_{pw}	Pipe water level (m)
h	Hardening parameter (–)
h_c	Thickness of colmation layer (m)
h_f	Freshwater hydraulic head (m)
h_s	Saltwater hydraulic head (m)
$\triangle h_I$	Specific enthalpy of fusion (J)
I	Ionic strength (–)
I_1	First principal invariant of the stress tensor (Pa)
I_r	Friction slope (–)

I_s	Bottom slope (–)
J_2	Second principal invariant of the deviatoric stress tensor (Pa^2)
J_3	Third principal invariant of the deviatoric stress tensor (Pa^3)
k	Residual stiffness parameter (–)
k	Permeability tensor (m^2)
k	Permeability tensor ($m\ s^{-1}$)
K	Intrinsic permeability (m^2)
k_c	Swelling/recompression index (–)
K_{eq}	Equilibrium constant (–)
K_M	Maxwell bulk modulus (–)
K_n	Normal stiffness ($Pa\ m^{-1}$)
K_s	Shear stiffness ($Pa\ m^{-1}$)
$K_r el$	Relative permeability (m^2)
k_s	Saturated hydraulic conductivity ($m\ d^{-1}$)
k^+	Dissolution rate constant ($mol\ m^{-2}\ s^{-1}$)
k°	Reaction rate constant ($mol\ m^{-2}\ s^{-1}$)
M	Kinetic coefficient ($mm^2\ N^{-1}\ s^{-1}$)
M	Slope of critical state line (–)
M_w	Molecular mass of water vapor (18.016 g mol^{-1})
n	Porosity ($m^3\ m^{-3}$)
n	Van Genuchten parameter (–)
m	Van Genuchten parameter (–)
p	Pressure ($kg\ s^{-1}\ m^{-1}$)
p	Pressure (Pa)
P	Load (Pa)
P_c	Capillary pressure (Pa)
p_s	Mean stress (Pa)
p_{scn}	Isotropic pre-consolidation pressure (Pa)
Pe	Péclet number (–)
q	Source/sink term (–)
q	Shear stress (Pa)
q	Heat source (W)
q	Darcy velocity ($m\ s^{-1}$)
q	Darcy velocity vector ($m\ s^{-1}$)
Q	Ion activity product (–)
Q_{leak}	Leakage flow ($m^3\ s^{-1}$)
q_T	Heat flux through unit area ($W\ m^{-2}$)
R	Universal gas constant (8.31432 J $mol^{-1}\ K^{-1}$)
R_c	Contact area ratio (–)
Ra	Rayleigh number (–)
Ra_{crit}	Critical Rayleigh number (–)
s	Soil suction (kPa)
S	Saturation (–)
S	Storage (1 Pa^{-1})
S_e	Effective saturation (–)

S_{max}	Maximum water saturation (–)
S_r	Residual saturation (–)
S_r	Residual water saturation (–)
SA	Reactive surface area (m^2)
t	Time (s)
T	Temperature (K)
T	Absolute temperature (K)
T_c	Top temperature (cold) (K)
T_h	Bottom temperature (hot) (K)
T_{init}	Initial temperature (K)
T_v	Approximation temperature (K)
u	Displacement (m)
$\mathbf{u}$	Displacement vector (m)
v	Velocity (m s^{-1})
V	Volume (m^3)
V_m	Molar volume (m^3 mol^{-1})
V_u	Cell volume (m^3)
w	Margules parameter (J mol^{-1})
X	Molar fraction (–)
Z	Ionic integer charge (–)
Z	Charge (–)

Indices

$\parallel$	Co-linear direction
$\perp$	Orthogonal direction
e	Efficient value
f	Fluid
f	Fracture
w	Water
s	Solid
0	Reference value

Operators

div, $\nabla\cdot$	Divergence operator
grad, ∇	Nabla, gradient operator
tr	Trace

Part I
Introduction

Chapter 1
Introduction

Olaf Kolditz, Thomas Nagel and Hua Shao

1.1 Recent Developments in THMC Research

Olaf Kolditz, Thomas Nagel, Hua Shao

In this section we discuss recent literature in thermo-hydro-mechanical-chemical (THMC) analysis particularly in fractured-porous media. Figure 1.1 depicts the quantitative development of Scopus listed publications in the field and highlights the increasing interest in THMC research.

THMC research is mainly related to geoscientific and environmental applications based on porous media approaches. A few works are dedicated to material research.

1.2 Events

In 2017 several important events took place related to THMC research such as the Mont Terri Technical Meeting 2017 in Porrentruy (Switzerland) and the DECO-VALEX 25th Anniversary in Stockholm (Sweden) (see Sect. 1.2).

O. Kolditz (✉) · T. Nagel · H. Shao
Helmholtz Centre for Environmental Research - UFZ, Leipzig, Germany
e-mail: Olaf.kolditz@ufz.de

O. Kolditz
Technische Universität Dresden, Dresden, Germany

T. Nagel
Trinity College Dublin, Dublin, Ireland

H. Shao
BGR - Federal Institute for Geosciences and Natural Resources, Hanover, Germany

© Springer International Publishing AG 2018
O. Kolditz et al. (eds.), *Thermo-Hydro-Mechanical-Chemical Processes in Fractured Porous Media: Modelling and Benchmarking*,
Terrestrial Environmental Sciences, https://doi.org/10.1007/978-3-319-68225-9_1

3

Porrentruy: The Mont Terri Technical Meeting (TM-35) from 08–09.02.2017 was held for annual review of selected ongoing experiments but also for discussion of new experiments which will be designed within the framework of rock lab extension later on this year. This meeting was also dedicated to celebrate the 20th anniversary of the Mont Terri Underground Research Laboratory (URL) being an excellent place of international applied research in clay rocks world-wide for over 2 decades. Due to this occasion, a Special Issue was published in the *Swiss Journal of Geosciences* "Mont Terri rock laboratory, 20 years of research: introduction, site characteristics and overview of experiments" Bossart et al. (2017) compiling recent research works from the project participants.

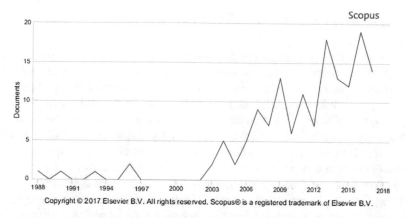

Fig. 1.1   Increase of THMC papers over years   (source: Scopus)

Stockholm: The 3rd DECOVALEX-2019 workshop from 25–28.04.2017 in Sweden's capital was more than the ordinary status meeting with progress reports about the seven individual tasks of the project phase 7 ...

... it was also dedicated to the celebration of the DECOVALEX's 25th anniversary. To this purpose the pioneers of DECOVALEX - DEvelopment of COupled models and their VALidation against EXperiments - were invited for a panel discussion to explain the unusual success story of this project (Ove Stephansson, Ivars Neretnieks, Johan Andersson, Ki-Bok Min, Chin-Fu Tsang, from left to right).

1.3   Literature Review

A literature review in Scopus (2016–early 2018) yielded the following topics in THMC analyses:

- General and review works (Sect. 1.3.1),
- Nuclear and chemo-toxic waste management, landfills Sect. 1.3.2,
- Geological CO2 storage (Sect. 1.3.3),
- Geothermal energy systems (Sect. 1.3.4),
- Reservoir exploitation and drilling (Sect. 1.3.5).

Nuclear waste management is the most frequent and intensive discussed topic in recent literature.

1.3.1 General and Review Works

Wang et al. (2017) provide a comprehensive literature review on the simulation techniques for describing flow processes in shale and tight gas reservoirs. "The capabilities of existing reservoir simulation tools are discussed in terms of numerical methods (finite difference, finite element, explicit/implicit scheme, sequentially and fully coupled schemes), fluid flow behavior (Darcy and non-Darcy flow, desorption, Klinkenberg effect and gas slip flow, transitional flow, Knudsen diffusion), reservoir rock properties (pore size distribution, fractures, geomechanics), coupling schemes, modeling scale, and computational efficiency." The authors also elaborate on pros and cons of pore-scale modeling, e.g. using lattice-Boltzmann approaches.

Wang (2017) present a fully coupled THMC model for fracture opening and closure by explicitly accounting for stress concentration on aperture surface, stress-activated mineral dissolution, pressure solution at contacting asperities, and channel flow dynamics. They showed that "a tangential surface stress created by a far-field compressive normal stress may play an important role in controlling fracture aperture evolution in a stressed geologic medium."

THMC simulation is also important for the understanding of fundamental Earth system processes such as lithospheric deformation, e.g. for analysis of Episodic Tremor and Slip (ETS) events. Veveakis et al. (2017) consider THMC instabilities triggered by the fluid release reactions in fault zones. "Data from ETS sequences in subduction zones reveal a geophysically tractable temporal evolution with no access to the fault zone." Alevizos et al. (2017) present a fundamental theoretical analysis on the important lithosphere deformation mechanism of creep enhanced by fluid-release reactions. "This mechanism features surprisingly rich dynamics stemming from the feedback between deformation induced fluid release through mineral breakdown reactions (dissolution) and fluid cementation into a solid matrix (precipitation) when the tectonic forces are locally relaxed." They show that TM feedback processes in the temperature and pressure evolution and the resulting feedback between fluid flow and mechanical deformation (HM processes) result in a highly dynamic system.

Faoro et al. (2016) present a fundamental study concerning permeability changes in fractured rocks due to THMC processes. "These experiments examine the influence of thermally and mechanically activated dissolution of minerals on the mechanical (stress/strain) and transport (permeability) responses of fractures." The data base relies on heated ($25\,^{\circ}$C up to $150\,^{\circ}$C) flow-through experiments on fractured core samples of Westerly granite. The measured efflux of dissolved mineral mass provides a record of the net mass removal, which is correlated with observed changes in relative hydraulic fracture aperture. The authors argue "that at low temperature and high stresses, mechanical crushing of the asperities and the production of gouge explain the permeability decrease although most of the permeability is recoverable as the stress is released. While at high temperature, the permeability changes are governed by mechanical deformation as well as chemical processes, in particular, we infer dissolution of minerals adjacent to the fracture and precipitation of kaolinite."

springer.com  springer.com

DECOVALEX— DEvelopment of Coupled
models and their VALidation against EXperiments

The DECOVALEX project is an international research and model comparison collaboration, initiated in 1992, for advancing the understanding and modeling of coupled thermo-hydro-mechanical-chemical (THMC) processes in geological systems. Prediction of these coupled effects is an essential part of the performance and safety assessment of geologic disposal systems for radioactive waste and spent nuclear fuel, and also for a range of sub-surface engineering activities. The project has been conducted by research teams supported by a large number of radioactive-waste-management organizations and regulatory authorities. Research teams work collaboratively on selected modeling cases, followed by comparative assessment of model results. This work has yielded in-depth knowledge of coupled THM and THMC processes associated with nuclear waste repositories and wider geo-engineering applications, as well as the suitability of numerical simulation models for quantitative analysis.

Topical Collection
DECOVALEX 2015
Guest editors:
Jens T. Birkholzer,
Alexander E. Bond,
John A. Hudson, Lanru Jing,
Hua Shao and Olaf Kolditz

Environmental
Earth Sciences

• Simulation of coupled thermal hydraulic mechanical
  and chemical processes
• Transport in fractured rock
• Modelling of the SEALEX experiment

Fig. 1.2 DECOVALEX tasks explore different complexity of T-H-M/C exercises based on experimental evidence. Fully coupled problems have been investigated e.g. by Bond et al. (2017, 2016), Pfeiffer et al. (2016), Pan et al. (2016), McDermott et al. (2015)

Rutqvist (2016) provides an overview of TOUGH-based geomechanics models and summarizes the history of TOUGH-FLAC3D developments in the past 15 years for various geoengineering applications such as "geologic CO2 sequestration, enhanced geothermal systems, unconventional hydrocarbon production, and most recently, related to reservoir stimulation and injection-induced seismicity."

1.3.2 Nuclear Waste Management

The scientific results of the previous **DECOVALEX**-2015 phase have been published recently as a Thematic Issue in *Environmental Earth Sciences*.[1] The volume contains 18 research papers describing details of Tasks A, B, C1, C2 as well as summarizing the model comparison exercises (Fig. 1.2).

A short introduction to the ongoing **DECOVALEX**-2019 can be found in Sect. 1.5 and through the related website.[2]

"Geologic repositories for radioactive waste are designed as multi-barrier disposal systems that perform a number of functions including the long-term isolation and containment of waste from the human environment, and the attenuation of radionuclides released to the subsurface. The rock laboratory at **Mont Terri** (canton Jura,

---

[1] https://link.springer.com/journal/12665/topicalCollection/AC_4afd5d3151e292e32fb5583c8a0d4b9a/page/1.

[2] www.decovalex.org.

Switzerland) in the Opalinus Clay plays an important role in the development of such repositories." Bossart et al. (2017) summarized the experimental results gained in the last 20 years to study the possible evolution of a repository and investigate processes closely related to the safety functions of a repository hosted in a clay rock. "At the same time, these experiments have increased the general knowledge of the complex behaviour of argillaceous formations in response to coupled hydrological, mechanical, thermal, chemical, and biological processes". Numerous research aspects are covered in a Special Issue of the *Swiss Journal of Geosciences*: bentonite buffer emplacement, high-pH concrete-clay interaction experiments, anaerobic steel corrosion with hydrogen formation, depletion of hydrogen by microbial activity, and finally, release of radionuclides into the bentonite buffer and the Opalinus Clay barrier. "The research at Mont Terri carried out in the last 20 years provides valuable information on repository evolution and strong arguments for a sound safety case for a repository in argillaceous formations."

Bernier et al. (2017) pointed out the key THMC processes that might influence radionuclide transport in a disposal system and its surrounding environment, considering the dynamic nature of these processes. These THMC processes have a potential impact on safety; it is important "to identify and to understand them properly when developing a disposal concept to ensure compliance with relevant safety requirements."

Lu et al. (2017a) and Lu and Fall (2017) developed a coupled thermo-hydro-mechanical-chemical (THMC)-visco-plastic cap model to characterize the behavior of cementing mine backfill material under blast loading. "The model is coupled to a Perzyna type of visco-plastic model with a modified smooth surface cap envelope and a variable bulk modulus, in order to reasonably capture the nonlinear and rate-dependent behaviors of the cemented tailings backfill under blast loading." The proposed model allows for a better understanding of hydrating cemented backfill under blasting conditions, and also practical risk management of backfill structures associated with dynamic environments. Ghirian and Fall (2016) conducted experimental work using a pressure cell apparatus to study the long-term hydro-mechanical behaviour of cemented paste backfill (CPB).

Bente et al. (2017) deal with a model for time-dependent compaction of porous materials with applications to degradation-induced settlements in municipal solid waste landfills. The Theory of Porous Media is used as continuum mechanical framework; for kinematic description, large strain continuum mechanics is applied.

Nishimura (2016) investigated compacted bentonite as an important component of engineered barrier systems. They present a new method of determining creep behavior and failure of compacted bentonite under constant (or maintained) high relative humidity. THMC modeling is suggested as an appropriate analysis tool.

Abed et al. (2016) started modelling coupled THMC behaviour of unsaturated bentonite relying on the Barcelona Basic Model (BBM). "As an alternative, BBM is used alongside the Krähn's model which assumes that bentonite re-saturation is mainly driven by water vapour diffusion." Both methods have been compared with laboratory experiments based on X-ray tomography and show similar results.

Yasuhara et al. (2016b) developed a THMC numerical model to examine the long-term change in permeability of the porous sedimentary rocks (quartz-rich), in particular, the chemo-mechanical process of the pressure solution is incorporated. "The model predictions clearly showed a significant influence of the pressure dissolution on the change in permeability with time" and, therefore, are important consideration for long-term high-level radioactive waste deposition (Yasuhara et al. 2016a).

GMZ-Na-bentonite is a selected buffer material for the preliminary concept of HLW repository in China. Liu et al. (2016) investigate THMC processes for the China-Mock-up facility. "Stress evolution of the compacted bentonite may be influenced by several mechanisms, including gravity, thermal expansion induced by high temperature, and the swelling pressure generated by bentonite saturation." The experimental results and achievements obtained from the THMC experiments provide important insight into GMZ-bentonite under realistic HLW repository conditions and the design.

1.3.3 Geological CO2 Storage

Li et al. (2016a): "Carbon dioxide (CO_2) capture and storage (CCS) is considered widely as one of promising options for CO2 emissions reduction, especially for those countries with coal-dominant energy mix like China. Injecting and storing a huge volume of CO2 in deep formations are likely to cause a series of geomechanical issues, including ground surface uplift, damage of caprock integrity, and fault reactivation." They present results from the Shenhua CCS demonstration project in Ordos Basin, China, which is the largest full-chain saline aquifer storage project of CO2 in Asia - a combination of CCS and overlying coal seam mining. Interferometric synthetic aperture radar (InSAR) technology was used for subsidence monitoring. THMC modeling was mainly used for geomechanical stability analysis. Li et al. (2016b) provide a comprehensive review of numerical approaches for analyzing the geomechanical effects induced by CO2 geological storage. They introduce theoretical aspects and classify numerical simulation methods for THMC analyses.

Zhang et al. (2016a, b) propose a sequentially coupled computational THMC framework for a variety of geo-applications such as CO2 geo-sequestration (CCS) and Engineered Geothermal Systems (EGS). Zhang and Wu (2016) present a practical reactive transport example with complex chemical compositions (Fig. 1.3).

1.3.4 Geothermal Energy Systems

Blöcher et al. (2016) investigate the THMC behaviour of a research well doublet consisting of the injection well E GrSk 3/90 and the production well Gt GrSk 4/05 A(2) in the deep geothermal reservoir of Groß Schönebeck (north of Berlin, Germany)

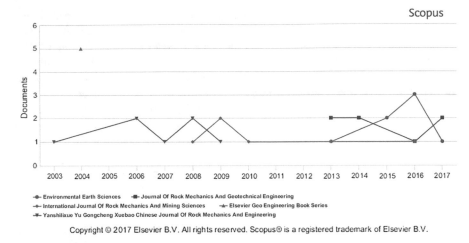

Fig. 1.3 THMC bibliography by journals (source: SCOPUS)

which is located in the Lower Permian of the North German Basin (NGB). The
authors use THMC analysis in order to explain the decrease of productivity and
identified five possible reasons: wellbore fill, wellbore skin, the sustainability of
induced fractures, two phase flow and compartmentalisation.

1.3.5 Reservoir Exploitation and Drilling

Kanfar et al. (2017) derived the governing equations for anisotropic poro-chemo-
thermo-elasticity to simulate the drilling of an inclined borehole problem in a transversely isotropic rock. A finite element method based numerical model is constructed
to estimate the pore pressure, temperature, solute concentration and stress distribution. The model is used to assess time-dependent wellbore stability during and after
drilling operations.

1.3.6 Continuous Workflows

To support THM/CB analysis for complex, real-world applications, complete workflows have been developed in order to combine data integration, process simulation,
and data analytics steps into continuous analysis based on observation and facilitate
experimental design (Fig. 1.4).

Related OpenGeoSys (OGS) developments are highlighted in the enlarged boxes.
These workflows have been implemented for hydrological, geothermal (shallow and

Fig. 1.4 Continuous
workflow concept

deep), energy storage as well as geotechnical applications and are embedded into
VR (Virtual Reality) environments.

1.4 Bibliography

We use Scopus biometrics tools to provide an actual THMC bibliography on journals
(Fig. 1.3), countries, affiliations, and authors (Fig. 1.5, Accessed 31.07.2017).

1.5 DECOVALEX-2019

The international collaborative project DECOVALEX is one of the most successful
projects dealing with the development and validation of THMC models and numerical
codes.[3] For that purpose, laboratory and field experiments are employed for a better
process understanding in geological and geotechnical materials along with their
interactions in deep geological repositories for radioactive waste. The project was
established in 1992; an account of its 25-year history is given elsewhere in this book.

In the current phase of the project DECOVALEX-2019, launched in April 2016,
material inhomogeneity plays an important role. Fractured rock is considered as
strong inhomogeneous, even heterogeneous medium. It is also important to consider
the porous media used as geological and geotechnical barrier, e.g., bentonite, as
inhomogeneous materials. During conventional modelling, a REV (representative
elementary volume) concept is applied by defining, usually implicitly, the small-
est volume over which a value obtained by measurement may represent the whole
system or at least a homogenized area in a statistical sense. Based on the REV
strategy, different mathematical approaches such as random number generation, stochastic treatment, have been developed to characterize the material inhomogeneity
by generating statistically representative distributions of physical parameters at the
microscale, e.g., porosity or permeability.

---

[3] www.decovalex.org.

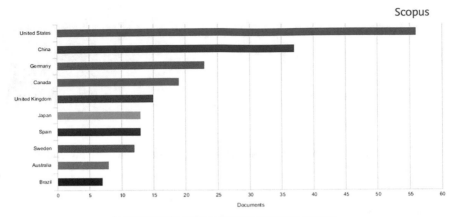

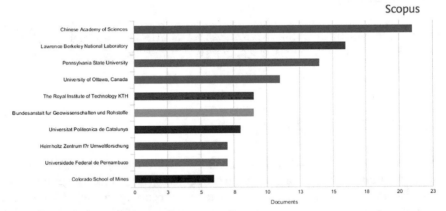

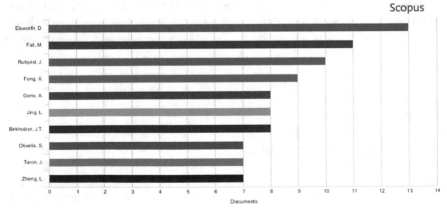

Fig. 1.5 THMC bibliography by countries, affiliations, and authors (source: Scopus)

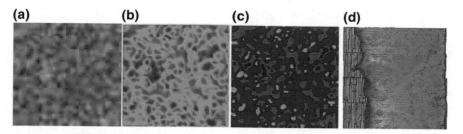

Fig. 1.6 Example of a pore-scale simulation workflow. **a** optical microscopy image of a porous medium, **b** contrast threshold analysis, **c** derived finite element mesh with different material groups and **d** pressure distribution and flow along channels calculated by OGS

To observe and understand the microscopic behavior, several technical methods have been developed in the fields of optical mineralogy and petrography and widely used in the analysis of rock samples. Examples include laboratory thin sectioning, CT (Computer Tomography) scanning, X-ray-based approaches and recently SEM (Scanning Electron Microscopy). With the help of these methods, materials previously considered as "homogeneous" may now be observed as exhibiting a pronounced inhomogeneous microstructure. Different mineral compositions in a "homogenous" material possess different physical and chemical properties and their reactions can now be observed in a more direct manner at the microscopic level. The difference between states before and after an experiment may therefore be qualitatively analysed.

Fully coupled THMC(B) models and the corresponding numerical codes have been developed following a macroscopic approach. Relevant physical-chemical processes and their interactions can be simulated reasonably well in comparison with many experimental measurements. In the course of the rapid development of computer technology and computational techniques, especially parallel high-performance computing, large-scale and long-term simulation has become a practical possibility. Consequently, down-scale modelling, such as pore-scale simulations for the direct simulation of coupled processes occurring in the microstructure of an (in-) homogeneous medium, has also become possible (Fig. 1.6).

The process understanding gained by detailed process analyses at the microscale has to be transferred back to the macroscale at which analyses and predictions relevant for the evolution of the deep geological repository have to be made. Therefore, down-scaling approaches have to be complemented by their reverse counterparts, i.e. by up-scaling or coarse-graining techniques. This step is crucial for rendering the new-found knowledge productive for safety cases and large-scale analyses. The importance of this cross-scale process analytics approach is reflected in the fact that several tasks explicitly study THMC processes at different scales in the current DECOVALEX phase.

Part II
Single Processes

Chapter 2
H Processes

Tao Chen, Jobst Maßmann and Peter Vogel

Steady Flow in a Rectangle

Peter Vogel and Jobst Maßmann

This section presents a set of examples which are standard practice in groundwater hydraulics. We focus on the closed form solutions. The associated simulation exercises have been checked by OGS; they may serve as verification test. Throughout this section we are concerned with the evaluation of pressure distributions. For the underlying theory of groundwater movement see Freeze and Cherry (1979), for more advanced examples see Polubarinova-Kochina (1962).

2.1 Toth's Box

This exercise has been adopted from Toth (1962). Given length $L = 600$ m and thickness $H = 300$ m the domain represents the cuboid $[0, L] \times [0, H] \times [0, H]$. It is discretized by $500 \times 1 \times 250$ equally sized hexahedral elements. An isotropic permeability of 10^{-12} m^2 is assumed for the material. Liquid viscosity is 1 mPa·s and gravity is neglected via zero liquid density. The pressure distribution $p_0 x / L$ ($p_0 = 10^5$ Pa) represents the groundwater table along the top face $z = H$, all other faces are no-flow faces by default. The simulation (Table 2.1) comprises one timestep to establish the steady-state pressure distribution $p(x, z)$ (Fig. 2.1).

T. Chen (✉)
RWTH, E.ON Energy Research Center, RWTH Aachen University, Aachen, Germany
e-mail: TChen@eonerc.rwth-aachen.de

J. Maßmann · P. Vogel
BGR, Federal Institute for Geosciences and Natural Resources, Hanover, Germany

© Springer International Publishing AG 2018
O. Kolditz et al. (eds.), *Thermo-Hydro-Mechanical-Chemical Processes in Fractured Porous Media: Modelling and Benchmarking*,
Terrestrial Environmental Sciences, https://doi.org/10.1007/978-3-319-68225-9_2

Fig. 2.1 Pressure distribution

Table 2.1 Benchmark deposit (https://docs.opengeosys.org/books/bmb-5)

BM code	Author	Code	Files	CTest
BMB5-2.1	Peter Vogel	OGS-5.XX (github link)	Available	ToDo

(https://oc.ufz.de/index.php/s/nlph7bhfDkj6tC7)

For incompressible liquids Darcy's law and continuity equation yield the Laplace equation as the governing equation describing the steady-state pressure distribution. Due to the setup the problem does not depend on the y-coordinate and the Laplace equation becomes

$$\frac{\partial^2 p}{\partial x^2} + \frac{\partial^2 p}{\partial z^2} = 0. \tag{2.1.1}$$

The specified boundary conditions read along the top face $z = H$

$$p(x, H) = p_0 \frac{x}{L} \qquad \text{for } 0 \leq x \leq L, \tag{2.1.2}$$

along the no-flow face $z = 0$

$$\frac{\partial p}{\partial z}(x, 0) = 0 \qquad \text{for } 0 \leq x \leq L, \tag{2.1.3}$$

and along the no-flow faces $x = 0$ and $x = L$

$$\frac{\partial p}{\partial x}(0, z) = \frac{\partial p}{\partial x}(L, z) = 0 \qquad \text{for } 0 \leq z \leq H. \tag{2.1.4}$$

This boundary value problem will be solved by separation of variables. Assuming a product solution

$$p(x, z) = F(x)G(z) \tag{2.1.5}$$

the Laplace equation gives

$$\frac{1}{F}\frac{d^2 F}{dx^2} = -\frac{1}{G}\frac{d^2 G}{dz^2}. \tag{2.1.6}$$

Since the left hand side depends only on x and the right hand side depends only on z, both sides are equal to some constant value $-\omega^2$. Thus the Laplace equation separates into two ordinary differential equations

$$\frac{d^2 F}{dx^2} = -\omega^2 F, \tag{2.1.7}$$

$$\frac{d^2 G}{dz^2} = \omega^2 G \tag{2.1.8}$$

with the general solutions

$$F(x) = C_1 \cos(\omega x) + C_2 \sin(\omega x), \tag{2.1.9}$$
$$G(z) = C_3 \cosh(\omega z) + C_4 \sinh(\omega z). \tag{2.1.10}$$

This yields

$$p(x, z) = A \cos(\omega x) \cosh(\omega z) + B \sin(\omega x) \cosh(\omega z)$$
$$+ C \cos(\omega x) \sinh(\omega z) + D \sin(\omega x) \sinh(\omega z). \tag{2.1.11}$$

The free constants A, B, C, D, and the eigenvalue ω will be determined from the boundary conditions. The no-flow boundary condition along the bottom face $z = 0$

$$\frac{\partial p}{\partial z}(x, 0) = 0 = \omega[C \cos(\omega x) + D \sin(\omega x)] \tag{2.1.12}$$

is satisfied by

$$C = D = 0, \tag{2.1.13}$$

and the no-flow boundary condition along the face $x = 0$

$$\frac{\partial p}{\partial x}(0, z) = 0 = \omega B \cosh(\omega z) \tag{2.1.14}$$

is satisfied by

$$B = 0. \tag{2.1.15}$$

The no-flow boundary condition along the face $x = L$

$$\frac{\partial p}{\partial x}(L, z) = 0 = \omega A \sin(\omega L) \cosh(\omega z) \tag{2.1.16}$$

yields the eigenvalues

$$\omega_n = \frac{n\pi}{L} \quad \text{for } n = 0, 1, 2, \ldots \tag{2.1.17}$$

and the associated solutions

$$p_n(x, z) = A_n \cosh\left(n\pi\frac{z}{L}\right) \cos\left(n\pi\frac{x}{L}\right) \quad \text{for } n = 0, 1, 2, \ldots . \tag{2.1.18}$$

The pressure distribution takes the form

$$p(x, z) = A_0 + \sum_{n=1}^{\infty} A_n \cosh\left(n\pi\frac{z}{L}\right) \cos\left(n\pi\frac{x}{L}\right). \tag{2.1.19}$$

The specified boundary condition along the top face $z = H$ yields the remaining constants $A_0, A_1, A_2, \ldots$.

$$p(x, H) = p_0\frac{x}{L} = A_0 + \sum_{n=1}^{\infty} A_n \cosh\left(n\pi\frac{H}{L}\right) \cos\left(n\pi\frac{x}{L}\right)$$

$$= \frac{a_0}{2} + \sum_{n=1}^{\infty} a_n \cos\left(n\pi\frac{x}{L}\right) \tag{2.1.20}$$

with the cosine series expansion of the prescribed surface pressure $p_0 x/L$ on the last line. The Fourier coefficients $a_0, a_1, a_2, \ldots$ read

$$a_0 = \frac{2}{L} \int_0^L p_0\frac{x}{L} dx = p_0, \tag{2.1.21}$$

and for $n = 1, 2, \ldots$

$$a_n = \frac{2}{L} \int_0^L p_0\frac{x}{L} \cos\left(n\pi\frac{x}{L}\right) dx = \frac{2p_0}{(n\pi)^2}[(-1)^n - 1]. \tag{2.1.22}$$

Comparing coefficients gives the constants $A_0, A_1, A_2, \ldots$ and the pressure distribution $p(x, z)$ becomes

$$p(x, z) = p_0\left(\frac{1}{2} - \frac{4}{\pi^2} \sum_{n=1}^{\infty} \frac{\cos\left[(2n-1)\pi\frac{x}{L}\right]}{(2n-1)^2} \frac{\cosh\left[(2n-1)\pi\frac{z}{L}\right]}{\cosh\left[(2n-1)\pi\frac{H}{L}\right]}\right). \tag{2.1.23}$$

Table 2.2 Benchmark deposit (https://docs.opengeosys.org/books/bmb-5)

BM code	Author	Code	Files	CTest
BMB5-2.2	Peter Vogel	OGS-5.XX (github link)	Available	ToDo

(https://oc.ufz.de/index.php/s/nlph7bhfDkj6tC7)

The series thus obtained is uniformly absolutely-convergent on the entire domain. It satisfies the Laplace equation and the boundary conditions, hence, the pressure distribution $p(x, z)$ is the solution of the boundary value problem.

2.2 Flow Under a Dam

Given length $L = 60\,\text{m}$ and thickness $H = 30\,\text{m}$ the domain represents the cuboid $[0, L] \times [0, H] \times [0, H]$. It is discretized by $600 \times 1 \times 300$ equally sized hexahedral elements. An isotropic permeability of $10^{-15}\,\text{m}^2$ is assumed for the material. Liquid viscosity is $1\,\text{mPa·s}$ and gravity is neglected via zero liquid density. A prescribed pressure distribution $p_0 f(x)$ ($p_0 = 10^5\,\text{Pa}$) prevails along the top face $z = H$. It represents the groundwater table in the vicinity of the dam and is specified below. The bottom $z = 0$ and the lateral faces are no-flow faces by default. The simulation (Table 2.2) comprises one timestep to establish the steady-state pressure distribution $p(x, z)$ (Fig. 2.2).

For incompressible liquids Darcy's law and continuity equation yield the Laplace equation as the governing equation describing the steady-state pressure distribution.

Fig. 2.2 Pressure distribution

Due to the setup the problem does not depend on the y-coordinate and the Laplace equation becomes

$$\frac{\partial^2 p}{\partial x^2} + \frac{\partial^2 p}{\partial z^2} = 0. \tag{2.2.1}$$

The specified boundary conditions read along the top face $z = H$

$$p(x, H) = p_0 f(x) = \begin{cases} p_0 & \text{for } 0 \le x \le \frac{L}{3}, \\ p_0 \left(2 - \frac{3x}{L}\right) & \text{for } \frac{L}{3} \le x \le \frac{2L}{3}, \\ 0 & \text{for } \frac{2L}{3} \le x \le L, \end{cases} \tag{2.2.2}$$

along the bottom face $z = 0$

$$\frac{\partial p}{\partial z}(x, 0) = 0 \quad \text{for } 0 \le x \le L, \tag{2.2.3}$$

and along the no-flow faces $x = 0$ and $x = L$

$$\frac{\partial p}{\partial x}(0, z) = \frac{\partial p}{\partial x}(L, z) = 0 \quad \text{for } 0 \le z \le H. \tag{2.2.4}$$

Similar to the previous example the pressure distribution is represented by a series of harmonic functions

$$p(x, z) = A_0 + \sum_{n=1}^{\infty} A_n \cosh\left(n\pi \frac{z}{L}\right) \cos\left(n\pi \frac{x}{L}\right), \tag{2.2.5}$$

which satisfies the boundary conditions along the no-flow faces $x = 0$, $x = L$, and $z = 0$. The specified boundary condition along the top face $z = H$ yields the remaining constants $A_0, A_1, A_2, \ldots$.

$$p(x, H) = p_0 f(x) = A_0 + \sum_{n=1}^{\infty} A_n \cosh\left(n\pi \frac{H}{L}\right) \cos\left(n\pi \frac{x}{L}\right)$$

$$= \frac{a_0}{2} + \sum_{n=1}^{\infty} a_n \cos\left(n\pi \frac{x}{L}\right) \tag{2.2.6}$$

with the cosine series expansion of the prescribed surface pressure $p_0 f(x)$ on the last line. The Fourier coefficients $a_0, a_1, a_2, \ldots$ read

$$a_0 = \frac{2}{L} \int_0^L p_0 f(x) dx = p_0, \tag{2.2.7}$$

and for $n = 1, 2, \ldots$

$$a_n = \frac{2}{L} \int_0^L p_0 f(x) \cos\left(n\pi \frac{x}{L}\right) dx = \frac{12 p_0}{(n\pi)^2} \sin \frac{n\pi}{2} \sin \frac{n\pi}{6}. \qquad (2.2.8)$$

Comparing coefficients gives the constants A_0, A_1, A_2, ... and the pressure distribution $p(x, z)$ becomes

$$p(x, z) = p_0 \left(\frac{1}{2} + \frac{12}{\pi^2} \sum_{n=1}^{\infty} \frac{\cos\left[(2n-1)\pi \frac{x}{L}\right]}{(2n-1)^2} \frac{\cosh\left[(2n-1)\pi \frac{z}{L}\right]}{\cosh\left[(2n-1)\pi \frac{H}{L}\right]} \right.$$
$$\left. \times \sin \frac{(2n-1)\pi}{2} \sin \frac{(2n-1)\pi}{6} \right). \qquad (2.2.9)$$

The series thus obtained is uniformly absolutely-convergent on the entire domain. It satisfies the Laplace equation and the boundary conditions, hence, the pressure distribution $p(x, z)$ is the solution of the boundary value problem.

2.3 A Gallery of Disposal Wells

Given length $L = 500$ m and thickness $H = 50$ m the domain represents the cuboid $[0, L] \times [0, L] \times [0, H]$. It is discretized by $400 \times 400 \times 1$ equally sized hexahedral elements. An isotropic permeability of 10^{-12} m^2 is assumed for the material. Liquid viscosity is 1 mPa·s and gravity is neglected via zero liquid density. The bottom $z = 0$ and the top face $z = H$ are no-flow faces by default. A prescribed pressure distribution $p_0 f(y)$ ($p_0 = 10^5$ Pa) prevails at the lateral face $x = L$. It represents the well gallery and is specified below. Zero pressure has been assigned along the remaining lateral faces. The simulation (Table 2.3) comprises one timestep to establish the steady-state pressure distribution $p(x, y)$ (Fig. 2.3).

For incompressible liquids Darcy's law and continuity equation yield the Laplace equation as the governing equation describing the steady-state pressure distribution. Due to the setup the problem does not depend on the z-coordinate and the Laplace equation becomes

$$\frac{\partial^2 p}{\partial x^2} + \frac{\partial^2 p}{\partial y^2} = 0. \qquad (2.3.1)$$

The specified boundary conditions read along the faces $y = 0$ and $y = L$

Table 2.3 Benchmark deposit (https://docs.opengeosys.org/books/bmb-5)

BM code	Author	Code	Files	CTest
BMB5-2.3	Peter Vogel	OGS-5.XX (github link)	Available	ToDo

(https://oc.ufz.de/index.php/s/nlph7bhfDkj6tC7)

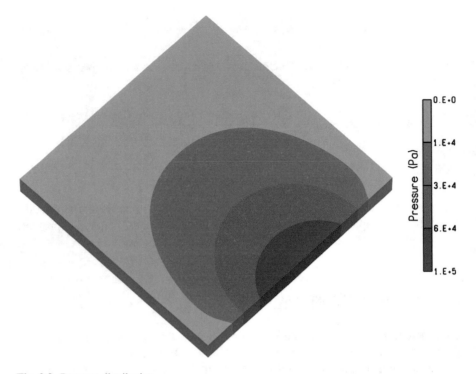

Fig. 2.3 Pressure distribution

$$p(x, 0) = p(x, L) = 0 \quad \text{for } 0 \le x \le L, \tag{2.3.2}$$

along the face $x = 0$

$$p(0, y) = 0 \quad \text{for } 0 \le y \le L, \tag{2.3.3}$$

and along the face $x = L$

$$p(L, y) = p_0 f(y) = \begin{cases} 0 & \text{for } 0 \le y \le \dfrac{L}{10}, \\[2mm] p_0 \left(\dfrac{10}{3L} y - \dfrac{1}{3} \right) & \text{for } \dfrac{L}{10} \le y \le \dfrac{4L}{10}, \\[2mm] p_0 & \text{for } \dfrac{4L}{10} \le y \le \dfrac{6L}{10}, \\[2mm] p_0 \left(3 - \dfrac{10}{3L} y \right) & \text{for } \dfrac{6L}{10} \le y \le \dfrac{9L}{10}, \\[2mm] 0 & \text{for } \dfrac{9L}{10} \le y \le L. \end{cases} \tag{2.3.4}$$

This boundary value problem will be solved by separation of variables. Assuming a product solution

$$p(x, y) = F(x)G(y) \tag{2.3.5}$$

the Laplace equation gives

$$-\frac{1}{F}\frac{d^2 F}{dx^2} = \frac{1}{G}\frac{d^2 G}{dy^2}. \tag{2.3.6}$$

Since the left hand side depends only on x and the right hand side depends only on y, both sides are equal to some constant value $-\omega^2$. Thus the Laplace equation separates into two ordinary differential equations

$$\frac{d^2 F}{dx^2} = \omega^2 F, \tag{2.3.7}$$

$$\frac{d^2 G}{dy^2} = -\omega^2 G \tag{2.3.8}$$

with the general solutions

$$F(x) = C_1 \sinh(\omega x) + C_2 \cosh(\omega x), \tag{2.3.9}$$
$$G(y) = C_3 \cos(\omega y) + C_4 \sin(\omega y). \tag{2.3.10}$$

This yields

$$\begin{aligned} p(x, y) = {} & A \sinh(\omega x) \cos(\omega y) + B \sinh(\omega x) \sin(\omega y) \\ & + C \cosh(\omega x) \cos(\omega y) + D \cosh(\omega x) \sin(\omega y). \end{aligned} \tag{2.3.11}$$

The free constants A, B, C, D, and the eigenvalue ω will be determined from the boundary conditions. Zero pressure along the face $x = 0$

$$p(0, y) = 0 = C \cos(\omega y) + D \sin(\omega y) \tag{2.3.12}$$

is satisfied by

$$C = D = 0, \tag{2.3.13}$$

and zero pressure along the face $y = 0$

$$p(x, 0) = 0 = A \sinh(\omega x) \tag{2.3.14}$$

is satisfied by

$$A = 0. \tag{2.3.15}$$

Zero pressure along the face $y = L$

$$p(x, L) = 0 = B \sinh(\omega x) \sin(\omega L) \tag{2.3.16}$$

yields the eigenvalues

$$\omega_n = \frac{n\pi}{L} \qquad \text{for } n = 1, 2, \ldots \tag{2.3.17}$$

and the associated solutions

$$p_n(x, y) = B_n \sinh\left(n\pi\frac{x}{L}\right) \sin\left(n\pi\frac{y}{L}\right) \qquad \text{for } n = 1, 2, \ldots . \tag{2.3.18}$$

The pressure distribution takes the form

$$p(x, y) = \sum_{n=1}^{\infty} B_n \sinh\left(n\pi\frac{x}{L}\right) \sin\left(n\pi\frac{y}{L}\right). \tag{2.3.19}$$

The specified boundary condition along the face $x = L$ yields the remaining constants $B_1, B_2, \ldots$.

$$p(L, y) = p_0 f(y) = \sum_{n=1}^{\infty} B_n \sinh(n\pi) \sin\left(n\pi\frac{y}{L}\right)$$

$$= \sum_{n=1}^{\infty} b_n \sin\left(n\pi\frac{y}{L}\right) \tag{2.3.20}$$

with the sine series expansion of the prescribed pressure $p_0 f(y)$ on the last line. The Fourier coefficients $b_1, b_2, \ldots$ read for $n = 1, 2, \ldots$

$$b_n = \frac{2}{L} \int_0^L p_0 f(y) \sin\left(n\pi\frac{y}{L}\right) dy$$

$$= p_0 \frac{80}{3(n\pi)^2} \sin\frac{n\pi}{2} \sin\frac{n\pi}{4} \sin\frac{3n\pi}{20}. \tag{2.3.21}$$

Comparing coefficients gives the constants $B_1, B_2, \ldots$ and the pressure distribution $p(x, y)$ becomes

$$p(x, y) = p_0 \frac{80}{3\pi^2} \sum_{n=1}^{\infty} \frac{\sinh\left[(2n-1)\pi\frac{x}{L}\right]}{\sinh\left[(2n-1)\pi\right]} \frac{\sin\left[(2n-1)\pi\frac{y}{L}\right]}{(2n-1)^2}$$

$$\times \sin\frac{(2n-1)\pi}{2} \sin\frac{(2n-1)\pi}{4} \sin\frac{3(2n-1)\pi}{20}. \tag{2.3.22}$$

The series thus obtained is uniformly absolutely-convergent on the entire domain. It satisfies the Laplace equation and the boundary conditions, hence, the pressure distribution $p(x, y)$ is the solution of the boundary value problem.

Table 2.4 Benchmark deposit (https://docs.opengeosys.org/books/bmb-5)

BM code	Author	Code	Files	CTest
BMB5-2.4	Peter Vogel	OGS-5.XX (github link)	Available	ToDo

(https://oc.ufz.de/index.php/s/nlph7bhfDkj6tC7)

2.4 A Catchment

Given length $L = 500$ m and thickness $H = 25$ m the domain represents the cuboid $[0, L/2] \times [0, L] \times [0, H]$. It is discretized by $250 \times 500 \times 1$ equally sized hexahedral elements. An isotropic permeability of 10^{-12} m^2 is assumed for the material. Liquid viscosity is 1 mPa·s and gravity is neglected via zero liquid density. The bottom $z = 0$ and the top face $z = H$ are no-flow faces by default. The constant pressure $p_0 = 10^5$ Pa prevails along the lateral faces $x = 0$, $y = 0$, and $y = L$. A prescribed pressure distribution $p_0(1 - f(y))$ represents outflow along the lateral face $x = L/2$ and is specified below. The simulation (Table 2.4) comprises one timestep to establish the steady-state pressure distribution $p(x, y)$ (Fig. 2.4).

For incompressible liquids Darcy's law and continuity equation yield the Laplace equation as the governing equation describing the steady-state pressure distribution.

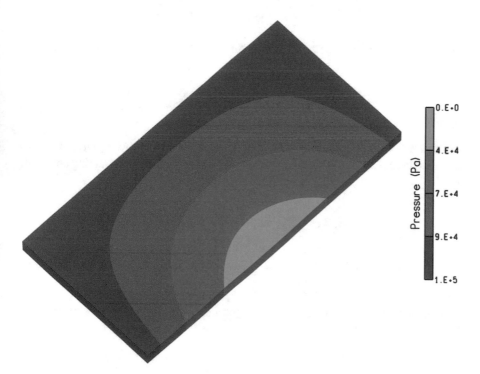

Fig. 2.4 Pressure distribution

Due to the setup the problem does not depend on the z-coordinate and the Laplace equation becomes

$$\frac{\partial^2 p}{\partial x^2} + \frac{\partial^2 p}{\partial y^2} = 0. \tag{2.4.1}$$

The specified boundary conditions read along the faces $y = 0$ and $y = L$

$$p(x, 0) = p(x, L) = p_0 \quad \text{for } 0 \leq x \leq \frac{L}{2}, \tag{2.4.2}$$

along the face $x = 0$

$$p(0, y) = p_0 \quad \text{for } 0 \leq y \leq L, \tag{2.4.3}$$

and along the face $x = L/2$

$$p\left(\frac{L}{2}, y\right) = p_0(1 - f(y)) \quad \text{for } 0 \leq y \leq L, \tag{2.4.4}$$

where $f(y)$ is given by

$$f(y) = \begin{cases} \frac{y}{L/2} & \text{for } 0 \leq y \leq \frac{L}{2}, \\ 2 - \frac{y}{L/2} & \text{for } \frac{L}{2} \leq y \leq L. \end{cases} \tag{2.4.5}$$

Similar to the previous example the pressure distribution is represented by a series of harmonic functions

$$p(x, y) = p_0 - \sum_{n=1}^{\infty} B_n \sinh\left(n\pi\frac{x}{L}\right) \sin\left(n\pi\frac{y}{L}\right), \tag{2.4.6}$$

which satisfies the boundary conditions along the faces $y = 0$, $y = L$, and $x = 0$. The specified boundary condition along the face $x = L/2$ yields the remaining constants $B_1, B_2, \ldots$.

$$p\left(\frac{L}{2}, y\right) = p_0(1 - f(y)) = p_0 - \sum_{n=1}^{\infty} B_n \sinh\left(\frac{n\pi}{2}\right) \sin\left(n\pi\frac{y}{L}\right)$$

$$= p_0 - \sum_{n=1}^{\infty} b_n \sin\left(n\pi\frac{y}{L}\right) \tag{2.4.7}$$

with the sine series expansion of $p_0 f(y)$ on the last line. The Fourier coefficients $b_1, b_2, \ldots$ read for $n = 1, 2, \ldots$

$$b_n = \frac{2}{L} \int_0^L p_0 f(y) \sin\left(n\pi\frac{y}{L}\right) dy = \frac{8p_0}{(n\pi)^2} \sin\frac{n\pi}{2}. \tag{2.4.8}$$

Comparing coefficients gives the constants B_1, B_2, ... and the pressure distribution $p(x, y)$ becomes

$$p(x, y) = p_0 - \frac{8p_0}{\pi^2} \sum_{n=1}^{\infty} \frac{\sinh\left[(2n-1)\pi\frac{x}{L}\right]}{\sinh\left[(2n-1)\frac{\pi}{2}\right]} \frac{\sin\left[(2n-1)\pi\frac{y}{L}\right]}{(2n-1)^2} \sin\frac{(2n-1)\pi}{2}.$$

$$(2.4.9)$$

The series thus obtained is uniformly absolutely-convergent on the entire domain. It satisfies the Laplace equation and the boundary conditions, hence, the pressure distribution $p(x, y)$ is the solution of the boundary value problem.

2.5 A Gallery of Recharge Wells

Given length $L = 500\,\text{m}$ and thickness $H = 50\,\text{m}$ the domain represents the cuboid $[0, L] \times [0, L] \times [0, H]$. It is discretized by $200 \times 200 \times 1$ equally sized hexahedral elements. An isotropic permeability $k = 10^{-12}\,\text{m}^2$ is assumed for the material. Liquid viscosity is $\mu = 1\,\text{mPa·s}$ and gravity is neglected via zero liquid density. The constant pressure $p_0 = 10^5\,\text{Pa}$ prevails at the face $x = 0$, the prescribed specific discharge $q_0 y/L$ ($q_0 = -3 \cdot 10^{-7}\,\text{m/s}$) represents the well gallery along the face $x = L$. The remaining faces are no-flow faces by default. The simulation (Table 2.5) comprises one timestep to establish the steady-state pressure distribution $p(x, y)$ (Fig. 2.5).

For incompressible liquids Darcy's law and continuity equation yield the Laplace equation as the governing equation describing the steady-state pressure distribution. Due to the setup the problem does not depend on the z-coordinate and the Laplace equation becomes

$$\frac{\partial^2 p}{\partial x^2} + \frac{\partial^2 p}{\partial y^2} = 0.$$

$$(2.5.1)$$

The specified boundary conditions read along the no-flow faces $y = 0$ and $y = L$

$$\frac{\partial p}{\partial y}(x, 0) = \frac{\partial p}{\partial y}(x, L) = 0 \qquad \text{for } 0 \leq x \leq L,$$

$$(2.5.2)$$

and along the specified pressure boundary $x = 0$

$$p(0, y) = p_0 \qquad \text{for } 0 \leq y \leq L.$$

$$(2.5.3)$$

Table 2.5 Benchmark deposit (https://docs.opengeosys.org/books/bmb-5)

BM code	Author	Code	Files	CTest
BMB5-2.5	Peter Vogel	OGS-5.XX (github link)	Available	ToDo

(https://oc.ufz.de/index.php/s/nlph7bhfDkj6tC7)

Fig. 2.5 Pressure distribution

Darcy's law yields the boundary condition associated with the well gallery

$$\frac{\partial p}{\partial x}(L, y) = q_0 \frac{\mu}{k} \frac{y}{L} \quad \text{for } 0 \le y \le L. \tag{2.5.4}$$

Similar to the previous examples the pressure distribution is represented by a series of harmonic functions

$$p(x, y) = p_0 + A_0 x + \sum_{n=1}^{\infty} A_n \sinh\left(n\pi \frac{x}{L}\right) \cos\left(n\pi \frac{y}{L}\right), \tag{2.5.5}$$

which satisfies the boundary conditions along the faces $y = 0$, $y = L$, and $x = 0$. The specified boundary condition along the face $x = L$ yields the remaining constants $A_0, A_1, A_2, \ldots$.

$$\frac{\partial p}{\partial x}(L, y) = q_0 \frac{\mu}{k} \frac{y}{L} = A_0 + \sum_{n=1}^{\infty} A_n \frac{n\pi}{L} \cosh(n\pi) \cos\left(n\pi \frac{y}{L}\right)$$

$$= \frac{a_0}{2} + \sum_{n=1}^{\infty} a_n \cos\left(n\pi \frac{y}{L}\right) \tag{2.5.6}$$

with the cosine series expansion of the imposed boundary condition on the last line. The Fourier coefficients $a_0, a_1, a_2, \ldots$ read

$$a_0 = \frac{2}{L} \int_0^L q_0 \frac{\mu}{k} \frac{y}{L} dy = q_0 \frac{\mu}{k}, \qquad (2.5.7)$$

and for $n = 1, 2, \ldots$

$$a_n = \frac{2}{L} \int_0^L q_0 \frac{\mu}{k} \frac{y}{L} \cos\left(n\pi \frac{y}{L}\right) dy = \frac{\mu}{k} \frac{2q_0}{(\pi n)^2} [(-1)^n - 1]. \qquad (2.5.8)$$

Comparing coefficients gives the constants $A_0, A_1, A_2, \ldots$ and the pressure distribution $p(x, y)$ becomes

$$p(x, y) = p_0 + \frac{q_0 \mu L}{2k} \left(\frac{x}{L} - \frac{8}{\pi^3} \sum_{n=1}^{\infty} \frac{\sinh\left[(2n-1)\pi \frac{x}{L}\right] \cos\left[(2n-1)\pi \frac{y}{L}\right]}{\cosh[(2n-1)\pi]} \frac{}{(2n-1)^3} \right). \qquad (2.5.9)$$

The series thus obtained is uniformly absolutely-convergent on the entire domain. It satisfies the Laplace equation and the boundary conditions, hence, the pressure distribution $p(x, y)$ is the solution of the boundary value problem.

2.6   H Processes in Stochastic Discrete Fracture Networks

Tao Chen

2.6.1   Problem Definition

This benchmark simulates flow through a stochastically generated discrete fracture model. The dimensions of the model are 500 m × 500 m × 150 m (Fig. 2.6). The fractures are created by FracMan (Dershowitz et al. 1998) based on the fracture data from the Soultz-sous-Forêts site (Massart et al. 2010; Sausse et al. 2010). There are 40 fractures in the model with multiple-scale sizes. The hydraulic conductivities of the fractures and the rock matrix are 1.17 m/s and 9.79×10^{-9} m/s, respectively. The linear boundary conditions are applied to the model along the y-axis with the hydraulic gradient of 1.

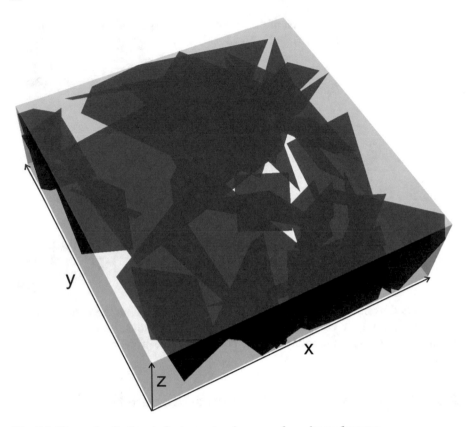

Fig. 2.6 The stochastic discrete fracture network, grey surfaces denote fractures

2.6.2 Input Files

The model is defined by input files. They are listed in the following with explanations.

MMP: Medium Properties

Listing 2.1 Medium property file: Stoch_Frac_Flow.mmp

```
#MEDIUM_PROPERTIES
 $GEOMETRY_DIMENSION
   3
 $PERMEABILITY_TENSOR
   ISOTROPIC   9.7904e-09
#MEDIUM_PROPERTIES
 $GEOMETRY_DIMENSION
   2
 $PERMEABILITY_TENSOR
   ISOTROPIC   1.1748
#STOP
```

MSH: Mesh

The meshes for the stochastic fracture model are created with Hypermesh (Altair 2016). The rock matrix meshes and the fracture meshes are represented by tetrahedrons and triangles, respectively.

Listing 2.2   Mesh file: Stoch_Frac_Flow.msh

```
#FEM_MSH
$NODES
32552
0  -0.000118  440.187897  0.000059
1  -0.000098  446.348785  15.173040
2  -0.000077  452.534393  30.406719
3  -0.000056  458.720001  45.640388
4  -0.000037  464.705688  60.381859
5  -0.000020  469.761414  72.832741
6  -0.000000  482.000092  73.765808
7  -0.000000  491.930298  74.522881

[..]

32550 0.000000  280.569489  49.522678
32551 0.000000  242.460602  14.133650
$ELEMENTS
202263
0  0  tet  27048  2941   5985   2943
1  0  tet  19182  27930  8712   8710
2  0  tet  19203  19204  8101   8100
3  0  tet  8352   8344   26593  26665
4  0  tet  15154  14069  16322  16321
5  0  tet  12557  12571  12099  31785
6  0  tet  12239  15554  12252  12219
7  0  tet  19662  19392  26947  26962

[..]

158448  0  tet  13084  16522  16549  16523
158449  0  tet  14017  31899  14152  14016
158450  0  tet  12985  31632  12971  31680
158451  1  tri  6244   6245   6246
158452  1  tri  1886   1887   6278
158453  1  tri  6278   1887   6271

[..]

202261  1  tri  7555   24957  27274
202262  1  tri  27274  24957  22678
#STOP
```

MFP: Fluid Properties

Listing 2.3 Fluid property file: Stoch_Frac_Flow.mfp

```
#FLUID_PROPERTIES
 $FLUID_TYPE
   LIQUID
 $DENSITY
   1 998.0
 $VISCOSITY
   1 1e-3
#STOP
```

BC: Boundary Condition

The linear boundary conditions are applied to the model surface. The hydraulic head values for nodes are defined directly by using the file bc_y.tex.

Listing 2.4 Boundary condition file: Stoch_Frac_Flow.bc

```
#BOUNDARY_CONDITION
 $PCS_TYPE
   GROUNDWATER_FLOW
 $PRIMARY_VARIABLE
   HEAD
 $DIS_TYPE
   DIRECT bc_y.tex
#STOP
```

The constant hydraulic head is defined on four surfaces: two x-z surfaces and two y-z surfaces which contain 6378 nodes in total. The first column in the file indicates sequence numbers of the nodes which are identical with those in the mesh file. The second column represents the hydraulic head values for the corresponding nodes.

Listing 2.5 Hydraulic head file for the linear boundary conditions: bc_y.tex

```
0      59.8121033
1      53.6512146
2      47.4656067
3      41.2799988
4      35.2943115
5      30.2385864
6      17.9999084
7      8.06970215
8      0.00000000
9      0.00000000

[..]

6589    500.000000
6590    500.000000
8108    4.64291382
8109    10.1155090
8110    15.5880127
8111    20.8116150
8112    25.6043091
9912    500.000000
9913    500.000000
9914    500.000000
```

```
10864     0.00000000
10865     0.00000000

[..]

32549     176.505402
32550     219.430511
32551     257.539398
```

PCS: Process

Listing 2.6 Process file: Stoch_Frac_Flow.pcs

```
#PROCESS
 $PCS_TYPE
    GROUNDWATER_FLOW
#STOP
```

NUM: Numerical Properties

Listing 2.7 Numeric parameters file: Stoch_Frac_Flow.num

```
#NUMERICS
 $PCS_TYPE
    GROUNDWATER_FLOW
 $LINEAR_SOLVER
; method error_tolerance  max_iterations  theta precond  storage
    2       1 1.e-014         1000          1.0     1        2
#STOP
```

OUT: Output

Listing 2.8 Output file: Stoch_Frac_Flow.out

```
#OUTPUT
 $PCS_TYPE
    GROUNDWATER_FLOW
 $NOD_VALUES
    HEAD
    VELOCITY_Y1
 $GEO_TYPE
    DOMAIN
 $DAT_TYPE
    VTK
 $TIM_TYPE
    STEADY
#STOP
```

2.6.3   Results

Using the input files listed above, we can simulate flow through a stochastically generated discrete fracture model by running OGS5 (Table 2.6). The vtk file is generated when the simulation is finished. The benchmark model illustrates how fractures influence the hydraulic head distribution of the model which act as preferable

Table 2.6 Benchmark deposit (https://docs.opengeosys.org/books/bmb-5)

BM code	Author	Code	Files	CTest
BMB5-2.6	Tao Chen	OGS-5.7.0 (Mac)	Available	ToDo

(https://oc.ufz.de/index.php/s/nlph7bhfDkj6tC7)

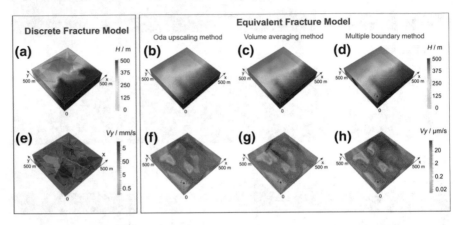

Fig. 2.7 Hydraulic head and Darcy's velocity in the y-direction for the stochastically generated discrete fracture model (**a, e**) and for the corresponding equivalent fracture models based on different upscaling methods (**b–d**) and (**f–h**) (Chen 2017)

paths for fluid flow (Fig. 2.7). Furthermore, by comparing with equivalent fracture models, which are based on different upscaling methods, it demonstrates that the hydraulic head distributions in the discrete fracture model and in the equivalent fracture model are quite similar. However, for Darcy's velocity, it can be represented explicitly according to the fracture geometry and has a higher magnitude than that in equivalent fracture models.

Chapter 3
M Processes

Xing-Yuan Miao, Jobst Maßmann, Thomas Nagel, Dmitri Naumov,
Francesco Parisio and Peter Vogel

Bending of Plates

Peter Vogel and Jobst Maßmann

This section presents problems on bending of elastic plates. Most of the material is based on ideas outlined by Timoshenko and Goodier (1951), the last exercise of this section has been adopted from Woinowsky-Krieger (1933). We focus on the closed form solutions. The associated simulation exercises have been checked by OGS; they may serve as verification tests. For the underlying theory of linear elasticity see Gurtin (1972).

3.1 A Thick Plate Undergoes Compression

Given length $2L = 20$ m and thickness $H = 2$ m the domain represents the rectangular plate $[-L, L] \times [-L, L] \times [-H, 0]$. It is discretized by $20 \times 20 \times 4$ equally sized hexahedral elements. The material has been selected elastic with Young's modulus $E = 12{,}600$ MPa and Poisson's ratio $\nu = 0.2$, gravity is neglected via zero material density. Fixities have been prescribed in the interior of the domain with zero x-displacement along the plane $x = 0$, zero y-displacement along the plane $y = 0$, and zero z-displacement at the origin. Specified loads $P(x, y, z)$ prevail along the surface

X.-Y. Miao (✉) · T. Nagel · D. Naumov · F. Parisio
UFZ, Helmholtz Centre for Environmental Research, Leipzig, Germany
e-mail: xing-yuan.miao@ufz.de

X.-Y. Miao
TU Dresden, Dresden, Germany

J. Maßmann · P. Vogel
BGR, Federal Institute for Geosciences and Natural Resources, Hanover, Germany

T. Nagel
Trinity College Dublin, Dublin, Ireland

© Springer International Publishing AG 2018
O. Kolditz et al. (eds.), *Thermo-Hydro-Mechanical-Chemical Processes
in Fractured Porous Media: Modelling and Benchmarking*,
Terrestrial Environmental Sciences, https://doi.org/10.1007/978-3-319-68225-9_3

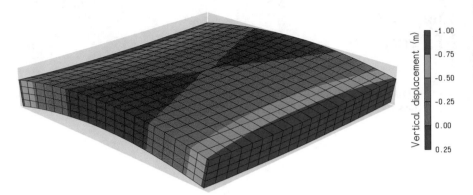

Fig. 3.1 Deformed mesh, vertical displacements

of the domain. With the aid of the stress $\sigma_0 = 500\,\text{MPa}$ these loads read along the faces $x = L$ and $x = -L$

$$P(L, y, z) = \frac{\sigma_0}{H} \begin{pmatrix} z \\ 0 \\ 0 \end{pmatrix}, \quad P(-L, y, z) = -\frac{\sigma_0}{H} \begin{pmatrix} z \\ 0 \\ 0 \end{pmatrix}, \quad (3.1.1)$$

and the remaining faces are free of load by default. The simulation (Table 3.1) comprises one timestep to establish the stresses, the strains, and the displacement vectorr (u_x, u_y, u_z) (Fig. 3.1).

The stress tensor

$$\sigma = \begin{pmatrix} \sigma_{11} & \sigma_{12} & \sigma_{13} \\ \sigma_{12} & \sigma_{22} & \sigma_{23} \\ \sigma_{13} & \sigma_{23} & \sigma_{33} \end{pmatrix} = \frac{\sigma_0}{H} \begin{pmatrix} z & 0 & 0 \\ 0 & 0 & 0 \\ 0 & 0 & 0 \end{pmatrix} \quad (3.1.2)$$

satisfies the equation of mechanical equilibrium

$$\text{div } \sigma = 0 \quad (3.1.3)$$

as well as the specified surface loads because along the faces $x = L$ and $x = -L$

$$\sigma \begin{pmatrix} 1 \\ 0 \\ 0 \end{pmatrix} = -\sigma \begin{pmatrix} -1 \\ 0 \\ 0 \end{pmatrix} = \begin{pmatrix} \sigma_{11} \\ \sigma_{12} \\ \sigma_{13} \end{pmatrix} = \frac{\sigma_0}{H} \begin{pmatrix} z \\ 0 \\ 0 \end{pmatrix}, \quad (3.1.4)$$

and similarly for the remaining faces. Hooke's law reads for the strains

$$\epsilon_{11} = \frac{\partial u_x}{\partial x} = \frac{1}{E}[\sigma_{11} - \nu(\sigma_{22} + \sigma_{33})] = \frac{\sigma_0}{EH}z, \tag{3.1.5}$$

$$\epsilon_{22} = \frac{\partial u_y}{\partial y} = \frac{1}{E}[\sigma_{22} - \nu(\sigma_{11} + \sigma_{33})] = -\nu\frac{\sigma_0}{EH}z, \tag{3.1.6}$$

$$\epsilon_{33} = \frac{\partial u_z}{\partial z} = \frac{1}{E}[\sigma_{33} - \nu(\sigma_{11} + \sigma_{22})] = -\nu\frac{\sigma_0}{EH}z, \tag{3.1.7}$$

$$\epsilon_{12} = \frac{\partial u_x}{\partial y} + \frac{\partial u_y}{\partial x} = \frac{2(1+\nu)}{E}\sigma_{12} = 0, \tag{3.1.8}$$

$$\epsilon_{13} = \frac{\partial u_x}{\partial z} + \frac{\partial u_z}{\partial x} = \frac{2(1+\nu)}{E}\sigma_{13} = 0, \tag{3.1.9}$$

$$\epsilon_{23} = \frac{\partial u_y}{\partial z} + \frac{\partial u_z}{\partial y} = \frac{2(1+\nu)}{E}\sigma_{23} = 0, \tag{3.1.10}$$

which constitute a set of equations for the partial derivatives of the displacements. Integrating the strains ϵ_{11}, ϵ_{22}, and ϵ_{33} gives the displacement vector (u_x, u_y, u_z)

$$u_x(x, y, z) = \frac{\sigma_0}{EH}zx + f(y, z), \tag{3.1.11}$$

$$u_y(x, y, z) = -\nu\frac{\sigma_0}{EH}zy + g(x, z), \tag{3.1.12}$$

$$u_z(x, y, z) = -\nu\frac{\sigma_0}{EH}\frac{1}{2}z^2 + h(x, y), \tag{3.1.13}$$

where $f(y, z)$, $g(x, z)$, and $h(x, y)$ have to be determined from the shear strains and the specified fixities. Due to the x-fixities along the plane $x = 0$

$$f(y, z) = 0, \tag{3.1.14}$$

and due to the y-fixities along the plane $y = 0$

$$g(x, z) = 0. \tag{3.1.15}$$

Then

$$\epsilon_{12} = \frac{\partial u_x}{\partial y} + \frac{\partial u_y}{\partial x} = 0, \tag{3.1.16}$$

$$\epsilon_{13} = \frac{\partial u_x}{\partial z} + \frac{\partial u_z}{\partial x} = \frac{\sigma_0}{EH}x + \frac{\partial h}{\partial x} = 0, \tag{3.1.17}$$

$$\epsilon_{23} = \frac{\partial u_y}{\partial z} + \frac{\partial u_z}{\partial y} = -\nu\frac{\sigma_0}{EH}y + \frac{\partial h}{\partial y} = 0, \tag{3.1.18}$$

Table 3.1 Benchmark deposit (https://docs.opengeosys.org/books/bmb-5)

BM code	Author	Code	Files	CTest
BMB5-3.1	Peter Vogel	OGS-5	Available	OK

(https://oc.ufz.de/index.php/s/nlph7bhfDkj6tC7)

hence,

$$\frac{\partial h}{\partial x} = -\frac{\sigma_0}{EH} x, \tag{3.1.19}$$

$$\frac{\partial h}{\partial y} = \nu \frac{\sigma_0}{EH} y, \tag{3.1.20}$$

and therefore,

$$h(x, y) = -\frac{\sigma_0}{EH} \frac{1}{2} x^2 + \nu \frac{\sigma_0}{EH} \frac{1}{2} y^2 + C. \tag{3.1.21}$$

The z-fixity at the origin yields the free constant C. The displacement vector (u_x, u_y, u_z) becomes

$$u_x(x, z) = \frac{\sigma_0}{EH} xz, \tag{3.1.22}$$

$$u_y(y, z) = -\nu \frac{\sigma_0}{EH} yz, \tag{3.1.23}$$

$$u_z(x, y, z) = -\frac{\sigma_0}{2EH} [x^2 + \nu(z^2 - y^2)]. \tag{3.1.24}$$

3.2 A Thick Plate Undergoes Tension

Given length $2L = 20$ m and thickness $H = 2$ m the domain represents the rectangular plate $[-L, L] \times [-L, L] \times [-H, 0]$. It is discretized by $20 \times 20 \times 4$ equally sized hexahedral elements. The material has been selected elastic with Young's modulus $E = 12,600$ MPa and Poisson's ratio $\nu = 0.2$, gravity is neglected via zero material density. Fixities have been prescribed in the interior of the domain with zero x-displacement along the plane $x = 0$, zero y-displacement along the plane $y = 0$, and zero z-displacement at the origin. Specified loads $P(x, y, z)$ prevail along the surface of the domain. With the aid of the stress $\sigma_0 = 500$ MPa these loads read along the faces $y = L$ and $y = -L$

$$P(x, L, z) = -\frac{\sigma_0}{H} \begin{pmatrix} 0 \\ z \\ 0 \end{pmatrix}, \quad P(x, -L, z) = \frac{\sigma_0}{H} \begin{pmatrix} 0 \\ z \\ 0 \end{pmatrix}, \tag{3.2.1}$$

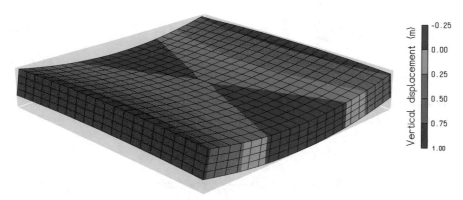

Fig. 3.2 Deformed mesh, vertical displacements

and the remaining faces are free of load by default. The simulation (Table 3.2) comprises one timestep to establish the stresses, the strains, and the displacement vector (u_x, u_y, u_z) (Fig. 3.2).

The stress tensor

$$\sigma = \begin{pmatrix} \sigma_{11} & \sigma_{12} & \sigma_{13} \\ \sigma_{12} & \sigma_{22} & \sigma_{23} \\ \sigma_{13} & \sigma_{23} & \sigma_{33} \end{pmatrix} = -\frac{\sigma_0}{H} \begin{pmatrix} 0 & 0 & 0 \\ 0 & z & 0 \\ 0 & 0 & 0 \end{pmatrix} \tag{3.2.2}$$

satisfies the equation of mechanical equilibrium

$$\text{div } \sigma = 0 \tag{3.2.3}$$

as well as the specified surface loads because along the faces $y = L$ and $y = -L$

$$\sigma \begin{pmatrix} 0 \\ 1 \\ 0 \end{pmatrix} = -\sigma \begin{pmatrix} 0 \\ -1 \\ 0 \end{pmatrix} = \begin{pmatrix} \sigma_{12} \\ \sigma_{22} \\ \sigma_{23} \end{pmatrix} = -\frac{\sigma_0}{H} \begin{pmatrix} 0 \\ z \\ 0 \end{pmatrix}, \tag{3.2.4}$$

and similarly for the remaining faces. Hooke's law reads for the strains

$$\epsilon_{11} = \frac{\partial u_x}{\partial x} = \frac{1}{E}[\sigma_{11} - \nu(\sigma_{22} + \sigma_{33})] = \nu\frac{\sigma_0}{EH}z, \tag{3.2.5}$$

$$\epsilon_{22} = \frac{\partial u_y}{\partial y} = \frac{1}{E}[\sigma_{22} - \nu(\sigma_{11} + \sigma_{33})] = -\frac{\sigma_0}{EH}z, \tag{3.2.6}$$

$$\epsilon_{33} = \frac{\partial u_z}{\partial z} = \frac{1}{E}[\sigma_{33} - \nu(\sigma_{11} + \sigma_{22})] = \nu\frac{\sigma_0}{EH}z, \tag{3.2.7}$$

$$\epsilon_{12} = \frac{\partial u_x}{\partial y} + \frac{\partial u_y}{\partial x} = \frac{2(1+\nu)}{E}\sigma_{12} = 0, \tag{3.2.8}$$

$$\epsilon_{13} = \frac{\partial u_x}{\partial z} + \frac{\partial u_z}{\partial x} = \frac{2(1+\nu)}{E}\sigma_{13} = 0, \tag{3.2.9}$$

$$\epsilon_{23} = \frac{\partial u_y}{\partial z} + \frac{\partial u_z}{\partial y} = \frac{2(1+\nu)}{E}\sigma_{23} = 0, \tag{3.2.10}$$

which constitute a set of equations for the partial derivatives of the displacements. Integrating the strains ϵ_{11}, ϵ_{22}, and ϵ_{33} gives the displacement vector (u_x, u_y, u_z)

$$u_x(x, y, z) = \nu\frac{\sigma_0}{EH}zx + f(y, z), \tag{3.2.11}$$

$$u_y(x, y, z) = -\frac{\sigma_0}{EH}zy + g(x, z), \tag{3.2.12}$$

$$u_z(x, y, z) = \nu\frac{\sigma_0}{EH}\frac{1}{2}z^2 + h(x, y), \tag{3.2.13}$$

where $f(y, z)$, $g(x, z)$, and $h(x, y)$ have to be determined from the shear strains and the specified fixities. Due to the x-fixities along the plane $x = 0$

$$f(y, z) = 0, \tag{3.2.14}$$

and due to the y-fixities along the plane $y = 0$

$$g(x, z) = 0. \tag{3.2.15}$$

Then

$$\epsilon_{12} = \frac{\partial u_x}{\partial y} + \frac{\partial u_y}{\partial x} = 0, \tag{3.2.16}$$

$$\epsilon_{13} = \frac{\partial u_x}{\partial z} + \frac{\partial u_z}{\partial x} = \nu\frac{\sigma_0}{EH}x + \frac{\partial h}{\partial x} = 0, \tag{3.2.17}$$

$$\epsilon_{23} = \frac{\partial u_y}{\partial z} + \frac{\partial u_z}{\partial y} = -\frac{\sigma_0}{EH}y + \frac{\partial h}{\partial y} = 0, \tag{3.2.18}$$

hence,

$$\frac{\partial h}{\partial x} = -\nu\frac{\sigma_0}{EH}x, \tag{3.2.19}$$

$$\frac{\partial h}{\partial y} = \frac{\sigma_0}{EH}y, \tag{3.2.20}$$

and therefore,

$$h(x, y) = -\nu\frac{\sigma_0}{EH}\frac{1}{2}x^2 + \frac{\sigma_0}{EH}\frac{1}{2}y^2 + C. \tag{3.2.21}$$

Table 3.2 Benchmark deposit (https://docs.opengeosys.org/books/bmb-5)

BM code	Author	Code	Files	CTest
BMB5-3.2	Peter Vogel	OGS-5	Available	ToDo

(https://oc.ufz.de/index.php/s/nlph7bhfDkj6tC7)

The z-fixity at the origin yields the free constant C. The displacement vector (u_x, u_y, u_z) becomes

$$u_x(x, z) = \nu \frac{\sigma_0}{EH} xz, \tag{3.2.22}$$

$$u_y(y, z) = -\frac{\sigma_0}{EH} yz, \tag{3.2.23}$$

$$u_z(x, y, z) = \frac{\sigma_0}{2EH}[y^2 + \nu(z^2 - x^2)]. \tag{3.2.24}$$

3.3   A Thick Plate Undergoes Compression and Tension

Given length $2L = 20$ m and thickness $H = 2$ m the domain represents the rectangular plate $[-L, L] \times [-L, L] \times [-H, 0]$. It is discretized by $20 \times 20 \times 4$ equally sized hexahedral elements. The material has been selected elastic with Young's modulus $E = 15{,}000$ MPa and Poisson's ratio $\nu = 0.2$, gravity is neglected via zero material density. Fixities have been prescribed in the interior of the domain with zero x-displacement along the plane $x = 0$, zero y-displacement along the plane $y = 0$, and zero z-displacement at the origin. Specified loads $P(x, y, z)$ prevail along the lateral boundaries of the domain. With the aid of the stress $\sigma_0 = 500$ MPa these loads read along the faces $x = L$ and $x = -L$

$$P(L, y, z) = \frac{\sigma_0}{H} \begin{pmatrix} z \\ 0 \\ 0 \end{pmatrix}, \quad P(-L, y, z) = -\frac{\sigma_0}{H} \begin{pmatrix} z \\ 0 \\ 0 \end{pmatrix}, \tag{3.3.1}$$

and along the faces $y = L$ and $y = -L$

$$P(x, L, z) = -\frac{\sigma_0}{H} \begin{pmatrix} 0 \\ z \\ 0 \end{pmatrix}, \quad P(x, -L, z) = \frac{\sigma_0}{H} \begin{pmatrix} 0 \\ z \\ 0 \end{pmatrix}. \tag{3.3.2}$$

The top $z = 0$ and the bottom $z = -H$ are free of load by default. The simulation (Table 3.3) comprises one timestep to establish the stresses, the strains, and the displacement vector (u_x, u_y, u_z) (Fig. 3.3).

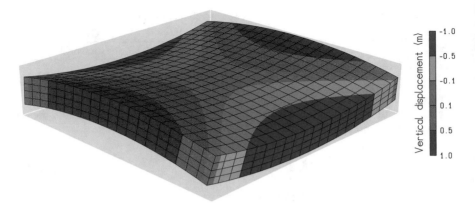

Fig. 3.3 Deformed mesh, vertical displacements

The stress tensor

$$\sigma = \begin{pmatrix} \sigma_{11} & \sigma_{12} & \sigma_{13} \\ \sigma_{12} & \sigma_{22} & \sigma_{23} \\ \sigma_{13} & \sigma_{23} & \sigma_{33} \end{pmatrix} = \frac{\sigma_0}{H} \begin{pmatrix} z & 0 & 0 \\ 0 & -z & 0 \\ 0 & 0 & 0 \end{pmatrix} \tag{3.3.3}$$

satisfies the equation of mechanical equilibrium

$$\text{div } \sigma = 0 \tag{3.3.4}$$

as well as the specified surface loads because along the faces $x = L$ and $x = -L$

$$\sigma \begin{pmatrix} 1 \\ 0 \\ 0 \end{pmatrix} = -\sigma \begin{pmatrix} -1 \\ 0 \\ 0 \end{pmatrix} = \begin{pmatrix} \sigma_{11} \\ \sigma_{12} \\ \sigma_{13} \end{pmatrix} = \frac{\sigma_0}{H} \begin{pmatrix} z \\ 0 \\ 0 \end{pmatrix}, \tag{3.3.5}$$

along the faces $y = L$ and $y = -L$

$$\sigma \begin{pmatrix} 0 \\ 1 \\ 0 \end{pmatrix} = -\sigma \begin{pmatrix} 0 \\ -1 \\ 0 \end{pmatrix} = \begin{pmatrix} \sigma_{12} \\ \sigma_{22} \\ \sigma_{23} \end{pmatrix} = -\frac{\sigma_0}{H} \begin{pmatrix} 0 \\ z \\ 0 \end{pmatrix}, \tag{3.3.6}$$

and similarly for the top face and the bottom face. Hooke's law reads for the strains

$$\epsilon_{11} = \frac{\partial u_x}{\partial x} = \frac{1}{E}[\sigma_{11} - \nu(\sigma_{22} + \sigma_{33})] = (1 + \nu)\frac{\sigma_0}{EH}z, \tag{3.3.7}$$

$$\epsilon_{22} = \frac{\partial u_y}{\partial y} = \frac{1}{E}[\sigma_{22} - \nu(\sigma_{11} + \sigma_{33})] = -(1 + \nu)\frac{\sigma_0}{EH}z, \tag{3.3.8}$$

$$\epsilon_{33} = \frac{\partial u_z}{\partial z} = \frac{1}{E}[\sigma_{33} - \nu(\sigma_{11} + \sigma_{22})] = 0, \tag{3.3.9}$$

$$\epsilon_{12} = \frac{\partial u_x}{\partial y} + \frac{\partial u_y}{\partial x} = \frac{2(1+\nu)}{E}\sigma_{12} = 0, \tag{3.3.10}$$

$$\epsilon_{13} = \frac{\partial u_x}{\partial z} + \frac{\partial u_z}{\partial x} = \frac{2(1+\nu)}{E}\sigma_{13} = 0, \tag{3.3.11}$$

$$\epsilon_{23} = \frac{\partial u_y}{\partial z} + \frac{\partial u_z}{\partial y} = \frac{2(1+\nu)}{E}\sigma_{23} = 0, \tag{3.3.12}$$

which constitute a set of equations for the partial derivatives of the displacements. Integrating the strains ϵ_{11}, ϵ_{22}, and ϵ_{33} gives the displacement vector (u_x, u_y, u_z)

$$u_x(x, y, z) = (1+\nu)\frac{\sigma_0}{EH}zx + f(y, z), \tag{3.3.13}$$

$$u_y(x, y, z) = -(1+\nu)\frac{\sigma_0}{EH}zy + g(x, z), \tag{3.3.14}$$

$$u_z(x, y, z) = h(x, y), \tag{3.3.15}$$

where $f(y, z)$, $g(x, z)$, and $h(x, y)$ have to be determined from the shear strains and the specified fixities. Due to the x-fixities along the plane $x = 0$

$$f(y, z) = 0, \tag{3.3.16}$$

and due to the y-fixities along the plane $y = 0$

$$g(x, z) = 0. \tag{3.3.17}$$

Then

$$\epsilon_{12} = \frac{\partial u_x}{\partial y} + \frac{\partial u_y}{\partial x} = 0, \tag{3.3.18}$$

$$\epsilon_{13} = \frac{\partial u_x}{\partial z} + \frac{\partial u_z}{\partial x} = (1+\nu)\frac{\sigma_0}{EH}x + \frac{\partial h}{\partial x} = 0, \tag{3.3.19}$$

$$\epsilon_{23} = \frac{\partial u_y}{\partial z} + \frac{\partial u_z}{\partial y} = -(1+\nu)\frac{\sigma_0}{EH}y + \frac{\partial h}{\partial y} = 0, \tag{3.3.20}$$

hence,

$$\frac{\partial h}{\partial x} = -(1+\nu)\frac{\sigma_0}{EH}x, \tag{3.3.21}$$

$$\frac{\partial h}{\partial y} = (1+\nu)\frac{\sigma_0}{EH}y, \tag{3.3.22}$$

Table 3.3 Benchmark deposit (https://docs.opengeosys.org/books/bmb-5)

BM code	Author	Code	Files	CTest
BMB5-3.3	Peter Vogel	OGS-5	Available	ToDo

(https://oc.ufz.de/index.php/s/nlph7bhfDkj6tC7)

and therefore,

$$h(x, y) = \frac{1+\nu}{2} \frac{\sigma_0}{EH}(y^2 - x^2) + C. \tag{3.3.23}$$

The z-fixity at the origin yields the free constant C. The displacement vector (u_x, u_y, u_z) becomes

$$u_x(x, z) = (1+\nu)\frac{\sigma_0}{EH}xz, \tag{3.3.24}$$

$$u_y(y, z) = -(1+\nu)\frac{\sigma_0}{EH}yz, \tag{3.3.25}$$

$$u_z(x, y) = \frac{1+\nu}{2} \frac{\sigma_0}{EH}(y^2 - x^2). \tag{3.3.26}$$

3.4   A Thick Plate Undergoes Tension and Shear

Given length $2L = 20$ m and thickness $H = 2$ m the domain represents the rectangular plate $[0, 2L] \times [-L, L] \times [-H, 0]$. It is discretized by $20 \times 20 \times 4$ equally sized hexahedral elements. The material has been selected elastic with Young's modulus $E = 12,600$ MPa and Poisson's ratio $\nu = 0.2$, gravity is neglected via zero material density. Fixities have been prescribed with zero x-displacement along the face $x = 0$, zero y-displacement along the plane $y = 0$, and zero z-displacement at the origin. Specified loads $P(x, y, z)$ prevail along the entire surface of the domain. With the aid of the stress $\sigma_0 = 500$ MPa these loads read along the faces $x = 2L$ and $x = 0$

$$P(2L, y, z) = -\sigma_0 \begin{pmatrix} 0 \\ 0 \\ 1 \end{pmatrix}, \quad P(0, y, z) = \sigma_0 \begin{pmatrix} 0 \\ 0 \\ 1 \end{pmatrix}, \tag{3.4.1}$$

along the faces $y = L$ and $y = -L$

$$P(x, L, z) = -\frac{\sigma_0}{H} \begin{pmatrix} 0 \\ z \\ 0 \end{pmatrix}, \quad P(x, -L, z) = \frac{\sigma_0}{H} \begin{pmatrix} 0 \\ z \\ 0 \end{pmatrix}, \tag{3.4.2}$$

and along the faces $z = 0$ and $z = -H$

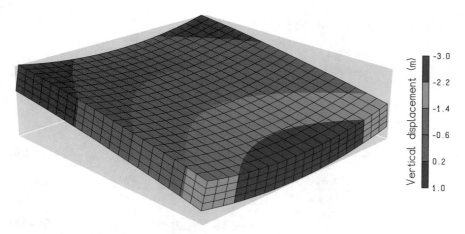

Fig. 3.4 Deformed mesh, vertical displacements

$$P(x, y, 0) = -\sigma_0 \begin{pmatrix} 1 \\ 0 \\ 0 \end{pmatrix}, \quad P(x, y, -H) = \sigma_0 \begin{pmatrix} 1 \\ 0 \\ 0 \end{pmatrix}. \tag{3.4.3}$$

The simulation (Table 3.4) comprises one timestep to establish the stresses, the strains, and the displacement vector (u_x, u_y, u_z) (Fig. 3.4).

The stress tensor

$$\boldsymbol{\sigma} = \begin{pmatrix} \sigma_{11} & \sigma_{12} & \sigma_{13} \\ \sigma_{12} & \sigma_{22} & \sigma_{23} \\ \sigma_{13} & \sigma_{23} & \sigma_{33} \end{pmatrix} = -\sigma_0 \begin{pmatrix} 0 & 0 & 1 \\ 0 & z/H & 0 \\ 1 & 0 & 0 \end{pmatrix} \tag{3.4.4}$$

satisfies the equation of mechanical equilibrium

$$\operatorname{div} \boldsymbol{\sigma} = 0 \tag{3.4.5}$$

as well as the specified surface loads because along the faces $x = 2L$ and $x = 0$

$$\boldsymbol{\sigma} \begin{pmatrix} 1 \\ 0 \\ 0 \end{pmatrix} = -\boldsymbol{\sigma} \begin{pmatrix} -1 \\ 0 \\ 0 \end{pmatrix} = \begin{pmatrix} \sigma_{11} \\ \sigma_{12} \\ \sigma_{13} \end{pmatrix} = -\sigma_0 \begin{pmatrix} 0 \\ 0 \\ 1 \end{pmatrix}, \tag{3.4.6}$$

along the faces $y = L$ and $y = -L$

$$\boldsymbol{\sigma} \begin{pmatrix} 0 \\ 1 \\ 0 \end{pmatrix} = -\boldsymbol{\sigma} \begin{pmatrix} 0 \\ -1 \\ 0 \end{pmatrix} = \begin{pmatrix} \sigma_{12} \\ \sigma_{22} \\ \sigma_{23} \end{pmatrix} = -\frac{\sigma_0}{H} \begin{pmatrix} 0 \\ z \\ 0 \end{pmatrix}, \tag{3.4.7}$$

and along the faces $z = 0$ and $z = -H$

$$\sigma \begin{pmatrix} 0 \\ 0 \\ 1 \end{pmatrix} = -\sigma \begin{pmatrix} 0 \\ 0 \\ -1 \end{pmatrix} = \begin{pmatrix} \sigma_{13} \\ \sigma_{23} \\ \sigma_{33} \end{pmatrix} = -\sigma_0 \begin{pmatrix} 1 \\ 0 \\ 0 \end{pmatrix}. \tag{3.4.8}$$

Hooke's law reads for the strains

$$\epsilon_{11} = \frac{\partial u_x}{\partial x} = \frac{1}{E}[\sigma_{11} - \nu(\sigma_{22} + \sigma_{33})] = \nu \frac{\sigma_0}{EH} z, \tag{3.4.9}$$

$$\epsilon_{22} = \frac{\partial u_y}{\partial y} = \frac{1}{E}[\sigma_{22} - \nu(\sigma_{11} + \sigma_{33})] = -\frac{\sigma_0}{EH} z, \tag{3.4.10}$$

$$\epsilon_{33} = \frac{\partial u_z}{\partial z} = \frac{1}{E}[\sigma_{33} - \nu(\sigma_{11} + \sigma_{22})] = \nu \frac{\sigma_0}{EH} z, \tag{3.4.11}$$

$$\epsilon_{12} = \frac{\partial u_x}{\partial y} + \frac{\partial u_y}{\partial x} = \frac{2(1+\nu)}{E}\sigma_{12} = 0, \tag{3.4.12}$$

$$\epsilon_{13} = \frac{\partial u_x}{\partial z} + \frac{\partial u_z}{\partial x} = \frac{2(1+\nu)}{E}\sigma_{13} = -\frac{2(1+\nu)}{E}\sigma_0, \tag{3.4.13}$$

$$\epsilon_{23} = \frac{\partial u_y}{\partial z} + \frac{\partial u_z}{\partial y} = \frac{2(1+\nu)}{E}\sigma_{23} = 0, \tag{3.4.14}$$

which constitute a set of equations for the partial derivatives of the displacements. Integrating the strains ϵ_{11}, ϵ_{22}, and ϵ_{33} gives the displacement vector (u_x, u_y, u_z)

$$u_x(x, y, z) = \nu \frac{\sigma_0}{EH} zx + f(y, z), \tag{3.4.15}$$

$$u_y(x, y, z) = -\frac{\sigma_0}{EH} zy + g(x, z), \tag{3.4.16}$$

$$u_z(x, y, z) = \nu \frac{\sigma_0}{EH} \frac{1}{2} z^2 + h(x, y), \tag{3.4.17}$$

where $f(y, z)$, $g(x, z)$, and $h(x, y)$ have to be determined from the shear strains and the specified fixities. Due to the x-fixities along the face $x = 0$

$$f(y, z) = 0, \tag{3.4.18}$$

and due to the y-fixities along the plane $y = 0$

$$g(x, z) = 0. \tag{3.4.19}$$

Then

$$\epsilon_{12} = \frac{\partial u_x}{\partial y} + \frac{\partial u_y}{\partial x} = 0, \tag{3.4.20}$$

Table 3.4 Benchmark deposit (https://docs.opengeosys.org/books/bmb-5)

BM code	Author	Code	Files	CTest
BMB5-3.4	Peter Vogel	OGS-5	Available	ToDo

(https://oc.ufz.de/index.php/s/nlph7bhfDkj6tC7)

$$\epsilon_{13} = \frac{\partial u_x}{\partial z} + \frac{\partial u_z}{\partial x} = \nu \frac{\sigma_0}{EH}x + \frac{\partial h}{\partial x} = -\frac{2(1+\nu)}{E}\sigma_0, \tag{3.4.21}$$

$$\epsilon_{23} = \frac{\partial u_y}{\partial z} + \frac{\partial u_z}{\partial y} = -\frac{\sigma_0}{EH}y + \frac{\partial h}{\partial y} = 0, \tag{3.4.22}$$

hence,

$$\frac{\partial h}{\partial x} = -\frac{2(1+\nu)}{E}\sigma_0 - \nu \frac{\sigma_0}{EH}x, \tag{3.4.23}$$

$$\frac{\partial h}{\partial y} = \frac{\sigma_0}{EH}y, \tag{3.4.24}$$

and therefore,

$$h(x, y) = -\frac{2(1+\nu)}{E}\sigma_0 x - \nu \frac{\sigma_0}{EH}\frac{1}{2}x^2 + \frac{\sigma_0}{EH}\frac{1}{2}y^2 + C. \tag{3.4.25}$$

The z-fixity at the origin yields the free constant C. The displacement vector (u_x, u_y, u_z) becomes

$$u_x(x, z) = \nu \frac{\sigma_0}{EH}xz, \tag{3.4.26}$$

$$u_y(y, z) = -\frac{\sigma_0}{EH}yz, \tag{3.4.27}$$

$$u_z(x, y, z) = \frac{\sigma_0}{2EH}[y^2 + \nu(z^2 - x^2)] - \frac{2(1+\nu)}{E}\sigma_0 x. \tag{3.4.28}$$

3.5 A Thick Plate Undergoes Tension and Twist

Given length $2L = 20$ m and thickness $H = 2$ m the domain represents the rectangular plate $[0, 2L] \times [-L, L] \times [-H, 0]$. It is discretized by $20 \times 20 \times 4$ equally sized hexahedral elements. The material has been selected elastic with Young's modulus $E = 8,100$ MPa and Poisson's ratio $\nu = 0.2$, gravity is neglected via zero material density. Fixities have been prescribed with zero x-displacement along the face $x = 0$, zero y-displacement along the plane $y = 0$, and zero z-displacement at the origin. Specified loads $P(x, y, z)$ prevail along the entire surface of the domain. With the aid of the stress $\sigma_0 = 500$ MPa these loads read along the faces $x = 2L$ and $x = 0$

$$P(2L, y, z) = \frac{\sigma_0}{2L} \begin{pmatrix} 0 \\ 0 \\ y \end{pmatrix}, \quad P(0, y, z) = -\frac{\sigma_0}{2L} \begin{pmatrix} 0 \\ 0 \\ y \end{pmatrix}, \tag{3.5.1}$$

along the faces $y = L$ and $y = -L$

$$P(x, L, z) = \sigma_0 \begin{pmatrix} 0 \\ -z/H \\ x/(2L) \end{pmatrix}, \quad P(x, -L, z) = -\sigma_0 \begin{pmatrix} 0 \\ -z/H \\ x/(2L) \end{pmatrix}, \tag{3.5.2}$$

and along the faces $z = 0$ and $z = -H$

$$P(x, y, 0) = \frac{\sigma_0}{2L} \begin{pmatrix} y \\ x \\ 0 \end{pmatrix}, \quad P(x, y, -H) = -\frac{\sigma_0}{2L} \begin{pmatrix} y \\ x \\ 0 \end{pmatrix}. \tag{3.5.3}$$

The simulation (Table 3.5) comprises one timestep to establish the stresses, the strains, and the displacement vector (u_x, u_y, u_z) (Fig. 3.5).

The stress tensor

$$\boldsymbol{\sigma} = \begin{pmatrix} \sigma_{11} & \sigma_{12} & \sigma_{13} \\ \sigma_{12} & \sigma_{22} & \sigma_{23} \\ \sigma_{13} & \sigma_{23} & \sigma_{33} \end{pmatrix} = \sigma_0 \begin{pmatrix} 0 & 0 & y/(2L) \\ 0 & -z/H & x/(2L) \\ y/(2L) & x/(2L) & 0 \end{pmatrix} \tag{3.5.4}$$

satisfies the equation of mechanical equilibrium

$$\operatorname{div} \boldsymbol{\sigma} = 0 \tag{3.5.5}$$

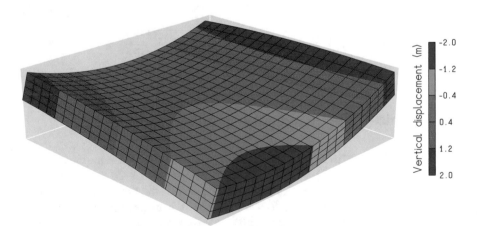

Fig. 3.5 Deformed mesh, vertical displacements

as well as the specified surface loads because along the faces $x = 2L$ and $x = 0$

$$\sigma \begin{pmatrix} 1 \\ 0 \\ 0 \end{pmatrix} = -\sigma \begin{pmatrix} -1 \\ 0 \\ 0 \end{pmatrix} = \begin{pmatrix} \sigma_{11} \\ \sigma_{12} \\ \sigma_{13} \end{pmatrix} = \frac{\sigma_0}{2L} \begin{pmatrix} 0 \\ 0 \\ y \end{pmatrix}, \tag{3.5.6}$$

along the faces $y = L$ and $y = -L$

$$\sigma \begin{pmatrix} 0 \\ 1 \\ 0 \end{pmatrix} = -\sigma \begin{pmatrix} 0 \\ -1 \\ 0 \end{pmatrix} = \begin{pmatrix} \sigma_{12} \\ \sigma_{22} \\ \sigma_{23} \end{pmatrix} = \sigma_0 \begin{pmatrix} 0 \\ -z/H \\ x/(2L) \end{pmatrix}, \tag{3.5.7}$$

and along the faces $z = 0$ and $z = -H$

$$\sigma \begin{pmatrix} 0 \\ 0 \\ 1 \end{pmatrix} = -\sigma \begin{pmatrix} 0 \\ 0 \\ -1 \end{pmatrix} = \begin{pmatrix} \sigma_{13} \\ \sigma_{23} \\ \sigma_{33} \end{pmatrix} = \frac{\sigma_0}{2L} \begin{pmatrix} y \\ x \\ 0 \end{pmatrix}. \tag{3.5.8}$$

Hooke's law reads for the strains

$$\epsilon_{11} = \frac{\partial u_x}{\partial x} = \frac{1}{E}[\sigma_{11} - \nu(\sigma_{22} + \sigma_{33})] = \nu \frac{\sigma_0}{EH} z, \tag{3.5.9}$$

$$\epsilon_{22} = \frac{\partial u_y}{\partial y} = \frac{1}{E}[\sigma_{22} - \nu(\sigma_{11} + \sigma_{33})] = -\frac{\sigma_0}{EH} z, \tag{3.5.10}$$

$$\epsilon_{33} = \frac{\partial u_z}{\partial z} = \frac{1}{E}[\sigma_{33} - \nu(\sigma_{11} + \sigma_{22})] = \nu \frac{\sigma_0}{EH} z, \tag{3.5.11}$$

$$\epsilon_{12} = \frac{\partial u_x}{\partial y} + \frac{\partial u_y}{\partial x} = \frac{2(1+\nu)}{E} \sigma_{12} = 0, \tag{3.5.12}$$

$$\epsilon_{13} = \frac{\partial u_x}{\partial z} + \frac{\partial u_z}{\partial x} = \frac{2(1+\nu)}{E} \sigma_{13} = \frac{1+\nu}{EL} \sigma_0 y, \tag{3.5.13}$$

$$\epsilon_{23} = \frac{\partial u_y}{\partial z} + \frac{\partial u_z}{\partial y} = \frac{2(1+\nu)}{E} \sigma_{23} = \frac{1+\nu}{EL} \sigma_0 x, \tag{3.5.14}$$

which constitute a set of equations for the partial derivatives of the displacements. Integrating the strains ϵ_{11}, ϵ_{22}, and ϵ_{33} gives the displacement vector (u_x, u_y, u_z)

$$u_x(x, y, z) = \nu \frac{\sigma_0}{EH} zx + f(y, z), \tag{3.5.15}$$

$$u_y(x, y, z) = -\frac{\sigma_0}{EH} zy + g(x, z), \tag{3.5.16}$$

$$u_z(x, y, z) = \nu \frac{\sigma_0}{EH} \frac{1}{2} z^2 + h(x, y), \tag{3.5.17}$$

where $f(y, z)$, $g(x, z)$, and $h(x, y)$ have to be determined from the shear strains and the specified fixities. Due to the x-fixities along the face $x = 0$

$$f(y, z) = 0, \tag{3.5.18}$$

and due to the y-fixities along the plane $y = 0$

$$g(x, z) = 0. \tag{3.5.19}$$

Then

$$\epsilon_{12} = \frac{\partial u_x}{\partial y} + \frac{\partial u_y}{\partial x} = 0, \tag{3.5.20}$$

$$\epsilon_{13} = \frac{\partial u_x}{\partial z} + \frac{\partial u_z}{\partial x} = \nu \frac{\sigma_0}{EH} x + \frac{\partial h}{\partial x} = \frac{1 + \nu}{EL} \sigma_0 y, \tag{3.5.21}$$

$$\epsilon_{23} = \frac{\partial u_y}{\partial z} + \frac{\partial u_z}{\partial y} = -\frac{\sigma_0}{EH} y + \frac{\partial h}{\partial y} = \frac{1 + \nu}{EL} \sigma_0 x, \tag{3.5.22}$$

hence,

$$\frac{\partial h}{\partial x} = \frac{1 + \nu}{EL} \sigma_0 y - \frac{\nu \sigma_0}{EH} x, \tag{3.5.23}$$

$$\frac{\partial h}{\partial y} = \frac{1 + \nu}{EL} \sigma_0 x + \frac{\sigma_0}{EH} y, \tag{3.5.24}$$

and therefore,

$$h(x, y) = \frac{1 + \nu}{EL} \sigma_0 xy - \frac{\nu \sigma_0}{2EH} x^2 + \frac{\sigma_0}{2EH} y^2 + C. \tag{3.5.25}$$

The z-fixity at the origin yields the free constant C. The displacement vector (u_x, u_y, u_z) becomes

$$u_x(x, z) = \nu \frac{\sigma_0}{EH} xz, \tag{3.5.26}$$

$$u_y(y, z) = -\frac{\sigma_0}{EH} yz, \tag{3.5.27}$$

$$u_z(x, y, z) = \frac{\sigma_0}{2EH} [y^2 + \nu(z^2 - x^2)] + \frac{1 + \nu}{EL} \sigma_0 xy. \tag{3.5.28}$$

3.6 A Double Fourier Series Representation

This exercise has been adopted from Woinowsky-Krieger (1933). Given length $L = 20$ m and thickness $2H = 4$ m the domain represents the rectangular plate $[0, L] \times [0, L] \times [-H, H]$. It is discretized by 96,000 hexahedral elements. The material has been selected elastic with Young's modulus $E = 15,000$ MPa and Poisson's ratio $\nu = 0.2$, gravity is neglected via zero material density. The bottom face is free of load by default. A compressive stress $P(x, y)$ prevails along the top face, fixities have been prescribed along the lateral boundaries of the domain; details are given below. The simulation (Table 3.6) comprises one timestep to establish the stresses, the strains, and the displacement vector (u_x, u_y, u_z) (Fig. 3.6).

The entire solution will be represented by double Fourier series, the associated form of the imposed boundary conditions will be given first. With the aid of the stress $\sigma_0 = -1$ MPa and denoting by

$$
\begin{aligned}
&S_1 \text{ the convex hull of} \left\{ \begin{array}{llll} (0,0), & (\frac{L}{4}, \frac{L}{4}), & (\frac{L}{4}, \frac{3L}{4}), & (0, L) \end{array} \right\}, \\
&S_2 \text{ the convex hull of} \left\{ \begin{array}{llll} (0,0), & (0, L), & (\frac{3L}{4}, \frac{L}{4}), & (\frac{L}{4}, \frac{L}{4}) \end{array} \right\}, \\
&S_3 \text{ the convex hull of} \left\{ \begin{array}{llll} (\frac{L}{4}, \frac{L}{4}), & (\frac{3L}{4}, \frac{L}{4}), & (\frac{3L}{4}, \frac{3L}{4}), & (\frac{L}{4}, \frac{3L}{4}) \end{array} \right\}, \\
&S_4 \text{ the convex hull of} \left\{ \begin{array}{llll} (L,0), & (L, L), & (\frac{3L}{4}, \frac{3L}{4}), & (\frac{3L}{4}, \frac{L}{4}) \end{array} \right\}, \\
&S_5 \text{ the convex hull of} \left\{ \begin{array}{llll} (\frac{L}{4}, \frac{3L}{4}), & (\frac{3L}{4}, \frac{3L}{4}), & (L, L), & (0, L) \end{array} \right\}
\end{aligned}
$$

the specified top load (Fig. 3.7) reads

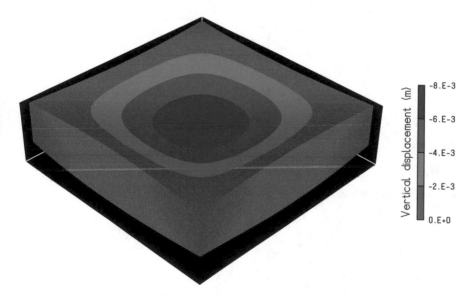

Fig. 3.6 Deformations scaled up, vertical displacements

Table 3.5 Benchmark deposit (https://docs.opengeosys.org/books/bmb-5)

BM code	Author	Code	Files	CTest
BMB5-3.5	Peter Vogel	OGS-5	Available	ToDo

(https://oc.ufz.de/index.php/s/nlph7bhfDkj6tC7)

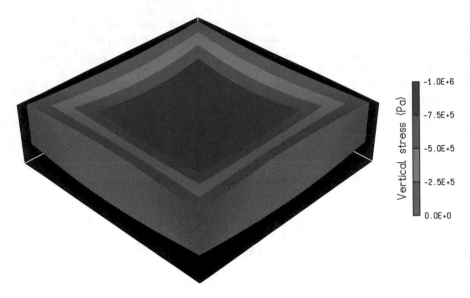

Fig. 3.7 Deformations scaled up, vertical stress

$$
P(x, y) = \begin{cases}
\sigma_0 \frac{4x}{L} & \text{on } S_1, \\[2mm]
\sigma_0 \frac{4y}{L} & \text{on } S_2, \\[2mm]
\sigma_0 & \text{on } S_3, \\[2mm]
\sigma_0 \frac{4}{L}(L - x) & \text{on } S_4, \\[2mm]
\sigma_0 \frac{4}{L}(L - y) & \text{on } S_5.
\end{cases}
\tag{3.6.1}
$$

The double Fourier series representation of $P(x, y)$ takes the form

$$
P(x, y) = \sum_{m=1}^{\infty} \sum_{n=1}^{\infty} P_{mn} \sin\left(\frac{m\pi}{L}x\right) \sin\left(\frac{n\pi}{L}y\right),
\tag{3.6.2}
$$

where for $m, n = 1, 2, \ldots$

$$
P_{mn} = \frac{4}{L^2} \int_0^L \int_0^L P(x, y) \sin\left(\frac{m\pi}{L}x\right) \sin\left(\frac{n\pi}{L}y\right) dx dy.
\tag{3.6.3}
$$

Now, with $P(x, y)$ as defined above, the integrals involved may be evaluated by elementary analytical methods. For $m, n = 1, 2, \ldots$ and with the aid of

$$
A_{mn} = \sigma_0 \frac{16}{mn\pi^2} \sin\left(\frac{m\pi}{2}\right) \sin\left(\frac{m\pi}{4}\right) \sin\left(\frac{n\pi}{2}\right) \sin\left(\frac{n\pi}{4}\right),
\tag{3.6.4}
$$

$$B_{mn} = \sigma_0 \frac{16}{\pi} \cos\left(\frac{m+n}{2}\pi\right) \cdot \left(\frac{\cos(n\pi) - 1}{n} + \frac{\cos(m\pi) - 1}{m}\right)$$
$$\cdot \left(\frac{\sin\left(\frac{m+n}{4}\pi\right)}{(m+n)^2\pi^2} + \frac{1}{4}\frac{\cos\left(\frac{m+n}{4}\pi\right)}{(m+n)\pi}\right), \qquad (3.6.5)$$

$$C_{mn} = \sigma_0 \frac{16}{\pi} \cos\left(\frac{m-n}{2}\pi\right) \cdot \left(\frac{\cos(n\pi) - 1}{n} - \frac{\cos(m\pi) - 1}{m}\right)$$
$$\cdot \left(\frac{\sin\left(\frac{m-n}{4}\pi\right)}{(m-n)^2\pi^2} + \frac{1}{4}\frac{\cos\left(\frac{m-n}{4}\pi\right)}{(m-n)\pi}\right) \qquad (3.6.6)$$

the Fourier coefficients P_{mn} become

$$P_{mn} = \begin{cases} A_{mn} + B_{mn} & \text{for } m = n, \\ A_{mn} + B_{mn} + C_{mn} & \text{for } m \neq n. \end{cases} \qquad (3.6.7)$$

Let

$$\sigma(x, y, z) = \begin{pmatrix} \sigma_{11}(x, y, z) & \sigma_{12}(x, y, z) & \sigma_{13}(x, y, z) \\ \sigma_{12}(x, y, z) & \sigma_{22}(x, y, z) & \sigma_{23}(x, y, z) \\ \sigma_{13}(x, y, z) & \sigma_{23}(x, y, z) & \sigma_{33}(x, y, z) \end{pmatrix} \qquad (3.6.8)$$

denote the stress tensor. The imposed boundary conditions read along the bottom face $z = -H$

$$\sigma_{13}(x, y, -H) = 0,$$
$$\sigma_{23}(x, y, -H) = 0, \qquad (3.6.9)$$
$$\sigma_{33}(x, y, -H) = 0,$$

along the top face $z = H$

$$\sigma_{13}(x, y, H) = 0,$$
$$\sigma_{23}(x, y, H) = 0, \qquad (3.6.10)$$
$$\sigma_{33}(x, y, H) = P(x, y) = \sum_{m=1}^{\infty}\sum_{n=1}^{\infty} P_{mn} \sin\left(\frac{m\pi}{L}x\right) \sin\left(\frac{n\pi}{L}y\right),$$

along the lateral faces $x = 0$ and $x = L$

$$\sigma_{11}(0, y, z) = \sigma_{11}(L, y, z) = 0,$$
$$u_y(0, y, z) = u_y(L, y, z) = 0, \qquad (3.6.11)$$
$$u_z(0, y, z) = u_z(L, y, z) = 0,$$

and along the lateral faces $y = 0$ and $y = L$

$$\sigma_{22}(x, 0, z) = \sigma_{22}(x, L, z) = 0,$$
$$u_x(x, 0, z) = u_x(x, L, z) = 0,$$
$$u_z(x, 0, z) = u_z(x, L, z) = 0. \tag{3.6.12}$$

The double Fourier series representations of the displacements

$$u_x(x, y, z) = -\sum_{m=1}^{\infty} \sum_{n=1}^{\infty} U_{mn}(z) \cos\left(\frac{m\pi}{L}x\right) \sin\left(\frac{n\pi}{L}y\right),$$

$$u_y(x, y, z) = -\sum_{m=1}^{\infty} \sum_{n=1}^{\infty} V_{mn}(z) \sin\left(\frac{m\pi}{L}x\right) \cos\left(\frac{n\pi}{L}y\right), \tag{3.6.13}$$

$$u_z(x, y, z) = \sum_{m=1}^{\infty} \sum_{n=1}^{\infty} W_{mn}(z) \sin\left(\frac{m\pi}{L}x\right) \sin\left(\frac{n\pi}{L}y\right)$$

satisfy the specified boundary conditions along the lateral faces of the domain. The displacements yield a double Fourier series representation of the volumetric strain

$$e = \frac{\partial u_x}{\partial x} + \frac{\partial u_y}{\partial y} + \frac{\partial u_z}{\partial z} \tag{3.6.14}$$

and the equations of mechanical equilibrium

$$\frac{\partial^2 u_x}{\partial x^2} + \frac{\partial^2 u_x}{\partial y^2} + \frac{\partial^2 u_x}{\partial z^2} + \frac{1}{1 - 2\nu}\frac{\partial e}{\partial x} = 0,$$

$$\frac{\partial^2 u_y}{\partial x^2} + \frac{\partial^2 u_y}{\partial y^2} + \frac{\partial^2 u_y}{\partial z^2} + \frac{1}{1 - 2\nu}\frac{\partial e}{\partial y} = 0, \tag{3.6.15}$$

$$\frac{\partial^2 u_z}{\partial x^2} + \frac{\partial^2 u_z}{\partial y^2} + \frac{\partial^2 u_z}{\partial z^2} + \frac{1}{1 - 2\nu}\frac{\partial e}{\partial z} = 0$$

give three ordinary differential equations for the Fourier coefficients $U_{mn}(z)$, $V_{mn}(z)$, and $W_{mn}(z)$. Introducing the notation

$$a_m = \frac{m\pi}{L} \qquad \text{for } m = 1, 2, \dots, \tag{3.6.16}$$

$$g_{mn} = \frac{\pi}{L}\sqrt{m^2 + n^2} \quad \text{for } m, n = 1, 2, \dots \tag{3.6.17}$$

these equations become

$$0 = (1 - 2\nu)\frac{\partial^2 U_{mn}}{\partial z^2} - [2a_m^2(1 - \nu) + a_n^2(1 - 2\nu)]U_{mn}$$

$$- a_m a_n V_{mn} - a_m \frac{\partial W_{mn}}{\partial z}, \tag{3.6.18}$$

$$0 = (1 - 2\nu) \frac{\partial^2 V_{mn}}{\partial z^2} - [2a_n^2(1 - \nu) + a_m^2(1 - 2\nu)]V_{mn}$$

$$- a_m a_n U_{mn} - a_n \frac{\partial W_{mn}}{\partial z}, \tag{3.6.19}$$

$$0 = a_m \frac{\partial U_{mn}}{\partial z} + a_n \frac{\partial V_{mn}}{\partial z} + 2(1 - \nu) \frac{\partial^2 W_{mn}}{\partial z^2} - (1 - 2\nu)g_{mn}^2 W_{mn}. \tag{3.6.20}$$

Then

$$0 = \frac{\partial^4}{\partial z^4}(a_m U_{mn} + a_n V_{mn}) - 2g_{mn}^2 \frac{\partial^2}{\partial z^2}(a_m U_{mn} + a_n V_{mn})$$

$$+ g_{mn}^4(a_m U_{mn} + a_n V_{mn}), \tag{3.6.21}$$

$$0 = \frac{\partial^2}{\partial z^2}(a_n U_{mn} - a_m V_{mn}) - g_{mn}^2(a_n U_{mn} - a_m V_{mn}) \tag{3.6.22}$$

and therefore,

$$a_m U_{mn} + a_n V_{mn} = \frac{1}{2Gg_{mn}^2}\Big[C1_{mn}\sinh(g_{mn}z)$$

$$+ C2_{mn}\cosh(g_{mn}z)$$

$$+ C3_{mn}z\cosh(g_{mn}z)$$

$$+ C4_{mn}z\sinh(g_{mn}z)\Big], \tag{3.6.23}$$

$$a_n U_{mn} - a_m V_{mn} = \frac{1}{2Gg_{mn}^2}\Big[C5_{mn}\sinh(g_{mn}z)$$

$$+ C6_{mn}\cosh(g_{mn}z)\Big], \tag{3.6.24}$$

where

$$G = \frac{E}{2(1 + \nu)} \tag{3.6.25}$$

is the shear modulus and the free constants $C1_{mn}$ to $C6_{mn}$ are still to be determined. For $m, n = 1, 2, \ldots$ the Fourier coefficients $U_{mn}(z)$, $V_{mn}(z)$, and $W_{mn}(z)$ become

$$U_{mn}(z) = \frac{1}{2G}\Big[(C1_{mn}a_m + C5_{mn}a_n)\sinh(g_{mn}z)$$

$$+ (C2_{mn}a_m + C6_{mn}a_n)\cosh(g_{mn}z)$$

$$+ C3_{mn}a_m z\cosh(g_{mn}z)$$

$$
+ C4_{mn}a_m z \sinh(g_{mn}z) \Big],
\tag{3.6.26}
$$

$$
\begin{aligned}
V_{mn}(z) = \frac{1}{2G}\Big[& (C1_{mn}a_n - C5_{mn}a_m)\sinh(g_{mn}z) \\
& + (C2_{mn}a_n - C6_{mn}a_m)\cosh(g_{mn}z) \\
& + C3_{mn}a_n z \cosh(g_{mn}z) \\
& + C4_{mn}a_n z \sinh(g_{mn}z) \Big],
\end{aligned}
\tag{3.6.27}
$$

$$
\begin{aligned}
W_{mn}(z) = \frac{1}{2G}\Big[& ((3-4\nu)C4_{mn} - C2_{mn}g_{mn})\sinh(g_{mn}z) \\
& + ((3-4\nu)C3_{mn} - C1_{mn}g_{mn})\cosh(g_{mn}z) \\
& - C4_{mn}g_{mn}z \cosh(g_{mn}z) \\
& - C3_{mn}g_{mn}z \sinh(g_{mn}z) \Big].
\end{aligned}
\tag{3.6.28}
$$

The double Fourier series representations of the displacements and hence, those of the strains and the stresses are thus established. The remaining constants $C1_{mn}$ to $C6_{mn}$ have to be determined from the series representations of the prescribed stresses along the top face and the bottom face of the plate. Woinowsky-Krieger (1933) gives details and intermediate results; the series representations of displacements and stresses finally become

$$
\begin{aligned}
u_x(x, y, z) = \sum_{m=1}^{\infty}\sum_{n=1}^{\infty} & \frac{P_{mn}}{2G}\cos(a_m x)\sin(a_n y) \\
\Big[& -\frac{-a_m g_{mn}H\sinh(g_{mn}H) + (1-2\nu)a_m\cosh(g_{mn}H)}{g_{mn}^2[\sinh(2g_{mn}H) - 2g_{mn}H]}\sinh(g_{mn}z) \\
& -\frac{a_m g_{mn}\cosh(g_{mn}H)}{g_{mn}^2[\sinh(2g_{mn}H) - 2g_{mn}H]}z\cosh(g_{mn}z) \\
& +\frac{a_m g_{mn}H\cosh(g_{mn}H) - (1-2\nu)a_m\sinh(g_{mn}H)}{g_{mn}^2[\sinh(2g_{mn}H) + 2g_{mn}H]}\cosh(g_{mn}z) \\
& -\frac{a_m g_{mn}\sinh(g_{mn}H)}{g_{mn}^2[\sinh(2g_{mn}H) + 2g_{mn}H]}z\sinh(g_{mn}z) \Big],
\end{aligned}
\tag{3.6.29}
$$

$$
\begin{aligned}
u_y(x, y, z) = \sum_{m=1}^{\infty}\sum_{n=1}^{\infty} & \frac{P_{mn}}{2G}\sin(a_m x)\cos(a_n y) \\
\Big[& -\frac{-a_n g_{mn}H\sinh(g_{mn}H) + (1-2\nu)a_n\cosh(g_{mn}H)}{g_{mn}^2[\sinh(2g_{mn}H) - 2g_{mn}H]}\sinh(g_{mn}z)
\end{aligned}
$$

$$-\frac{a_n g_{mn} \cosh(g_{mn} H)}{g_{mn}^2[\sinh(2g_{mn} H) - 2g_{mn} H]} z \cosh(g_{mn} z)$$

$$+\frac{a_n g_{mn} H \cosh(g_{mn} H) - (1 - 2\nu)a_n \sinh(g_{mn} H)}{g_{mn}^2[\sinh(2g_{mn} H) + 2g_{mn} H]} \cosh(g_{mn} z)$$

$$-\frac{a_n g_{mn} \sinh(g_{mn} H)}{g_{mn}^2[\sinh(2g_{mn} H) + 2g_{mn} H]} z \sinh(g_{mn} z) \Bigg], \qquad (3.6.30)$$

$$u_z(x, y, z) = \sum_{m=1}^{\infty} \sum_{n=1}^{\infty} \frac{P_{mn}}{2G} \sin(a_m x) \sin(a_n y)$$

$$\Bigg[\frac{g_{mn} H \sinh(g_{mn} H) + 2(1 - \nu) \cosh(g_{mn} H)}{g_{mn}[\sinh(2g_{mn} H) - 2g_{mn} H]} \cosh(g_{mn} z)$$

$$-\frac{g_{mn} \cosh(g_{mn} H)}{g_{mn}[\sinh(2g_{mn} H) - 2g_{mn} H]} z \sinh(g_{mn} z)$$

$$+\frac{g_{mn} H \cosh(g_{mn} H) + 2(1 - \nu) \sinh(g_{mn} H)}{g_{mn}[\sinh(2g_{mn} H) + 2g_{mn} H]} \sinh(g_{mn} z)$$

$$-\frac{g_{mn} \sinh(g_{mn} H)}{g_{mn}[\sinh(2g_{mn} H) + 2g_{mn} H]} z \cosh(g_{mn} z) \Bigg], \qquad (3.6.31)$$

$$\sigma_{11}(x, y, z) = \sum_{m=1}^{\infty} \sum_{n=1}^{\infty} P_{mn} \sin(a_m x) \sin(a_n y)$$

$$\Bigg[\frac{-a_m^2 g_{mn} H \sinh(g_{mn} H) + (a_m^2 + 2\nu a_n^2) \cosh(g_{mn} H)}{g_{mn}^2[\sinh(2g_{mn} H) - 2g_{mn} H]} \sinh(g_{mn} z)$$

$$+\frac{a_m^2 g_{mn} \cosh(g_{mn} H)}{g_{mn}^2[\sinh(2g_{mn} H) - 2g_{mn} H]} z \cosh(g_{mn} z)$$

$$-\frac{a_m^2 g_{mn} H \cosh(g_{mn} H) - (a_m^2 + 2\nu a_n^2) \sinh(g_{mn} H)}{g_{mn}^2[\sinh(2g_{mn} H) + 2g_{mn} H]} \cosh(g_{mn} z)$$

$$+\frac{a_m^2 g_{mn} \sinh(g_{mn} H)}{g_{mn}^2[\sinh(2g_{mn} H) + 2g_{mn} H]} z \sinh(g_{mn} z) \Bigg], \qquad (3.6.32)$$

$$\sigma_{22}(x, y, z) = \sum_{m=1}^{\infty} \sum_{n=1}^{\infty} P_{mn} \sin(a_m x) \sin(a_n y)$$

$$\Bigg[\frac{-a_n^2 g_{mn} H \sinh(g_{mn} H) + (a_n^2 + 2\nu a_m^2) \cosh(g_{mn} H)}{g_{mn}^2[\sinh(2g_{mn} H) - 2g_{mn} H]} \sinh(g_{mn} z)$$

$$+\frac{a_n^2 g_{mn} \cosh(g_{mn} H)}{g_{mn}^2[\sinh(2g_{mn} H) - 2g_{mn} H]} z \cosh(g_{mn} z)$$

$$-\frac{a_n^2 g_{mn} H \cosh(g_{mn} H) - (a_n^2 + 2\nu a_m^2) \sinh(g_{mn} H)}{g_{mn}^2[\sinh(2g_{mn} H) + 2g_{mn} H]} \cosh(g_{mn} z)$$

$$+\frac{a_n^2 g_{mn} \sinh(g_{mn} H)}{g_{mn}^2[\sinh(2g_{mn} H) + 2g_{mn} H]} z \sinh(g_{mn} z) \Bigg], \qquad (3.6.33)$$

$$\sigma_{33}(x, y, z) = \sum_{m=1}^{\infty} \sum_{n=1}^{\infty} P_{mn} \sin(a_m x) \sin(a_n y)$$

$$\left[\frac{g_{mn} H \sinh(g_{mn} H) + \cosh(g_{mn} H)}{\sinh(2g_{mn} H) - 2g_{mn} H} \sinh(g_{mn} z) \right.$$

$$- \frac{g_{mn} \cosh(g_{mn} H)}{\sinh(2g_{mn} H) - 2g_{mn} H} z \cosh(g_{mn} z)$$

$$+ \frac{g_{mn} H \cosh(g_{mn} H) + \sinh(g_{mn} H)}{\sinh(2g_{mn} H) + 2g_{mn} H} \cosh(g_{mn} z)$$

$$\left. - \frac{g_{mn} \sinh(g_{mn} H)}{\sinh(2g_{mn} H) + 2g_{mn} H} z \sinh(g_{mn} z) \right], \qquad (3.6.34)$$

$$\sigma_{12}(x, y, z) = \sum_{m=1}^{\infty} \sum_{n=1}^{\infty} P_{mn} \cos(a_m x) \cos(a_n y)$$

$$\left[\frac{a_m a_n}{g_{mn}^2} \frac{g_{mn} H \sinh(g_{mn} H) - (1 - 2\nu) \cosh(g_{mn} H)}{\sinh(2g_{mn} H) - 2g_{mn} H} \sinh(g_{mn} z) \right.$$

$$- \frac{a_m a_n}{g_{mn}^2} \frac{g_{mn} \cosh(g_{mn} H)}{\sinh(2g_{mn} H) - 2g_{mn} H} z \cosh(g_{mn} z)$$

$$+ \frac{a_m a_n}{g_{mn}^2} \frac{g_{mn} H \cosh(g_{mn} H) - (1 - 2\nu) \sinh(g_{mn} H)}{\sinh(2g_{mn} H) + 2g_{mn} H} \cosh(g_{mn} z)$$

$$\left. - \frac{a_m a_n}{g_{mn}^2} \frac{g_{mn} \sinh(g_{mn} H)}{\sinh(2g_{mn} H) + 2g_{mn} H} z \sinh(g_{mn} z) \right], \qquad (3.6.35)$$

$$\sigma_{13}(x, y, z) = \sum_{m=1}^{\infty} \sum_{n=1}^{\infty} P_{mn} \cos(a_m x) \sin(a_n y)$$

$$\left[\frac{a_m H \sinh(g_{mn} H)}{\sinh(2g_{mn} H) - 2g_{mn} H} \cosh(g_{mn} z) \right.$$

$$- \frac{a_m \cosh(g_{mn} H)}{\sinh(2g_{mn} H) - 2g_{mn} H} z \sinh(g_{mn} z)$$

$$+ \frac{a_m H \cosh(g_{mn} H)}{\sinh(2g_{mn} H) + 2g_{mn} H} \sinh(g_{mn} z)$$

$$\left. - \frac{a_m \sinh(g_{mn} H)}{\sinh(2g_{mn} H) + 2g_{mn} H} z \cosh(g_{mn} z) \right], \qquad (3.6.36)$$

$$\sigma_{23}(x, y, z) = \sum_{m=1}^{\infty} \sum_{n=1}^{\infty} P_{mn} \sin(a_m x) \cos(a_n y)$$

$$\left[\frac{a_n H \sinh(g_{mn} H)}{\sinh(2g_{mn} H) - 2g_{mn} H} \cosh(g_{mn} z) \right.$$

$$- \frac{a_n \cosh(g_{mn} H)}{\sinh(2g_{mn} H) - 2g_{mn} H} z \sinh(g_{mn} z)$$

Table 3.6 Benchmark deposit (https://docs.opengeosys.org/books/bmb-5)

BM code	Author	Code	Files	CTest
BMB5-3.6	Peter Vogel	OGS-5	Available	ToDo

(https://oc.ufz.de/index.php/s/nlph7bhfDkj6tC7)

$$\begin{aligned}
&+ \frac{a_n H \cosh(g_{mn} H)}{\sinh(2g_{mn} H) + 2g_{mn} H} \sinh(g_{mn} z) \\
&- \frac{a_n \sinh(g_{mn} H)}{\sinh(2g_{mn} H) + 2g_{mn} H} z \cosh(g_{mn} z) \Big].
\end{aligned} \qquad (3.6.37)$$

Hooke's law yields the strains once that the stresses are known.

$$\begin{aligned}
\epsilon_{11}(x, y, z) &= \frac{1}{E}[\sigma_{11}(x, y, z) - \nu(\sigma_{22}(x, y, z) + \sigma_{33}(x, y, z))], \\
\epsilon_{22}(x, y, z) &= \frac{1}{E}[\sigma_{22}(x, y, z) - \nu(\sigma_{11}(x, y, z) + \sigma_{33}(x, y, z))], \\
\epsilon_{33}(x, y, z) &= \frac{1}{E}[\sigma_{33}(x, y, z) - \nu(\sigma_{11}(x, y, z) + \sigma_{22}(x, y, z))], \\
\epsilon_{12}(x, y, z) &= \frac{1}{G}\sigma_{12}(x, y, z), \\
\epsilon_{13}(x, y, z) &= \frac{1}{G}\sigma_{13}(x, y, z), \\
\epsilon_{23}(x, y, z) &= \frac{1}{G}\sigma_{23}(x, y, z).
\end{aligned} \qquad (3.6.38)$$

The entire solution is thus established by double Fourier series representations. For the numerical evaluation of trigonometric series see Goertzel (1958).

Deformations Due to Gravity

Peter Vogel and Jobst Maßmann

Based on ideas outlined by Sokolnikoff (1956) we present problems on elastic bodies deforming by their own weight. We focus on the closed form solutions. The associated simulation exercises have been checked by OGS; they may serve as verification tests. For the underlying theory of linear elasticity see Gurtin (1972) or see above.

3.7   A Thick Plate, Bottom Loads

Given length $2L = 20$ m and thickness $H = 2$ m the domain represents the rectangular plate $[-L, L] \times [-L, L] \times [-H, 0]$. It is discretized by $20 \times 20 \times 4$ equally sized hexahedral elements. The material has been selected elastic with Young's modulus $E = 8{,}100$ MPa, Poisson's ratio $\nu = 0.25$, and density $\rho = 3058.104$ kg/m^3.

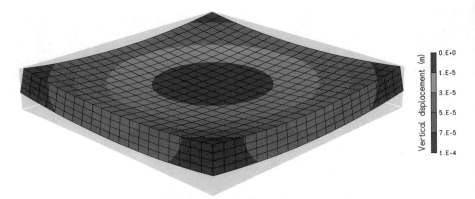

Fig. 3.8 Deformations scaled up, vertical displacements

Gravitational acceleration $g = 9.81$ m/s^2 is applied in negative z-direction by default. The top and the lateral faces of the domain are free of load, the vertical load $\rho g H$ is applied at the bottom. Fixities have been prescribed in the interior of the domain with zero x-displacement along the plane $x = 0$, zero y-displacement along the plane $y = 0$, and zero z-displacement at the origin. The simulation (Table 3.7) comprises one timestep to establish the stresses, the strains, and the displacement vector (u_x, u_y, u_z) (Fig. 3.8).

The stress tensor

$$\sigma = \begin{pmatrix} \sigma_{11} & \sigma_{12} & \sigma_{13} \\ \sigma_{12} & \sigma_{22} & \sigma_{23} \\ \sigma_{13} & \sigma_{23} & \sigma_{33} \end{pmatrix} = \rho g \begin{pmatrix} 0 & 0 & 0 \\ 0 & 0 & 0 \\ 0 & 0 & z \end{pmatrix} \tag{3.7.1}$$

satisfies the equation of mechanical equilibrium

$$\text{div } \sigma = (0, 0, \rho g) \tag{3.7.2}$$

as well as the specified surface loads because along the bottom face $z = -H$

$$\sigma \begin{pmatrix} 0 \\ 0 \\ -1 \end{pmatrix} = \rho g \begin{pmatrix} 0 & 0 & 0 \\ 0 & 0 & 0 \\ 0 & 0 & -H \end{pmatrix} \begin{pmatrix} 0 \\ 0 \\ -1 \end{pmatrix} = \rho g H \begin{pmatrix} 0 \\ 0 \\ 1 \end{pmatrix}, \tag{3.7.3}$$

along the top face $z = 0$

$$\sigma \begin{pmatrix} 0 \\ 0 \\ 1 \end{pmatrix} = \rho g \begin{pmatrix} 0 & 0 & 0 \\ 0 & 0 & 0 \\ 0 & 0 & 0 \end{pmatrix} \begin{pmatrix} 0 \\ 0 \\ 1 \end{pmatrix} = \begin{pmatrix} 0 \\ 0 \\ 0 \end{pmatrix}, \tag{3.7.4}$$

and similarly for the lateral faces. Hooke's law reads for the strains

$$\epsilon_{11} = \frac{\partial u_x}{\partial x} = \frac{1}{E}[\sigma_{11} - \nu(\sigma_{22} + \sigma_{33})] = -\frac{\nu}{E}\rho g z, \tag{3.7.5}$$

$$\epsilon_{22} = \frac{\partial u_y}{\partial y} = \frac{1}{E}[\sigma_{22} - \nu(\sigma_{11} + \sigma_{33})] = -\frac{\nu}{E}\rho g z, \tag{3.7.6}$$

$$\epsilon_{33} = \frac{\partial u_z}{\partial z} = \frac{1}{E}[\sigma_{33} - \nu(\sigma_{11} + \sigma_{22})] = \frac{1}{E}\rho g z, \tag{3.7.7}$$

$$\epsilon_{12} = \frac{\partial u_x}{\partial y} + \frac{\partial u_y}{\partial x} = \frac{2(1+\nu)}{E}\sigma_{12} = 0, \tag{3.7.8}$$

$$\epsilon_{13} = \frac{\partial u_x}{\partial z} + \frac{\partial u_z}{\partial x} = \frac{2(1+\nu)}{E}\sigma_{13} = 0, \tag{3.7.9}$$

$$\epsilon_{23} = \frac{\partial u_y}{\partial z} + \frac{\partial u_z}{\partial y} = \frac{2(1+\nu)}{E}\sigma_{23} = 0, \tag{3.7.10}$$

which constitute a set of equations for the partial derivatives of the displacements. Integrating the strains ϵ_{11}, ϵ_{22}, and ϵ_{33} with respect to the prescribed x- and y-fixities gives the displacement vector (u_x, u_y, u_z)

$$u_x(x, z) = -\frac{\nu}{E}\rho g z x, \tag{3.7.11}$$

$$u_y(y, z) = -\frac{\nu}{E}\rho g z y, \tag{3.7.12}$$

$$u_z(x, y, z) = \frac{1}{E}\rho g \frac{1}{2}z^2 + h(x, y), \tag{3.7.13}$$

where $h(x, y)$ has to be determined from the shear strains and the specified z-fixity. Then

$$\epsilon_{12} = \frac{\partial u_x}{\partial y} + \frac{\partial u_y}{\partial x} = 0, \tag{3.7.14}$$

$$\epsilon_{13} = \frac{\partial u_x}{\partial z} + \frac{\partial u_z}{\partial x} = -\frac{\nu}{E}\rho g x + \frac{\partial h}{\partial x} = 0, \tag{3.7.15}$$

$$\epsilon_{23} = \frac{\partial u_y}{\partial z} + \frac{\partial u_z}{\partial y} = -\frac{\nu}{E}\rho g y + \frac{\partial h}{\partial y} = 0, \tag{3.7.16}$$

hence,

$$\frac{\partial h}{\partial x} = \frac{\nu}{E}\rho g x, \tag{3.7.17}$$

$$\frac{\partial h}{\partial y} = \frac{\nu}{E}\rho g y, \tag{3.7.18}$$

and therefore,

$$h(x, y) = \frac{\nu}{2E}\rho g(x^2 + y^2) + C. \tag{3.7.19}$$

Table 3.7 Benchmark deposit (https://docs.opengeosys.org/books/bmb-5)

BM code	Author	Code	Files	CTest
BMB5-3.7	Peter Vogel	OGS-5	Available	ToDo

(https://oc.ufz.de/index.php/s/nlph7bhfDkj6tC7)

The z-fixity at the origin yields the free constant C. The displacement vector (u_x, u_y, u_z) becomes

$$u_x(x, z) = -\frac{\nu}{E}\rho gxz, \tag{3.7.20}$$

$$u_y(y, z) = -\frac{\nu}{E}\rho gyz, \tag{3.7.21}$$

$$u_z(x, y, z) = \frac{1}{2E}\rho g[z^2 + \nu(x^2 + y^2)]. \tag{3.7.22}$$

3.8 A Cuboid, Top Loads, Bottom Loads

The domain is a cuboid discretized by $10 \times 10 \times 10$ equally sized hexahedral elements. Given size $H = 10$ m the vertices of the front face $y = -H$ have the x-y-z-coordinates $(0, -H, H), (H, -H, 0), (0, -H, -H)$, and $(-H, -H, 0)$, respectively; the rear face is located on the $y = H$ level. The material has been selected elastic with Young's modulus $E = 5,000$ MPa, Poisson's ratio $\nu = 0.25$, and density $\rho = 2038.736$ kg/m^3. Gravitational acceleration $g = 9.81$ m/s^2 is applied in negative z-direction by default. Fixities have been prescribed in the interior of the domain with zero x-displacement along the plane $x = 0$, zero y-displacement along the plane $y = 0$, and zero z-displacement at the origin. Specified loads $P(x, y, z)$ prevail along the entire surface of the domain. These loads read along the faces $y = H$ and $y = -H$

$$P(x, H, z) = \begin{pmatrix} 0 \\ 0 \\ 0 \end{pmatrix}, \quad P(x, -H, z) = \begin{pmatrix} 0 \\ 0 \\ 0 \end{pmatrix}, \tag{3.8.1}$$

along the top face $x \geq 0, z \geq 0$

$$P(H - z, y, z) = \frac{1}{\sqrt{2}}\rho g \begin{pmatrix} 0 \\ 0 \\ z \end{pmatrix}, \tag{3.8.2}$$

along the top face $x \leq 0, z \geq 0$

$$P(-H + z, y, z) = \frac{1}{\sqrt{2}} \rho g \begin{pmatrix} 0 \\ 0 \\ z \end{pmatrix}, \tag{3.8.3}$$

along the bottom face $x \geq 0, z \leq 0$

$$P(H + z, y, z) = -\frac{1}{\sqrt{2}} \rho g \begin{pmatrix} 0 \\ 0 \\ z \end{pmatrix}, \tag{3.8.4}$$

and along the bottom face $x \leq 0, z \leq 0$

$$P(-H - z, y, z) = -\frac{1}{\sqrt{2}} \rho g \begin{pmatrix} 0 \\ 0 \\ z \end{pmatrix}. \tag{3.8.5}$$

The simulation (Table 3.8) comprises one timestep to evaluate the stresses, the strains, and the displacement vector (u_x, u_y, u_z) (Fig. 3.9).

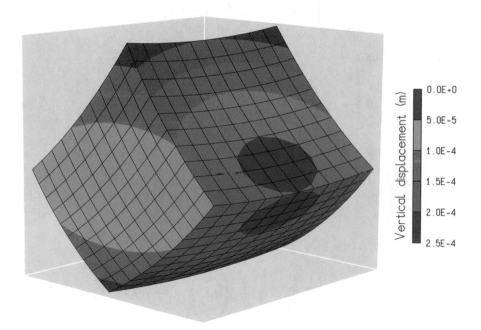

Fig. 3.9 Deformations scaled up, vertical displacements

The stress tensor

$$\sigma = \begin{pmatrix} \sigma_{11} & \sigma_{12} & \sigma_{13} \\ \sigma_{12} & \sigma_{22} & \sigma_{23} \\ \sigma_{13} & \sigma_{23} & \sigma_{33} \end{pmatrix} = \rho g \begin{pmatrix} 0 & 0 & 0 \\ 0 & 0 & 0 \\ 0 & 0 & z \end{pmatrix} \tag{3.8.6}$$

satisfies the equation of mechanical equilibrium

$$\text{div } \sigma = (0, 0, \rho g) \tag{3.8.7}$$

as well as the specified surface loads because along the face $y = H$

$$\sigma \begin{pmatrix} 0 \\ 1 \\ 0 \end{pmatrix} = \rho g \begin{pmatrix} 0 & 0 & 0 \\ 0 & 0 & 0 \\ 0 & 0 & z \end{pmatrix} \begin{pmatrix} 0 \\ 1 \\ 0 \end{pmatrix} = \begin{pmatrix} 0 \\ 0 \\ 0 \end{pmatrix}, \tag{3.8.8}$$

along the face $y = -H$

$$\sigma \begin{pmatrix} 0 \\ -1 \\ 0 \end{pmatrix} = \rho g \begin{pmatrix} 0 & 0 & 0 \\ 0 & 0 & 0 \\ 0 & 0 & z \end{pmatrix} \begin{pmatrix} 0 \\ -1 \\ 0 \end{pmatrix} = \begin{pmatrix} 0 \\ 0 \\ 0 \end{pmatrix}, \tag{3.8.9}$$

along the top face $x \geq 0, z \geq 0$

$$\sigma \frac{1}{\sqrt{2}} \begin{pmatrix} 1 \\ 0 \\ 1 \end{pmatrix} = \rho g \begin{pmatrix} 0 & 0 & 0 \\ 0 & 0 & 0 \\ 0 & 0 & z \end{pmatrix} \frac{1}{\sqrt{2}} \begin{pmatrix} 1 \\ 0 \\ 1 \end{pmatrix} = \frac{1}{\sqrt{2}} \rho g \begin{pmatrix} 0 \\ 0 \\ z \end{pmatrix}, \tag{3.8.10}$$

along the top face $x \leq 0, z \geq 0$

$$\sigma \frac{1}{\sqrt{2}} \begin{pmatrix} -1 \\ 0 \\ 1 \end{pmatrix} = \rho g \begin{pmatrix} 0 & 0 & 0 \\ 0 & 0 & 0 \\ 0 & 0 & z \end{pmatrix} \frac{1}{\sqrt{2}} \begin{pmatrix} -1 \\ 0 \\ 1 \end{pmatrix} = \frac{1}{\sqrt{2}} \rho g \begin{pmatrix} 0 \\ 0 \\ z \end{pmatrix}, \tag{3.8.11}$$

along the bottom face $x \geq 0, z \leq 0$

$$\sigma \frac{1}{\sqrt{2}} \begin{pmatrix} 1 \\ 0 \\ -1 \end{pmatrix} = \rho g \begin{pmatrix} 0 & 0 & 0 \\ 0 & 0 & 0 \\ 0 & 0 & z \end{pmatrix} \frac{1}{\sqrt{2}} \begin{pmatrix} 1 \\ 0 \\ -1 \end{pmatrix} = -\frac{1}{\sqrt{2}} \rho g \begin{pmatrix} 0 \\ 0 \\ z \end{pmatrix}, \tag{3.8.12}$$

and along the bottom face $x \leq 0, z \leq 0$

$$\sigma \frac{1}{\sqrt{2}} \begin{pmatrix} -1 \\ 0 \\ -1 \end{pmatrix} = \rho g \begin{pmatrix} 0 & 0 & 0 \\ 0 & 0 & 0 \\ 0 & 0 & z \end{pmatrix} \frac{1}{\sqrt{2}} \begin{pmatrix} -1 \\ 0 \\ -1 \end{pmatrix} = -\frac{1}{\sqrt{2}} \rho g \begin{pmatrix} 0 \\ 0 \\ z \end{pmatrix}. \tag{3.8.13}$$

Hooke's law reads for the strains

$$\epsilon_{11} = \frac{\partial u_x}{\partial x} = \frac{1}{E}[\sigma_{11} - \nu(\sigma_{22} + \sigma_{33})] = -\frac{\nu}{E}\rho gz, \tag{3.8.14}$$

$$\epsilon_{22} = \frac{\partial u_y}{\partial y} = \frac{1}{E}[\sigma_{22} - \nu(\sigma_{11} + \sigma_{33})] = -\frac{\nu}{E}\rho gz, \tag{3.8.15}$$

$$\epsilon_{33} = \frac{\partial u_z}{\partial z} = \frac{1}{E}[\sigma_{33} - \nu(\sigma_{11} + \sigma_{22})] = \frac{1}{E}\rho gz, \tag{3.8.16}$$

$$\epsilon_{12} = \frac{\partial u_x}{\partial y} + \frac{\partial u_y}{\partial x} = \frac{2(1+\nu)}{E}\sigma_{12} = 0, \tag{3.8.17}$$

$$\epsilon_{13} = \frac{\partial u_x}{\partial z} + \frac{\partial u_z}{\partial x} = \frac{2(1+\nu)}{E}\sigma_{13} = 0, \tag{3.8.18}$$

$$\epsilon_{23} = \frac{\partial u_y}{\partial z} + \frac{\partial u_z}{\partial y} = \frac{2(1+\nu)}{E}\sigma_{23} = 0, \tag{3.8.19}$$

which constitute a set of equations for the partial derivatives of the displacements. Integrating the strains ϵ_{11}, ϵ_{22}, and ϵ_{33} with respect to the prescribed x- and y-fixities gives the displacement vector (u_x, u_y, u_z)

$$u_x(x, z) = -\frac{\nu}{E}\rho gzx, \tag{3.8.20}$$

$$u_y(y, z) = -\frac{\nu}{E}\rho gzy, \tag{3.8.21}$$

$$u_z(x, y, z) = \frac{1}{E}\rho g\frac{1}{2}z^2 + h(x, y), \tag{3.8.22}$$

where $h(x, y)$ has to be determined from the shear strains and the specified z-fixity. Then

$$\epsilon_{12} = \frac{\partial u_x}{\partial y} + \frac{\partial u_y}{\partial x} = 0, \tag{3.8.23}$$

$$\epsilon_{13} = \frac{\partial u_x}{\partial z} + \frac{\partial u_z}{\partial x} = -\frac{\nu}{E}\rho gx + \frac{\partial h}{\partial x} = 0, \tag{3.8.24}$$

$$\epsilon_{23} = \frac{\partial u_y}{\partial z} + \frac{\partial u_z}{\partial y} = -\frac{\nu}{E}\rho gy + \frac{\partial h}{\partial y} = 0, \tag{3.8.25}$$

hence,

$$\frac{\partial h}{\partial x} = \frac{\nu}{E}\rho gx, \tag{3.8.26}$$

$$\frac{\partial h}{\partial y} = \frac{\nu}{E}\rho gy, \tag{3.8.27}$$

and therefore,

Table 3.8 Benchmark deposit (https://docs.opengeosys.org/books/bmb-5)

BM code	Author	Code	Files	CTest
BMB5-3.8	Peter Vogel	OGS-5	Available	ToDo

(https://oc.ufz.de/index.php/s/nlph7bhfDkj6tC7)

$$h(x, y) = \frac{\nu}{2E}\rho g(x^2 + y^2) + C. \tag{3.8.28}$$

The z-fixity at the origin yields the free constant C. The displacement vector (u_x, u_y, u_z) becomes

$$u_x(x, z) = -\frac{\nu}{E}\rho g x z, \tag{3.8.29}$$

$$u_y(y, z) = -\frac{\nu}{E}\rho g y z, \tag{3.8.30}$$

$$u_z(x, y, z) = \frac{1}{2E}\rho g[z^2 + \nu(x^2 + y^2)]. \tag{3.8.31}$$

3.9 A Cuboid, Bottom Loads

The domain is a cuboid discretized by $10 \times 10 \times 10$ equally sized hexahedral elements. Given size $H = 10$ m the vertices of the front face $y = -H$ have the x-y-z-coordinates $(0, -H, H)$, $(H, -H, 0)$, $(0, -H, -H)$, and $(-H, -H, 0)$, respectively; the rear face is located on the $y = H$ level. The material has been selected elastic with Young's modulus $E = 5{,}000$ MPa, Poisson's ratio $\nu = 0.25$, and density $\rho = 2038.736$ kg/m^3. Gravitational acceleration $g = 9.81$ m/s^2 is applied in negative z-direction by default. Fixities have been prescribed in the interior of the domain with zero x-displacement along the plane $x = 0$, zero y-displacement along the plane $y = 0$, and zero z-displacement at the origin. The top and the lateral faces are free of load by default. Specified loads $P(x, y, z)$ prevail along the bottom of the domain. These loads read along the bottom face $x \leq 0, z \leq 0$

$$P(x, y, -H - x) = \frac{1}{\sqrt{2}}\rho g H \begin{pmatrix} 1 \\ 0 \\ 1 \end{pmatrix} \tag{3.9.1}$$

and along the bottom face $x \geq 0, z \leq 0$

$$P(x, y, -H + x) = \frac{1}{\sqrt{2}}\rho g H \begin{pmatrix} -1 \\ 0 \\ 1 \end{pmatrix}. \tag{3.9.2}$$

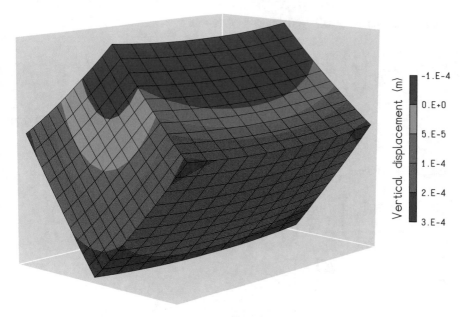

Fig. 3.10 Deformations scaled up, vertical displacements

The simulation (Table 3.9) comprises one timestep to evaluate the stresses, the strains, and the displacement vector (u_x, u_y, u_z) (Fig. 3.10).

The stress tensor

$$\boldsymbol{\sigma} = \begin{pmatrix} \sigma_{11} & \sigma_{12} & \sigma_{13} \\ \sigma_{12} & \sigma_{22} & \sigma_{23} \\ \sigma_{13} & \sigma_{23} & \sigma_{33} \end{pmatrix} = \frac{\rho g}{2} \begin{pmatrix} z - H & 0 & x \\ 0 & 0 & 0 \\ x & 0 & z - H \end{pmatrix} \tag{3.9.3}$$

satisfies the equation of mechanical equilibrium

$$\operatorname{div} \boldsymbol{\sigma} = (0, 0, \rho g) \tag{3.9.4}$$

as well as the specified surface loads because along the face $y = H$

$$\boldsymbol{\sigma} \begin{pmatrix} 0 \\ 1 \\ 0 \end{pmatrix} = \frac{\rho g}{2} \begin{pmatrix} z - H & 0 & x \\ 0 & 0 & 0 \\ x & 0 & z - H \end{pmatrix} \begin{pmatrix} 0 \\ 1 \\ 0 \end{pmatrix} = \begin{pmatrix} 0 \\ 0 \\ 0 \end{pmatrix}, \tag{3.9.5}$$

along the face $y = -H$

$$\boldsymbol{\sigma} \begin{pmatrix} 0 \\ -1 \\ 0 \end{pmatrix} = \frac{\rho g}{2} \begin{pmatrix} z - H & 0 & x \\ 0 & 0 & 0 \\ x & 0 & z - H \end{pmatrix} \begin{pmatrix} 0 \\ -1 \\ 0 \end{pmatrix} = \begin{pmatrix} 0 \\ 0 \\ 0 \end{pmatrix}, \tag{3.9.6}$$

along the top face $x \geq 0$, $z \geq 0$ where $z = H - x$

$$\sigma \frac{1}{\sqrt{2}} \begin{pmatrix} 1 \\ 0 \\ 1 \end{pmatrix} = \frac{\rho g}{2} \begin{pmatrix} H - x - H & 0 & x \\ 0 & 0 & 0 \\ x & 0 & H - x - H \end{pmatrix} \frac{1}{\sqrt{2}} \begin{pmatrix} 1 \\ 0 \\ 1 \end{pmatrix}$$
$$= \begin{pmatrix} 0 \\ 0 \\ 0 \end{pmatrix}, \qquad (3.9.7)$$

along the top face $x \leq 0$, $z \geq 0$ where $z = H + x$

$$\sigma \frac{1}{\sqrt{2}} \begin{pmatrix} -1 \\ 0 \\ 1 \end{pmatrix} = \frac{\rho g}{2} \begin{pmatrix} H + x - H & 0 & x \\ 0 & 0 & 0 \\ x & 0 & H + x - H \end{pmatrix} \frac{1}{\sqrt{2}} \begin{pmatrix} -1 \\ 0 \\ 1 \end{pmatrix}$$
$$= \begin{pmatrix} 0 \\ 0 \\ 0 \end{pmatrix}, \qquad (3.9.8)$$

along the bottom face $x \leq 0$, $z \leq 0$ where $z = -H - x$

$$\sigma \frac{1}{\sqrt{2}} \begin{pmatrix} -1 \\ 0 \\ -1 \end{pmatrix} = \frac{\rho g}{2} \begin{pmatrix} -H - x - H & 0 & x \\ 0 & 0 & 0 \\ x & 0 & -H - x - H \end{pmatrix} \frac{1}{\sqrt{2}} \begin{pmatrix} -1 \\ 0 \\ -1 \end{pmatrix}$$
$$= \frac{1}{\sqrt{2}} \rho g H \begin{pmatrix} 1 \\ 0 \\ 1 \end{pmatrix}, \qquad (3.9.9)$$

and along the bottom face $x \geq 0$, $z \leq 0$ where $z = -H + x$

$$\sigma \frac{1}{\sqrt{2}} \begin{pmatrix} 1 \\ 0 \\ -1 \end{pmatrix} = \frac{\rho g}{2} \begin{pmatrix} -H + x - H & 0 & x \\ 0 & 0 & 0 \\ x & 0 & -H + x - H \end{pmatrix} \frac{1}{\sqrt{2}} \begin{pmatrix} 1 \\ 0 \\ -1 \end{pmatrix}$$
$$= \frac{1}{\sqrt{2}} \rho g H \begin{pmatrix} -1 \\ 0 \\ 1 \end{pmatrix}. \qquad (3.9.10)$$

Hooke's law reads for the strains

$$\epsilon_{11} = \frac{\partial u_x}{\partial x} = \frac{1}{E} [\sigma_{11} - \nu(\sigma_{22} + \sigma_{33})] = \frac{1 - \nu}{2E} \rho g(z - H), \qquad (3.9.11)$$

$$\epsilon_{22} = \frac{\partial u_y}{\partial y} = \frac{1}{E} [\sigma_{22} - \nu(\sigma_{11} + \sigma_{33})] = -\frac{\nu}{E} \rho g(z - H), \qquad (3.9.12)$$

$$\epsilon_{33} = \frac{\partial u_z}{\partial z} = \frac{1}{E} [\sigma_{33} - \nu(\sigma_{11} + \sigma_{22})] = \frac{1 - \nu}{2E} \rho g(z - H), \qquad (3.9.13)$$

$$\epsilon_{12} = \frac{\partial u_x}{\partial y} + \frac{\partial u_y}{\partial x} = \frac{2(1+\nu)}{E}\sigma_{12} = 0, \tag{3.9.14}$$

$$\epsilon_{13} = \frac{\partial u_x}{\partial z} + \frac{\partial u_z}{\partial x} = \frac{2(1+\nu)}{E}\sigma_{13} = \frac{1+\nu}{E}\rho g x, \tag{3.9.15}$$

$$\epsilon_{23} = \frac{\partial u_y}{\partial z} + \frac{\partial u_z}{\partial y} = \frac{2(1+\nu)}{E}\sigma_{23} = 0, \tag{3.9.16}$$

which constitute a set of equations for the partial derivatives of the displacements. Integrating the strains ϵ_{11}, ϵ_{22}, and ϵ_{33} with respect to the prescribed x- and y-fixities gives the displacement vector (u_x, u_y, u_z)

$$u_x(x, z) = \frac{1-\nu}{2E}\rho g(z - H)x, \tag{3.9.17}$$

$$u_y(y, z) = -\frac{\nu}{E}\rho g(z - H)y, \tag{3.9.18}$$

$$u_z(x, y, z) = \frac{1-\nu}{2E}\rho g(\tfrac{1}{2}z^2 - Hz) + h(x, y), \tag{3.9.19}$$

where $h(x, y)$ has to be determined from the shear strains and the specified z-fixity. Then

$$\epsilon_{12} = \frac{\partial u_x}{\partial y} + \frac{\partial u_y}{\partial x} = 0, \tag{3.9.20}$$

$$\epsilon_{13} = \frac{\partial u_x}{\partial z} + \frac{\partial u_z}{\partial x} = \frac{1-\nu}{2E}\rho g x + \frac{\partial h}{\partial x} = \frac{1+\nu}{E}\rho g x, \tag{3.9.21}$$

$$\epsilon_{23} = \frac{\partial u_y}{\partial z} + \frac{\partial u_z}{\partial y} = -\frac{\nu}{E}\rho g y + \frac{\partial h}{\partial y} = 0, \tag{3.9.22}$$

hence,

$$\frac{\partial h}{\partial x} = \frac{1+3\nu}{2E}\rho g x, \tag{3.9.23}$$

$$\frac{\partial h}{\partial y} = \frac{\nu}{E}\rho g y, \tag{3.9.24}$$

and therefore,

$$h(x, y) = \frac{\nu}{2E}\rho g y^2 + \frac{1+3\nu}{4E}\rho g x^2 + C. \tag{3.9.25}$$

The z-fixity at the origin yields the free constant C. The displacement vector (u_x, u_y, u_z) becomes

Table 3.9 Benchmark deposit (https://docs.opengeosys.org/books/bmb-5)

BM code	Author	Code	Files	CTest
BMB5-3.9	Peter Vogel	OGS-5	Available	ToDo

(https://oc.ufz.de/index.php/s/nlph7bhfDkj6tC7)

$$u_x(x, z) = \frac{1 - \nu}{2E} \rho g x (z - H), \tag{3.9.26}$$

$$u_y(y, z) = -\frac{\nu}{E} \rho g y (z - H), \tag{3.9.27}$$

$$u_z(x, y, z) = \frac{1 - \nu}{2E} \rho g (\frac{z^2}{2} - Hz) + \frac{\nu}{2E} \rho g y^2 + \frac{1 + 3\nu}{4E} \rho g x^2. \tag{3.9.28}$$

3.10 A Cuboid, Top Loads

The domain is a cuboid discretized by $10 \times 10 \times 10$ equally sized hexahedral elements. Given size $H = 10$ m the vertices of the front face $y = -H$ have the x-y-z-coordinates $(0, -H, H)$, $(H, -H, 0)$, $(0, -H, -H)$, and $(-H, -H, 0)$, respectively; the rear face is located on the $y = H$ level. The material has been selected elastic with Young's modulus $E = 5,000$ MPa, Poisson's ratio $\nu = 0.25$, and density $\rho = 2038.736$ kg/m^3. Gravitational acceleration $g = 9.81$ m/s^2 is applied in negative z-direction by default. Fixities have been prescribed in the interior of the domain with zero x-displacement along the plane $x = 0$, zero y-displacement along the plane $y = 0$, and zero z-displacement at the origin. The bottom and the lateral faces are free of load by default. Specified loads $P(x, y, z)$ prevail along the top of the domain. These loads read along the top face $x \geq 0$, $z \geq 0$

$$P(x, y, H - x) = \frac{1}{\sqrt{2}} \rho g H \begin{pmatrix} 1 \\ 0 \\ 1 \end{pmatrix}, \tag{3.10.1}$$

and along the top face $x \leq 0$, $z \geq 0$

$$P(x, y, H + x) = \frac{1}{\sqrt{2}} \rho g H \begin{pmatrix} -1 \\ 0 \\ 1 \end{pmatrix}. \tag{3.10.2}$$

The simulation (Table 3.10) comprises one timestep to evaluate the stresses, the strains, and the displacement vector (u_x, u_y, u_z) (Fig. 3.11).

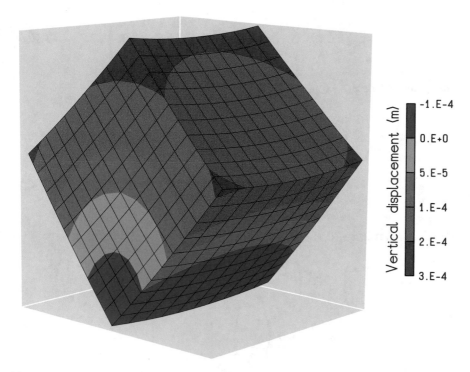

Fig. 3.11 Deformations scaled up, vertical displacements

The stress tensor

$$\sigma = \begin{pmatrix} \sigma_{11} & \sigma_{12} & \sigma_{13} \\ \sigma_{12} & \sigma_{22} & \sigma_{23} \\ \sigma_{13} & \sigma_{23} & \sigma_{33} \end{pmatrix} = \frac{\rho g}{2} \begin{pmatrix} z+H & 0 & x \\ 0 & 0 & 0 \\ x & 0 & z+H \end{pmatrix} \qquad (3.10.3)$$

satisfies the equation of mechanical equilibrium

$$\text{div } \sigma = (0, 0, \rho g) \qquad (3.10.4)$$

as well as the specified surface loads because along the face $y = H$

$$\sigma \begin{pmatrix} 0 \\ 1 \\ 0 \end{pmatrix} = \frac{\rho g}{2} \begin{pmatrix} z+H & 0 & x \\ 0 & 0 & 0 \\ x & 0 & z+H \end{pmatrix} \begin{pmatrix} 0 \\ 1 \\ 0 \end{pmatrix} = \begin{pmatrix} 0 \\ 0 \\ 0 \end{pmatrix}, \qquad (3.10.5)$$

along the face $y = -H$

$$\sigma \begin{pmatrix} 0 \\ -1 \\ 0 \end{pmatrix} = \frac{\rho g}{2} \begin{pmatrix} z+H & 0 & x \\ 0 & 0 & 0 \\ x & 0 & z+H \end{pmatrix} \begin{pmatrix} 0 \\ -1 \\ 0 \end{pmatrix} = \begin{pmatrix} 0 \\ 0 \\ 0 \end{pmatrix}, \quad (3.10.6)$$

along the top face $x \geq 0$, $z \geq 0$ where $z = H - x$

$$\sigma \frac{1}{\sqrt{2}} \begin{pmatrix} 1 \\ 0 \\ 1 \end{pmatrix} = \frac{\rho g}{2} \begin{pmatrix} H-x+H & 0 & x \\ 0 & 0 & 0 \\ x & 0 & H-x+H \end{pmatrix} \frac{1}{\sqrt{2}} \begin{pmatrix} 1 \\ 0 \\ 1 \end{pmatrix}$$

$$= \frac{1}{\sqrt{2}} \rho g H \begin{pmatrix} 1 \\ 0 \\ 1 \end{pmatrix}, \quad (3.10.7)$$

along the top face $x \leq 0$, $z \geq 0$ where $z = H + x$

$$\sigma \frac{1}{\sqrt{2}} \begin{pmatrix} -1 \\ 0 \\ 1 \end{pmatrix} = \frac{\rho g}{2} \begin{pmatrix} H+x+H & 0 & x \\ 0 & 0 & 0 \\ x & 0 & H+x+H \end{pmatrix} \frac{1}{\sqrt{2}} \begin{pmatrix} -1 \\ 0 \\ 1 \end{pmatrix}$$

$$= \frac{1}{\sqrt{2}} \rho g H \begin{pmatrix} -1 \\ 0 \\ 1 \end{pmatrix}, \quad (3.10.8)$$

along the bottom face $x \leq 0$, $z \leq 0$ where $z = -H - x$

$$\sigma \frac{1}{\sqrt{2}} \begin{pmatrix} -1 \\ 0 \\ -1 \end{pmatrix} = \frac{\rho g}{2} \begin{pmatrix} -H-x+H & 0 & x \\ 0 & 0 & 0 \\ x & 0 & -H-x+H \end{pmatrix} \frac{1}{\sqrt{2}} \begin{pmatrix} -1 \\ 0 \\ -1 \end{pmatrix}$$

$$= \begin{pmatrix} 0 \\ 0 \\ 0 \end{pmatrix}, \quad (3.10.9)$$

and along the bottom face $x \geq 0$, $z \leq 0$ where $z = -H + x$

$$\sigma \frac{1}{\sqrt{2}} \begin{pmatrix} 1 \\ 0 \\ -1 \end{pmatrix} = \frac{\rho g}{2} \begin{pmatrix} -H+x+H & 0 & x \\ 0 & 0 & 0 \\ x & 0 & -H+x+H \end{pmatrix} \frac{1}{\sqrt{2}} \begin{pmatrix} 1 \\ 0 \\ -1 \end{pmatrix}$$

$$= \begin{pmatrix} 0 \\ 0 \\ 0 \end{pmatrix}. \quad (3.10.10)$$

Hooke's law reads for the strains

$$\epsilon_{11} = \frac{\partial u_x}{\partial x} = \frac{1}{E}[\sigma_{11} - \nu(\sigma_{22} + \sigma_{33})] = \frac{1-\nu}{2E}\rho g(z + H), \quad (3.10.11)$$

$$\epsilon_{22} = \frac{\partial u_y}{\partial y} = \frac{1}{E}[\sigma_{22} - \nu(\sigma_{11} + \sigma_{33})] = -\frac{\nu}{E}\rho g(z + H), \quad (3.10.12)$$

$$\epsilon_{33} = \frac{\partial u_z}{\partial z} = \frac{1}{E}[\sigma_{33} - \nu(\sigma_{11} + \sigma_{22})] = \frac{1-\nu}{2E}\rho g(z + H), \quad (3.10.13)$$

$$\epsilon_{12} = \frac{\partial u_x}{\partial y} + \frac{\partial u_y}{\partial x} = \frac{2(1+\nu)}{E}\sigma_{12} = 0, \quad (3.10.14)$$

$$\epsilon_{13} = \frac{\partial u_x}{\partial z} + \frac{\partial u_z}{\partial x} = \frac{2(1+\nu)}{E}\sigma_{13} = \frac{1+\nu}{E}\rho g x, \quad (3.10.15)$$

$$\epsilon_{23} = \frac{\partial u_y}{\partial z} + \frac{\partial u_z}{\partial y} = \frac{2(1+\nu)}{E}\sigma_{23} = 0, \quad (3.10.16)$$

which constitute a set of equations for the partial derivatives of the displacements. Integrating the strains ϵ_{11}, ϵ_{22}, and ϵ_{33} with respect to the prescribed x- and y-fixities gives the displacement vector (u_x, u_y, u_z)

$$u_x(x, z) = \frac{1-\nu}{2E}\rho g(z + H)x, \quad (3.10.17)$$

$$u_y(y, z) = -\frac{\nu}{E}\rho g(z + H)y, \quad (3.10.18)$$

$$u_z(x, y, z) = \frac{1-\nu}{2E}\rho g(\frac{1}{2}z^2 + Hz) + h(x, y), \quad (3.10.19)$$

where $h(x, y)$ has to be determined from the shear strains and the specified z-fixity. Then

$$\epsilon_{12} = \frac{\partial u_x}{\partial y} + \frac{\partial u_y}{\partial x} = 0, \quad (3.10.20)$$

$$\epsilon_{13} = \frac{\partial u_x}{\partial z} + \frac{\partial u_z}{\partial x} = \frac{1-\nu}{2E}\rho g x + \frac{\partial h}{\partial x} = \frac{1+\nu}{E}\rho g x, \quad (3.10.21)$$

$$\epsilon_{23} = \frac{\partial u_y}{\partial z} + \frac{\partial u_z}{\partial y} = -\frac{\nu}{E}\rho g y + \frac{\partial h}{\partial y} = 0, \quad (3.10.22)$$

hence,

$$\frac{\partial h}{\partial x} = \frac{1+3\nu}{2E}\rho g x, \quad (3.10.23)$$

$$\frac{\partial h}{\partial y} = \frac{\nu}{E}\rho g y, \quad (3.10.24)$$

and therefore,

$$h(x, y) = \frac{\nu}{2E}\rho g y^2 + \frac{1+3\nu}{4E}\rho g x^2 + C. \quad (3.10.25)$$

Table 3.10 Benchmark deposit (https://docs.opengeosys.org/books/bmb-5)

BM code	Author	Code	Files	CTest
BMB5-3.10	Peter Vogel	OGS-5	Available	ToDo

(https://oc.ufz.de/index.php/s/nlph7bhfDkj6tC7)

The z-fixity at the origin yields the free constant C. The displacement vector (u_x, u_y, u_z) becomes

$$u_x(x, z) = \frac{1 - \nu}{2E} \rho g x (z + H), \tag{3.10.26}$$

$$u_y(y, z) = -\frac{\nu}{E} \rho g y (z + H), \tag{3.10.27}$$

$$u_z(x, y, z) = \frac{1 - \nu}{2E} \rho g (\frac{z^2}{2} + Hz) + \frac{\nu}{2E} \rho g y^2 + \frac{1 + 3\nu}{4E} \rho g x^2. \tag{3.10.28}$$

Anisotropy

Peter Vogel and Jobst Maßmann

This section presents problems on transversely isotropic bodies, most of the material has been adopted from Lekhnitskii (1981). We focus on the closed form solutions. The associated simulation exercises have been checked by OGS; they may serve as verification tests. For the underlying theory of anisotropic elasticity and for more advanced examples see Ting (1996).

3.11 A Cube Undergoes Uniform Compression, Anisotropy Parallel to X-Axis

Given size $2H = 2$ m the domain represents the cube $[-H, H] \times [-H, H] \times [-H, H]$. It is discretized by $4 \times 4 \times 4$ cubic elements. The material has been selected transversely isotropic, the axis of anisotropy is the x-axis. For the elastic parameters see Table 3.11, gravity is neglected via zero material density. Fixities have been prescribed in the interior of the domain with zero x-displacement along the plane $x = 0$, zero y-displacement along the plane $y = 0$, and zero z-displacement along the plane $z = 0$. The cube is deformed by normal forces uniformly distributed over the entire surface of the domain; $\sigma_0 = -1$ MPa is the constant surface pressure. The simulation (Table 3.12) comprises one timestep to evaluate the stresses, the strains, and the displacement vector (u_x, u_y, u_z) (Fig. 3.12).

The stress tensor

$$\sigma = \begin{pmatrix} \sigma_{11} & \sigma_{12} & \sigma_{13} \\ \sigma_{12} & \sigma_{22} & \sigma_{23} \\ \sigma_{13} & \sigma_{23} & \sigma_{33} \end{pmatrix} = \sigma_0 \begin{pmatrix} 1 & 0 & 0 \\ 0 & 1 & 0 \\ 0 & 0 & 1 \end{pmatrix} \tag{3.11.1}$$

Table 3.11 Elastic parameters of the transversely isotropic material (after Vietor et al. 2008)

Quantity	Meaning
$E_i = 7200$ MPa	Young's modulus parallel to the plane of isotropy
$E_a = 2800$ MPa	Young's modulus normal to the plane of isotropy
$\nu = 0.33$	Poisson's ratio within the plane of isotropy (load applied parallel to the plane of isotropy)
$\nu_{ia} = 0.24$	Poisson's ratio normal to the plane of isotropy (load applied parallel to the plane of isotropy)
$G_a = 1200$ MPa	Shear modulus normal to the plane of isotropy

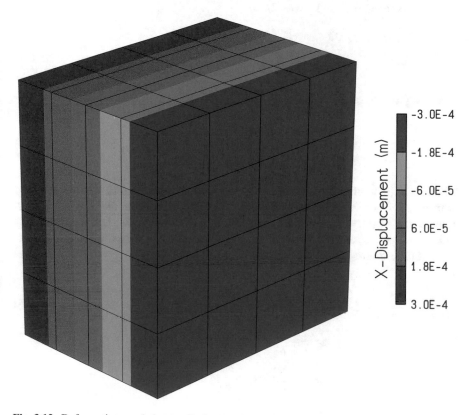

Fig. 3.12 Deformations scaled up, x-displacements

satisfies the equation of mechanical equilibrium

$$\text{div } \boldsymbol{\sigma} = 0 \tag{3.11.2}$$

as well as the specified surface loads because along the faces $x = H$ and $x = -H$

$$\boldsymbol{\sigma} \begin{pmatrix} 1 \\ 0 \\ 0 \end{pmatrix} = -\boldsymbol{\sigma} \begin{pmatrix} -1 \\ 0 \\ 0 \end{pmatrix} = \begin{pmatrix} \sigma_{11} \\ \sigma_{12} \\ \sigma_{13} \end{pmatrix} = \sigma_0 \begin{pmatrix} 1 \\ 0 \\ 0 \end{pmatrix}, \tag{3.11.3}$$

along the faces $y = H$ and $y = -H$

$$\boldsymbol{\sigma} \begin{pmatrix} 0 \\ 1 \\ 0 \end{pmatrix} = -\boldsymbol{\sigma} \begin{pmatrix} 0 \\ -1 \\ 0 \end{pmatrix} = \begin{pmatrix} \sigma_{12} \\ \sigma_{22} \\ \sigma_{23} \end{pmatrix} = \sigma_0 \begin{pmatrix} 0 \\ 1 \\ 0 \end{pmatrix}, \tag{3.11.4}$$

and along the faces $z = H$ and $z = -H$

$$\boldsymbol{\sigma} \begin{pmatrix} 0 \\ 0 \\ 1 \end{pmatrix} = -\boldsymbol{\sigma} \begin{pmatrix} 0 \\ 0 \\ 1 \end{pmatrix} = \begin{pmatrix} \sigma_{13} \\ \sigma_{23} \\ \sigma_{33} \end{pmatrix} = \sigma_0 \begin{pmatrix} 0 \\ 0 \\ 1 \end{pmatrix}. \tag{3.11.5}$$

The stress-strain relationships governing the present situation will next be obtained from those of an orthotropic material with principal axes equal to the coordinate axes. Introducing engineering constants, i.e. Young's moduli E_1, E_2, and E_3, Poisson's ratios $\nu_{12}, \nu_{13}, \ldots$, and shear moduli G_{12}, G_{13}, and G_{23} the generalized Hooke's law gives for the strains

$$\epsilon_{11} = \frac{1}{E_1}\sigma_{11} - \frac{\nu_{12}}{E_2}\sigma_{22} - \frac{\nu_{13}}{E_3}\sigma_{33}, \tag{3.11.6}$$

$$\epsilon_{22} = -\frac{\nu_{21}}{E_1}\sigma_{11} + \frac{1}{E_2}\sigma_{22} - \frac{\nu_{23}}{E_3}\sigma_{33}, \tag{3.11.7}$$

$$\epsilon_{33} = -\frac{\nu_{31}}{E_1}\sigma_{11} - \frac{\nu_{32}}{E_2}\sigma_{22} + \frac{1}{E_3}\sigma_{33}, \tag{3.11.8}$$

$$\epsilon_{12} = \frac{1}{G_{12}}\sigma_{12}, \tag{3.11.9}$$

$$\epsilon_{13} = \frac{1}{G_{13}}\sigma_{13}, \tag{3.11.10}$$

$$\epsilon_{23} = \frac{1}{G_{23}}\sigma_{23}. \tag{3.11.11}$$

Due to the underlying theory the following symmetry holds

$$\frac{\nu_{12}}{E_2} = \frac{\nu_{21}}{E_1},$$

(3.11.12)

$$\frac{\nu_{13}}{E_3} = \frac{\nu_{31}}{E_1},$$

(3.11.13)

$$\frac{\nu_{23}}{E_3} = \frac{\nu_{32}}{E_2}.$$

(3.11.14)

Therefore,

$$\epsilon_{11} = \frac{1}{E_1}\sigma_{11} - \frac{\nu_{12}}{E_2}\sigma_{22} - \frac{\nu_{13}}{E_3}\sigma_{33},$$

(3.11.15)

$$\epsilon_{22} = -\frac{\nu_{12}}{E_2}\sigma_{11} + \frac{1}{E_2}\sigma_{22} - \frac{\nu_{23}}{E_3}\sigma_{33},$$

(3.11.16)

$$\epsilon_{33} = -\frac{\nu_{13}}{E_3}\sigma_{11} - \frac{\nu_{23}}{E_3}\sigma_{22} + \frac{1}{E_3}\sigma_{33},$$

(3.11.17)

$$\epsilon_{12} = \frac{1}{G_{12}}\sigma_{12},$$

(3.11.18)

$$\epsilon_{13} = \frac{1}{G_{13}}\sigma_{13},$$

(3.11.19)

$$\epsilon_{23} = \frac{1}{G_{23}}\sigma_{23}.$$

(3.11.20)

Due to the transverse isotropy of the material

$$E_2 = E_3 = E_i,$$

(3.11.21)

$$E_1 = E_a,$$

(3.11.22)

$$\nu_{23} = \nu,$$

(3.11.23)

$$\nu_{12} = \nu_{13} = \nu_{ia},$$

(3.11.24)

$$G_{12} = G_{13} = G_a,$$

(3.11.25)

$$G_{23} = \frac{E_i}{2(1+\nu)}$$

(3.11.26)

and the constitutive equations become

$$\epsilon_{11} = \frac{1}{E_a}\sigma_{11} - \frac{\nu_{ia}}{E_i}\sigma_{22} - \frac{\nu_{ia}}{E_i}\sigma_{33},$$

(3.11.27)

$$\epsilon_{22} = -\frac{\nu_{ia}}{E_i}\sigma_{11} + \frac{1}{E_i}\sigma_{22} - \frac{\nu}{E_i}\sigma_{33},$$

(3.11.28)

$$\epsilon_{33} = -\frac{\nu_{ia}}{E_i}\sigma_{11} - \frac{\nu}{E_i}\sigma_{22} + \frac{1}{E_i}\sigma_{33},$$

(3.11.29)

Table 3.12 Benchmark deposit (https://docs.opengeosys.org/books/bmb-5)

BM code	Author	Code	Files	CTest
BMB5-3.11	Peter Vogel	OGS-5	Available	ToDo

(https://oc.ufz.de/index.php/s/nlph7bhfDkj6tC7)

$$\epsilon_{12} = \frac{1}{G_a}\sigma_{12}, \tag{3.11.30}$$

$$\epsilon_{13} = \frac{1}{G_a}\sigma_{13}, \tag{3.11.31}$$

$$\epsilon_{23} = \frac{2(1+\nu)}{E_i}\sigma_{23}. \tag{3.11.32}$$

Employing the stresses above the non-zero strains read

$$\epsilon_{11} = \frac{\partial u_x}{\partial x} = \left(\frac{1}{E_a} - 2\frac{\nu_{ia}}{E_i}\right)\sigma_0, \tag{3.11.33}$$

$$\epsilon_{22} = \frac{\partial u_y}{\partial y} = \frac{1-\nu-\nu_{ia}}{E_i}\sigma_0, \tag{3.11.34}$$

$$\epsilon_{33} = \frac{\partial u_z}{\partial z} = \frac{1-\nu-\nu_{ia}}{E_i}\sigma_0. \tag{3.11.35}$$

Integrating the strains with respect to the fixities at the coordinate planes yields the displacement vector (u_x, u_y, u_z)

$$u_x(x) = x\left(\frac{1}{E_a} - 2\frac{\nu_{ia}}{E_i}\right)\sigma_0, \tag{3.11.36}$$

$$u_y(y) = y\frac{1-\nu-\nu_{ia}}{E_i}\sigma_0, \tag{3.11.37}$$

$$u_z(z) = z\frac{1-\nu-\nu_{ia}}{E_i}\sigma_0. \tag{3.11.38}$$

The deformed cube takes the shape of a cuboid.

3.12   A Cuboid Undergoes Load Due to Gravity, Anisotropy Parallel to X-Axis

Given size $H = 30$ m the domain represents the cuboid $[0, 2H] \times [0, H] \times [0, H]$. It is discretized by an irregular mesh of hexahedral elements. The material has been selected transversely isotropic, the axis of anisotropy is the x-axis. For the elastic parameters see Table 3.11, density $\rho = 2450$ kg/m^3 has been assigned. Gravitational

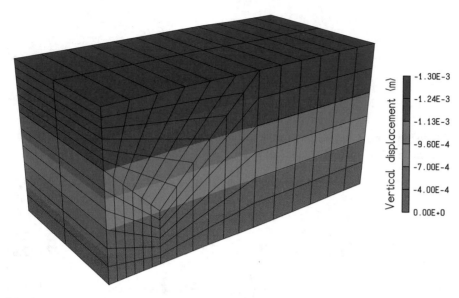

Fig. 3.13 Vertical displacements

acceleration $g = 9.81$ m/s^2 is applied in negative z-direction by default. The bottom $z = 0$ and the lateral faces are sliding planes, the top face is free. Gravity is the only load applied. The simulation (Table 3.13) comprises one timestep to establish the stresses, the strains, and the displacement vector (u_x, u_y, u_z) (Fig. 3.13).

Let σ denote the stress tensor. The equation of mechanical equilibrium

$$\text{div } \sigma = (0, 0, \rho g) \tag{3.12.1}$$

is satisfied by zero shear, if the horizontal stresses σ_{11} and σ_{22} are functions of the vertical coordinate z only and the vertical stress σ_{33} satisfies

$$\frac{\partial \sigma_{33}}{\partial z} = \rho g. \tag{3.12.2}$$

The face $z = H$ is free, hence, integration gives

$$\sigma_{33} = \rho(-g)(H - z). \tag{3.12.3}$$

The constitutive equations governing the present situation have already been outlined in the context of the previous example. The stress-strain relationships become

$$\epsilon_{11} = \frac{1}{E_a}\sigma_{11} - \frac{\nu_{ia}}{E_i}\sigma_{22} - \frac{\nu_{ia}}{E_i}\sigma_{33}, \qquad (3.12.4)$$

$$\epsilon_{22} = -\frac{\nu_{ia}}{E_i}\sigma_{11} + \frac{1}{E_i}\sigma_{22} - \frac{\nu}{E_i}\sigma_{33}, \qquad (3.12.5)$$

$$\epsilon_{33} = -\frac{\nu_{ia}}{E_i}\sigma_{11} - \frac{\nu}{E_i}\sigma_{22} + \frac{1}{E_i}\sigma_{33}. \qquad (3.12.6)$$

Due to the setup it may be assumed that there is no horizontal displacement anywhere and we have for the horizontal strains

$$\epsilon_{11} = \epsilon_{22} = 0. \qquad (3.12.7)$$

Then

$$0 = \frac{1}{E_a}\sigma_{11} - \frac{\nu_{ia}}{E_i}\sigma_{22} - \frac{\nu_{ia}}{E_i}\rho(-g)(H-z), \qquad (3.12.8)$$

$$0 = -\frac{\nu_{ia}}{E_i}\sigma_{11} + \frac{1}{E_i}\sigma_{22} - \frac{\nu}{E_i}\rho(-g)(H-z), \qquad (3.12.9)$$

$$\epsilon_{33} = -\frac{\nu_{ia}}{E_i}\sigma_{11} - \frac{\nu}{E_i}\sigma_{22} + \frac{1}{E_i}\rho(-g)(H-z). \qquad (3.12.10)$$

Solving for σ_{11}, σ_{22}, and the vertical strain ϵ_{33} yields

$$\sigma_{11} = \frac{(1+\nu)\nu_{ia}}{E_i/E_a - \nu_{ia}^2}\rho(-g)(H-z), \qquad (3.12.11)$$

$$\sigma_{22} = \left(\nu + \frac{(1+\nu)\nu_{ia}^2}{E_i/E_a - \nu_{ia}^2}\right)\rho(-g)(H-z), \qquad (3.12.12)$$

$$\epsilon_{33} = \frac{1}{E_i}\left(1 - \nu^2 - \frac{(1+\nu)^2\nu_{ia}^2}{E_i/E_a - \nu_{ia}^2}\right)\rho(-g)(H-z) \qquad (3.12.13)$$

in terms of the vertical coordinate. Integrating the strains with respect to the pre-scribed fixities gives the displacement vector (u_x, u_y, u_z)

$$u_x = u_y = 0, \qquad (3.12.14)$$

$$u_z(z) = \frac{1}{E_i}\left(1 - \nu^2 - \frac{(1+\nu)^2\nu_{ia}^2}{E_i/E_a - \nu_{ia}^2}\right)\rho(-g)\left(Hz - \frac{1}{2}z^2\right). \qquad (3.12.15)$$

Table 3.13 Benchmark deposit (https://docs.opengeosys.org/books/bmb-5)

BM code	Author	Code	Files	CTest
BMB5-3.12	Peter Vogel	OGS-5	Available	ToDo

(https://oc.ufz.de/index.php/s/nlph7bhfDkj6tC7)

3.13   A Thick Plate Undergoes Lateral Compression, Anisotropy Parallel to Y-Axis

Given length $2L = 20$ m and thickness $H = 2$ m the domain represents the rectangular plate $[-L, L] \times [-L, L] \times [-H, 0]$. It is discretized by $20 \times 20 \times 4$ equally sized hexahedral elements. The material has been selected transversely isotropic, the axis of anisotropy is the y-axis. For the elastic parameters see Table 3.11, gravity is neglected via zero material density. Fixities have been prescribed in the interior of the domain with zero x-displacement along the plane $x = 0$, zero y-displacement along the plane $y = 0$, and zero z-displacement at the origin. Specified loads $P(x, y, z)$ prevail along the lateral boundaries of the domain. With the aid of the stress $\sigma_0 = 1$ MPa these loads read along the faces $x = L$ and $x = -L$

$$P(L, y, z) = \frac{\sigma_0}{H} \begin{pmatrix} z \\ 0 \\ 0 \end{pmatrix}, \quad P(-L, y, z) = -\frac{\sigma_0}{H} \begin{pmatrix} z \\ 0 \\ 0 \end{pmatrix}, \tag{3.13.1}$$

and along the faces $y = L$ and $y = -L$

$$P(x, L, z) = \frac{\sigma_0}{H} \begin{pmatrix} 0 \\ z \\ 0 \end{pmatrix}, \quad P(x, -L, z) = -\frac{\sigma_0}{H} \begin{pmatrix} 0 \\ z \\ 0 \end{pmatrix}. \tag{3.13.2}$$

The top $z = 0$ and the bottom $z = -H$ are free of load by default. The simulation (Table 3.14) comprises one timestep to establish the stresses, the strains, and the displacement vector (u_x, u_y, u_z) (Fig. 3.14).

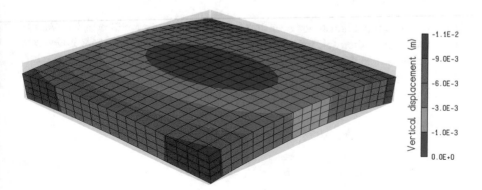

Vertical displacement (m)

-1.1E-2

-9.0E-3

-6.0E-3

-3.0E-3

-1.0E-3

0.0E+0

Fig. 3.14 Deformations scaled up, vertical displacements

The stress tensor

$$\boldsymbol{\sigma} = \begin{pmatrix} \sigma_{11} & \sigma_{12} & \sigma_{13} \\ \sigma_{12} & \sigma_{22} & \sigma_{23} \\ \sigma_{13} & \sigma_{23} & \sigma_{33} \end{pmatrix} = \frac{\sigma_0}{H} \begin{pmatrix} z & 0 & 0 \\ 0 & z & 0 \\ 0 & 0 & 0 \end{pmatrix} \qquad (3.13.3)$$

satisfies the equation of mechanical equilibrium

$$\text{div } \boldsymbol{\sigma} = 0 \qquad (3.13.4)$$

as well as the specified surface loads because along the faces $x = L$ and $x = -L$

$$\boldsymbol{\sigma} \begin{pmatrix} 1 \\ 0 \\ 0 \end{pmatrix} = -\boldsymbol{\sigma} \begin{pmatrix} -1 \\ 0 \\ 0 \end{pmatrix} = \begin{pmatrix} \sigma_{11} \\ \sigma_{12} \\ \sigma_{13} \end{pmatrix} = \frac{\sigma_0}{H} \begin{pmatrix} z \\ 0 \\ 0 \end{pmatrix}, \qquad (3.13.5)$$

along the faces $y = L$ and $y = -L$

$$\boldsymbol{\sigma} \begin{pmatrix} 0 \\ 1 \\ 0 \end{pmatrix} = -\boldsymbol{\sigma} \begin{pmatrix} 0 \\ -1 \\ 0 \end{pmatrix} = \begin{pmatrix} \sigma_{12} \\ \sigma_{22} \\ \sigma_{23} \end{pmatrix} = \frac{\sigma_0}{H} \begin{pmatrix} 0 \\ z \\ 0 \end{pmatrix}, \qquad (3.13.6)$$

and along the faces $z = 0$ and $z = -H$

$$\boldsymbol{\sigma} \begin{pmatrix} 0 \\ 0 \\ 1 \end{pmatrix} = -\boldsymbol{\sigma} \begin{pmatrix} 0 \\ 0 \\ -1 \end{pmatrix} = \begin{pmatrix} \sigma_{13} \\ \sigma_{23} \\ \sigma_{33} \end{pmatrix} = \begin{pmatrix} 0 \\ 0 \\ 0 \end{pmatrix}. \qquad (3.13.7)$$

The stress-strain relationships governing the present situation will next be obtained from those of an orthotropic material with principal axes equal to the coordinate axes. Introducing engineering constants, i.e. Young's moduli E_1, E_2, and E_3, Poisson's ratios $\nu_{12}, \nu_{13}, \ldots$, and shear moduli G_{12}, G_{13}, and G_{23} the generalized Hooke's law gives for the strains

$$\epsilon_{11} = \frac{1}{E_1}\sigma_{11} - \frac{\nu_{12}}{E_2}\sigma_{22} - \frac{\nu_{13}}{E_3}\sigma_{33}, \qquad (3.13.8)$$

$$\epsilon_{22} = -\frac{\nu_{21}}{E_1}\sigma_{11} + \frac{1}{E_2}\sigma_{22} - \frac{\nu_{23}}{E_3}\sigma_{33}, \qquad (3.13.9)$$

$$\epsilon_{33} = -\frac{\nu_{31}}{E_1}\sigma_{11} - \frac{\nu_{32}}{E_2}\sigma_{22} + \frac{1}{E_3}\sigma_{33}, \qquad (3.13.10)$$

$$\epsilon_{12} = \frac{1}{G_{12}} \sigma_{12}, \tag{3.13.11}$$

$$\epsilon_{13} = \frac{1}{G_{13}} \sigma_{13}, \tag{3.13.12}$$

$$\epsilon_{23} = \frac{1}{G_{23}} \sigma_{23}. \tag{3.13.13}$$

Due to the underlying theory the following symmetry holds

$$\frac{\nu_{12}}{E_2} = \frac{\nu_{21}}{E_1}, \tag{3.13.14}$$

$$\frac{\nu_{13}}{E_3} = \frac{\nu_{31}}{E_1}, \tag{3.13.15}$$

$$\frac{\nu_{23}}{E_3} = \frac{\nu_{32}}{E_2}. \tag{3.13.16}$$

Therefore,

$$\epsilon_{11} = \frac{1}{E_1} \sigma_{11} - \frac{\nu_{21}}{E_1} \sigma_{22} - \frac{\nu_{31}}{E_1} \sigma_{33}, \tag{3.13.17}$$

$$\epsilon_{22} = -\frac{\nu_{21}}{E_1} \sigma_{11} + \frac{1}{E_2} \sigma_{22} - \frac{\nu_{23}}{E_3} \sigma_{33}, \tag{3.13.18}$$

$$\epsilon_{33} = -\frac{\nu_{31}}{E_1} \sigma_{11} - \frac{\nu_{23}}{E_3} \sigma_{22} + \frac{1}{E_3} \sigma_{33}, \tag{3.13.19}$$

$$\epsilon_{12} = \frac{1}{G_{12}} \sigma_{12}, \tag{3.13.20}$$

$$\epsilon_{13} = \frac{1}{G_{13}} \sigma_{13}, \tag{3.13.21}$$

$$\epsilon_{23} = \frac{1}{G_{23}} \sigma_{23}. \tag{3.13.22}$$

Due to the transverse isotropy of the material

$$E_1 = E_3 = E_i, \tag{3.13.23}$$

$$E_2 = E_a, \tag{3.13.24}$$

$$\nu_{31} = \nu, \tag{3.13.25}$$

$$\nu_{21} = \nu_{23} = \nu_{ia}, \tag{3.13.26}$$

$$G_{12} = G_{23} = G_a, \tag{3.13.27}$$

$$G_{13} = \frac{E_i}{2(1 + \nu)} \tag{3.13.28}$$

and the constitutive equations become

$$\epsilon_{11} = \frac{1}{E_i}\sigma_{11} - \frac{\nu_{ia}}{E_i}\sigma_{22} - \frac{\nu}{E_i}\sigma_{33}, \tag{3.13.29}$$

$$\epsilon_{22} = -\frac{\nu_{ia}}{E_i}\sigma_{11} + \frac{1}{E_a}\sigma_{22} - \frac{\nu_{ia}}{E_i}\sigma_{33}, \tag{3.13.30}$$

$$\epsilon_{33} = -\frac{\nu}{E_i}\sigma_{11} - \frac{\nu_{ia}}{E_i}\sigma_{22} + \frac{1}{E_i}\sigma_{33}, \tag{3.13.31}$$

$$\epsilon_{12} = \frac{1}{G_a}\sigma_{12}, \tag{3.13.32}$$

$$\epsilon_{13} = \frac{2(1+\nu)}{E_i}\sigma_{13}, \tag{3.13.33}$$

$$\epsilon_{23} = \frac{1}{G_a}\sigma_{23}. \tag{3.13.34}$$

The stresses above yield the strains

$$\epsilon_{11} = \frac{\partial u_x}{\partial x} = \frac{1 - \nu_{ia}}{E_i}\frac{\sigma_0}{H}z, \tag{3.13.35}$$

$$\epsilon_{22} = \frac{\partial u_y}{\partial y} = \left(\frac{1}{E_a} - \frac{\nu_{ia}}{E_i}\right)\frac{\sigma_0}{H}z, \tag{3.13.36}$$

$$\epsilon_{33} = \frac{\partial u_z}{\partial z} = -\frac{\nu + \nu_{ia}}{E_i}\frac{\sigma_0}{H}z, \tag{3.13.37}$$

$$\epsilon_{12} = \frac{\partial u_x}{\partial y} + \frac{\partial u_y}{\partial x} = 0, \tag{3.13.38}$$

$$\epsilon_{13} = \frac{\partial u_x}{\partial z} + \frac{\partial u_z}{\partial x} = 0, \tag{3.13.39}$$

$$\epsilon_{23} = \frac{\partial u_y}{\partial z} + \frac{\partial u_z}{\partial y} = 0, \tag{3.13.40}$$

which constitute a set of equations for the partial derivatives of the displacements. Integrating the strains ϵ_{11}, ϵ_{22}, and ϵ_{33} with respect to the prescribed x- and y-fixities gives the displacement vector (u_x, u_y, u_z)

$$u_x(x, z) = \frac{1 - \nu_{ia}}{E_i}\frac{\sigma_0}{H}zx, \tag{3.13.41}$$

$$u_y(y, z) = \left(\frac{1}{E_a} - \frac{\nu_{ia}}{E_i}\right)\frac{\sigma_0}{H}zy, \tag{3.13.42}$$

$$u_z(x, y, z) = -\frac{\nu + \nu_{ia}}{E_i}\frac{\sigma_0}{2H}z^2 + h(x, y), \tag{3.13.43}$$

Table 3.14 Benchmark deposit (https://docs.opengeosys.org/books/bmb-5)

BM code	Author	Code	Files	CTest
BMB5-3.13	Peter Vogel	OGS-5	Available	ToDo

(https://oc.ufz.de/index.php/s/nlph7bhfDkj6tC7)

where $h(x, y)$ has to be determined from the shear strains and the specified z-fixity. Then

$$\epsilon_{12} = \frac{\partial u_x}{\partial y} + \frac{\partial u_y}{\partial x} = 0, \tag{3.13.44}$$

$$\epsilon_{13} = \frac{\partial u_x}{\partial z} + \frac{\partial u_z}{\partial x} = \frac{1 - \nu_{ia}}{E_i} \frac{\sigma_0}{H} x + \frac{\partial h}{\partial x} = 0, \tag{3.13.45}$$

$$\epsilon_{23} = \frac{\partial u_y}{\partial z} + \frac{\partial u_z}{\partial y} = \left(\frac{1}{E_a} - \frac{\nu_{ia}}{E_i} \right) \frac{\sigma_0}{H} y + \frac{\partial h}{\partial y} = 0, \tag{3.13.46}$$

hence,

$$\frac{\partial h}{\partial x} = -\frac{1 - \nu_{ia}}{E_i} \frac{\sigma_0}{H} x, \tag{3.13.47}$$

$$\frac{\partial h}{\partial y} = -\left(\frac{1}{E_a} - \frac{\nu_{ia}}{E_i} \right) \frac{\sigma_0}{H} y, \tag{3.13.48}$$

and therefore,

$$h(x, y) = -\frac{\sigma_0}{2H} \left[\frac{1 - \nu_{ia}}{E_i} x^2 + \left(\frac{1}{E_a} - \frac{\nu_{ia}}{E_i} \right) y^2 \right] + C. \tag{3.13.49}$$

The z-fixity at the origin yields the free constant C. The displacement vector (u_x, u_y, u_z) becomes

$$u_x(x, z) = \frac{1 - \nu_{ia}}{E_i} \frac{\sigma_0}{H} xz, \tag{3.13.50}$$

$$u_y(y, z) = \left(\frac{1}{E_a} - \frac{\nu_{ia}}{E_i} \right) \frac{\sigma_0}{H} yz, \tag{3.13.51}$$

$$u_z(x, y, z) = -\frac{\sigma_0}{2H} \left[\frac{1 - \nu_{ia}}{E_i} x^2 + \left(\frac{1}{E_a} - \frac{\nu_{ia}}{E_i} \right) y^2 + \frac{\nu + \nu_{ia}}{E_i} z^2 \right]. \tag{3.13.52}$$

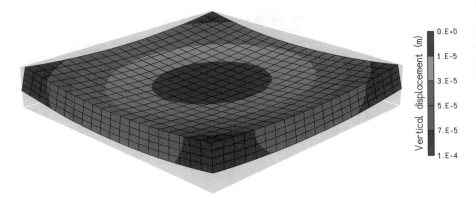

Fig. 3.15 Deformations scaled up, vertical displacements

3.14 A Thick Plate Deforms Under Its Own Weight, Anisotropy Parallel to Y-Axis

Given length $2L = 20$ m and thickness $H = 2$ m the domain represents the rectangular plate $[-L, L] \times [-L, L] \times [-H, 0]$. It is discretized by $20 \times 20 \times 4$ equally sized hexahedral elements. The material has been selected transversely isotropic, the axis of anisotropy is the y-axis. For the elastic parameters see Table 3.11, density $\rho = 2450$ kg/m^3 has been assigned. Gravitational acceleration $g = 9.81$ m/s^2 is applied in negative z-direction by default. The top and the lateral faces of the domain are free of load, the vertical load $\rho g H$ is applied at the bottom. Fixities have been prescribed in the interior of the domain with zero x-displacement along the plane $x = 0$, zero y-displacement along the plane $y = 0$, and zero z-displacement at the origin. The simulation (Table 3.15) comprises one timestep to establish the stresses, the strains, and the displacement vector (u_x, u_y, u_z) (Fig. 3.15).

The stress tensor

$$\boldsymbol{\sigma} = \begin{pmatrix} \sigma_{11} & \sigma_{12} & \sigma_{13} \\ \sigma_{12} & \sigma_{22} & \sigma_{23} \\ \sigma_{13} & \sigma_{23} & \sigma_{33} \end{pmatrix} = \rho g \begin{pmatrix} 0 & 0 & 0 \\ 0 & 0 & 0 \\ 0 & 0 & z \end{pmatrix} \tag{3.14.1}$$

satisfies the equation of mechanical equilibrium

$$\operatorname{div} \boldsymbol{\sigma} = (0, 0, \rho g) \tag{3.14.2}$$

as well as the specified surface loads because along the bottom face $z = -H$

$$\boldsymbol{\sigma} \begin{pmatrix} 0 \\ 0 \\ -1 \end{pmatrix} = \rho g \begin{pmatrix} 0 & 0 & 0 \\ 0 & 0 & 0 \\ 0 & 0 & -H \end{pmatrix} \begin{pmatrix} 0 \\ 0 \\ -1 \end{pmatrix} = \rho g H \begin{pmatrix} 0 \\ 0 \\ 1 \end{pmatrix}, \tag{3.14.3}$$

along the top face $z = 0$

$$\sigma \begin{pmatrix} 0 \\ 0 \\ 1 \end{pmatrix} = \rho g \begin{pmatrix} 0 & 0 & 0 \\ 0 & 0 & 0 \\ 0 & 0 & 0 \end{pmatrix} \begin{pmatrix} 0 \\ 0 \\ 1 \end{pmatrix} = \begin{pmatrix} 0 \\ 0 \\ 0 \end{pmatrix}, \tag{3.14.4}$$

and similarly for the lateral faces. The constitutive equations governing the present situation have already been outlined in the context of the previous example. The stress-strain relationships become

$$\epsilon_{11} = \frac{\partial u_x}{\partial x} = \frac{1}{E_i}\sigma_{11} - \frac{\nu_{ia}}{E_i}\sigma_{22} - \frac{\nu}{E_i}\sigma_{33} = -\frac{\nu}{E_i}\rho g z, \tag{3.14.5}$$

$$\epsilon_{22} = \frac{\partial u_y}{\partial y} = -\frac{\nu_{ia}}{E_i}\sigma_{11} + \frac{1}{E_a}\sigma_{22} - \frac{\nu_{ia}}{E_i}\sigma_{33} = -\frac{\nu_{ia}}{E_i}\rho g z, \tag{3.14.6}$$

$$\epsilon_{33} = \frac{\partial u_z}{\partial z} = -\frac{\nu}{E_i}\sigma_{11} - \frac{\nu_{ia}}{E_i}\sigma_{22} + \frac{1}{E_i}\sigma_{33} = \frac{1}{E_i}\rho g z, \tag{3.14.7}$$

$$\epsilon_{12} = \frac{\partial u_x}{\partial y} + \frac{\partial u_y}{\partial x} = \frac{1}{G_a}\sigma_{12} = 0, \tag{3.14.8}$$

$$\epsilon_{13} = \frac{\partial u_x}{\partial z} + \frac{\partial u_z}{\partial x} = \frac{2(1+\nu)}{E_i}\sigma_{13} = 0, \tag{3.14.9}$$

$$\epsilon_{23} = \frac{\partial u_y}{\partial z} + \frac{\partial u_z}{\partial y} = \frac{1}{G_a}\sigma_{23} = 0 \tag{3.14.10}$$

and constitute a set of equations for the partial derivatives of the displacements. Integrating the strains ϵ_{11}, ϵ_{22}, and ϵ_{33} with respect to the prescribed x- and y-fixities gives the displacement vector (u_x, u_y, u_z)

$$u_x(x, z) = -\frac{\nu}{E_i}\rho g z x, \tag{3.14.11}$$

$$u_y(y, z) = -\frac{\nu_{ia}}{E_i}\rho g z y, \tag{3.14.12}$$

$$u_z(x, y, z) = \frac{1}{E_i}\rho g \frac{1}{2}z^2 + h(x, y), \tag{3.14.13}$$

where $h(x, y)$ has to be determined from the shear strains and the specified z-fixity. Then

$$\epsilon_{12} = \frac{\partial u_x}{\partial y} + \frac{\partial u_y}{\partial x} = 0, \tag{3.14.14}$$

$$\epsilon_{13} = \frac{\partial u_x}{\partial z} + \frac{\partial u_z}{\partial x} = -\frac{\nu}{E_i}\rho g x + \frac{\partial h}{\partial x} = 0, \tag{3.14.15}$$

Table 3.15 Benchmark deposit (https://docs.opengeosys.org/books/bmb-5)

BM code	Author	Code	Files	CTest
BMB5-3.14	Peter Vogel	OGS-5	Available	ToDo

(https://oc.ufz.de/index.php/s/nlph7bhfDkj6tC7)

$$\epsilon_{23} = \frac{\partial u_y}{\partial z} + \frac{\partial u_z}{\partial y} = -\frac{\nu_{ia}}{E_i} \rho g y + \frac{\partial h}{\partial y} = 0, \tag{3.14.16}$$

hence,

$$\frac{\partial h}{\partial x} = \frac{\nu}{E_i} \rho g x, \tag{3.14.17}$$

$$\frac{\partial h}{\partial y} = \frac{\nu_{ia}}{E_i} \rho g y, \tag{3.14.18}$$

and therefore,

$$h(x, y) = \frac{\rho g}{2E_i} (\nu x^2 + \nu_{ia} y^2) + C. \tag{3.14.19}$$

The z-fixity at the origin yields the free constant C. The displacement vector (u_x, u_y, u_z) becomes

$$u_x(x, z) = -\frac{\nu}{E_i} \rho g x z, \tag{3.14.20}$$

$$u_y(y, z) = -\frac{\nu_{ia}}{E_i} \rho g y z, \tag{3.14.21}$$

$$u_z(x, y, z) = \frac{\rho g}{2E_i} (\nu x^2 + \nu_{ia} y^2 + z^2). \tag{3.14.22}$$

3.15 A Thick Plate Undergoes Shear, Anisotropy Parallel to Z-Axis

Given length $L = 10$ m and thickness $H = 2$ m the domain represents the rectangular plate $[0, L] \times [0, L] \times [0, H]$. It is discretized by $8 \times 8 \times 2$ equally sized hexahedral elements. The material has been selected transversely isotropic, the axis of anisotropy is the z-axis. For the elastic parameters see Table 3.11, gravity is neglected via zero material density. The face $x = 0$ is entirely fixed, specified loads $P(x, y, z)$ prevail along the remaining faces. With the aid of the stress $\sigma_0 = 1$ MPa these loads read along the face $x = L$

$$P(L, y, z) = \sigma_0 \begin{pmatrix} 0 \\ 1 \\ 1 \end{pmatrix}, \tag{3.15.1}$$

along the faces $y = L$ and $y = 0$

$$P(x, L, z) = \sigma_0 \begin{pmatrix} 1 \\ 0 \\ 0 \end{pmatrix}, \quad P(x, 0, z) = -\sigma_0 \begin{pmatrix} 1 \\ 0 \\ 0 \end{pmatrix}, \tag{3.15.2}$$

and along the faces $z = H$ and $z = 0$

$$P(x, y, H) = \sigma_0 \begin{pmatrix} 1 \\ 0 \\ 0 \end{pmatrix}, \quad P(x, y, 0) = -\sigma_0 \begin{pmatrix} 1 \\ 0 \\ 0 \end{pmatrix}. \tag{3.15.3}$$

The simulation (Table 3.16) comprises one timestep to evaluate the stresses, the strains, and the displacement vector (u_x, u_y, u_z) (Fig. 3.16).

The stress tensor

$$\sigma = \begin{pmatrix} \sigma_{11} & \sigma_{12} & \sigma_{13} \\ \sigma_{12} & \sigma_{22} & \sigma_{23} \\ \sigma_{13} & \sigma_{23} & \sigma_{33} \end{pmatrix} = \sigma_0 \begin{pmatrix} 0 & 1 & 1 \\ 1 & 0 & 0 \\ 1 & 0 & 0 \end{pmatrix} \tag{3.15.4}$$

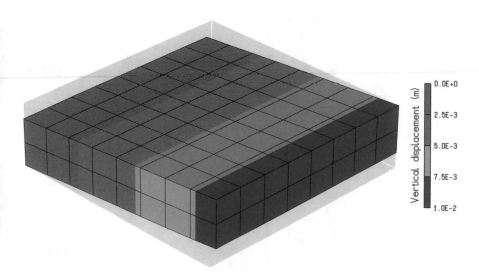

Fig. 3.16 Deformations scaled up, vertical displacements

satisfies the equation of mechanical equilibrium

$$\text{div } \sigma = 0 \tag{3.15.5}$$

as well as the specified surface loads because along the face $x = L$

$$\sigma \begin{pmatrix} 1 \\ 0 \\ 0 \end{pmatrix} = \begin{pmatrix} \sigma_{11} \\ \sigma_{12} \\ \sigma_{13} \end{pmatrix} = \sigma_0 \begin{pmatrix} 0 \\ 1 \\ 1 \end{pmatrix}, \tag{3.15.6}$$

along the faces $y = L$ and $y = 0$

$$\sigma \begin{pmatrix} 0 \\ 1 \\ 0 \end{pmatrix} = -\sigma \begin{pmatrix} 0 \\ -1 \\ 0 \end{pmatrix} = \begin{pmatrix} \sigma_{12} \\ \sigma_{22} \\ \sigma_{23} \end{pmatrix} = \sigma_0 \begin{pmatrix} 1 \\ 0 \\ 0 \end{pmatrix}, \tag{3.15.7}$$

and along the faces $z = H$ and $z = 0$

$$\sigma \begin{pmatrix} 0 \\ 0 \\ 1 \end{pmatrix} = -\sigma \begin{pmatrix} 0 \\ 0 \\ -1 \end{pmatrix} = \begin{pmatrix} \sigma_{13} \\ \sigma_{23} \\ \sigma_{33} \end{pmatrix} = \sigma_0 \begin{pmatrix} 1 \\ 0 \\ 0 \end{pmatrix}. \tag{3.15.8}$$

The stress-strain relationships governing the present situation will next be obtained from those of an orthotropic material with principal axes equal to the coordinate axes. Introducing engineering constants, i.e. Young's moduli E_1, E_2, and E_3, Poisson's ratios $\nu_{12}, \nu_{13}, \ldots$, and shear moduli G_{12}, G_{13}, and G_{23} the generalized Hooke's law gives for the strains

$$\epsilon_{11} = \frac{1}{E_1}\sigma_{11} - \frac{\nu_{12}}{E_2}\sigma_{22} - \frac{\nu_{13}}{E_3}\sigma_{33}, \tag{3.15.9}$$

$$\epsilon_{22} = -\frac{\nu_{21}}{E_1}\sigma_{11} + \frac{1}{E_2}\sigma_{22} - \frac{\nu_{23}}{E_3}\sigma_{33}, \tag{3.15.10}$$

$$\epsilon_{33} = -\frac{\nu_{31}}{E_1}\sigma_{11} - \frac{\nu_{32}}{E_2}\sigma_{22} + \frac{1}{E_3}\sigma_{33}, \tag{3.15.11}$$

$$\epsilon_{12} = \frac{1}{G_{12}}\sigma_{12}, \tag{3.15.12}$$

$$\epsilon_{13} = \frac{1}{G_{13}}\sigma_{13}, \tag{3.15.13}$$

$$\epsilon_{23} = \frac{1}{G_{23}}\sigma_{23}. \tag{3.15.14}$$

Due to the underlying theory the following symmetry holds

$$\frac{\nu_{12}}{E_2} = \frac{\nu_{21}}{E_1}, \tag{3.15.15}$$

$$\frac{\nu_{13}}{E_3} = \frac{\nu_{31}}{E_1}, \tag{3.15.16}$$

$$\frac{\nu_{23}}{E_3} = \frac{\nu_{32}}{E_2}. \tag{3.15.17}$$

Therefore,

$$\epsilon_{11} = \frac{1}{E_1}\sigma_{11} - \frac{\nu_{21}}{E_1}\sigma_{22} - \frac{\nu_{31}}{E_1}\sigma_{33}, \tag{3.15.18}$$

$$\epsilon_{22} = -\frac{\nu_{21}}{E_1}\sigma_{11} + \frac{1}{E_2}\sigma_{22} - \frac{\nu_{32}}{E_2}\sigma_{33}, \tag{3.15.19}$$

$$\epsilon_{33} = -\frac{\nu_{31}}{E_1}\sigma_{11} - \frac{\nu_{32}}{E_2}\sigma_{22} + \frac{1}{E_3}\sigma_{33}, \tag{3.15.20}$$

$$\epsilon_{12} = \frac{1}{G_{12}}\sigma_{12}, \tag{3.15.21}$$

$$\epsilon_{13} = \frac{1}{G_{13}}\sigma_{13}, \tag{3.15.22}$$

$$\epsilon_{23} = \frac{1}{G_{23}}\sigma_{23}. \tag{3.15.23}$$

Due to the transverse isotropy of the material

$$E_1 = E_2 = E_i, \tag{3.15.24}$$

$$E_3 = E_a, \tag{3.15.25}$$

$$\nu_{21} = \nu, \tag{3.15.26}$$

$$\nu_{31} = \nu_{32} = \nu_{ia}, \tag{3.15.27}$$

$$G_{12} = \frac{E_i}{2(1+\nu)}, \tag{3.15.28}$$

$$G_{13} = G_{23} = G_a \tag{3.15.29}$$

and the constitutive equations become

$$\epsilon_{11} = \frac{1}{E_i}\sigma_{11} - \frac{\nu}{E_i}\sigma_{22} - \frac{\nu_{ia}}{E_i}\sigma_{33}, \tag{3.15.30}$$

$$\epsilon_{22} = -\frac{\nu}{E_i}\sigma_{11} + \frac{1}{E_i}\sigma_{22} - \frac{\nu_{ia}}{E_i}\sigma_{33}, \tag{3.15.31}$$

$$\epsilon_{33} = -\frac{\nu_{ia}}{E_i}\sigma_{11} - \frac{\nu_{ia}}{E_i}\sigma_{22} + \frac{1}{E_a}\sigma_{33}, \tag{3.15.32}$$

$$\epsilon_{12} = \frac{2(1+\nu)}{E_i}\sigma_{12}, \qquad (3.15.33)$$

$$\epsilon_{13} = \frac{1}{G_a}\sigma_{13}, \qquad (3.15.34)$$

$$\epsilon_{23} = \frac{1}{G_a}\sigma_{23}. \qquad (3.15.35)$$

Employing the stresses above

$$\epsilon_{11} = \epsilon_{22} = \epsilon_{33} = \frac{\partial u_x}{\partial x} = \frac{\partial u_y}{\partial y} = \frac{\partial u_z}{\partial z} = 0, \qquad (3.15.36)$$

$$\epsilon_{12} = \frac{\partial u_x}{\partial y} + \frac{\partial u_y}{\partial x} = \frac{2(1+\nu)}{E_i}\sigma_0, \qquad (3.15.37)$$

$$\epsilon_{13} = \frac{\partial u_x}{\partial z} + \frac{\partial u_z}{\partial x} = \frac{1}{G_a}\sigma_0, \qquad (3.15.38)$$

$$\epsilon_{23} = \frac{\partial u_y}{\partial z} + \frac{\partial u_z}{\partial y} = 0. \qquad (3.15.39)$$

The displacements (u_x, u_y, u_z) will next be obtained from the strains.

$$\frac{\partial u_x}{\partial x} = \frac{\partial u_y}{\partial y} = \frac{\partial u_z}{\partial z} = 0 \qquad (3.15.40)$$

yield

$$u_x = u_x(y, z), \qquad (3.15.41)$$
$$u_y = u_y(x, z), \qquad (3.15.42)$$
$$u_z = u_z(x, y), \qquad (3.15.43)$$

thus reducing the number of independent variables. Because $u_x(y, z)$ does not depend on x and the face $x = 0$ is entirely fixed

$$u_x(y, z) = 0. \qquad (3.15.44)$$

Then

$$\epsilon_{12} = \frac{\partial u_x}{\partial y} + \frac{\partial u_y}{\partial x} = \frac{\partial}{\partial x}u_y(x, z) = \frac{2(1+\nu)}{E_i}\sigma_0, \qquad (3.15.45)$$

and due to the fixities at the face $x = 0$

$$u_y(x) = x\frac{2(1+\nu)}{E_i}\sigma_0. \qquad (3.15.46)$$

Table 3.16 Benchmark deposit (https://docs.opengeosys.org/books/bmb-5)

BM code	Author	Code	Files	CTest
BMB5-3.15	Peter Vogel	OGS-5	Available	ToDo

(https://oc.ufz.de/index.php/s/nlph7bhfDkj6tC7)

Because

$$\epsilon_{13} = \frac{\partial u_x}{\partial z} + \frac{\partial u_z}{\partial x} = \frac{\partial}{\partial x} u_z(x, y) = \frac{1}{G_a}\sigma_0, \qquad (3.15.47)$$

the fixities at the face $x = 0$ yield

$$u_z(x) = x\frac{1}{G_a}\sigma_0, \qquad (3.15.48)$$

and the above $u_y(x)$ and $u_z(x)$ also satisfy $\epsilon_{23} = 0$. The displacement vector (u_x, u_y, u_z) becomes

$$u_x = 0, \qquad (3.15.49)$$

$$u_y(x) = x\frac{2(1 + \nu)}{E_i}\sigma_0, \qquad (3.15.50)$$

$$u_z(x) = x\frac{1}{G_a}\sigma_0. \qquad (3.15.51)$$

Due to the applied load the plate undergoes shear in the x-y-plane and in the x-z-plane.

3.16   A Cuboid Deformes Under Its Own Weight, Anisotropy Parallel to Z-Axis

The domain is a cuboid discretized by $10 \times 10 \times 10$ equally sized hexahedral elements. Given size $H = 10$ m the vertices of the front face $y = -H$ have the x-y-z-coordinates $(0, -H, H), (H, -H, 0), (0, -H, -H)$, and $(-H, -H, 0)$, respectively; the rear face is located on the $y = H$ level. The material has been selected transversely isotropic, the axis of anisotropy is the z-axis. For the elastic parameters see Table 3.11, density $\rho = 2450$ kg/m^3 has been assigned. Gravitational acceleration $g = 9.81$ m/s^2 is applied in negative z-direction by default. Fixities have been prescribed in the interior of the domain with zero x-displacement along the plane $x = 0$, zero y-displacement along the plane $y = 0$, and zero z-displacement at the origin. The bottom and the lateral faces are free of load by default. Specified loads $P(x, y, z)$ prevail along the top faces of the domain. These loads read along the top face $x \geq 0$, $z \geq 0$

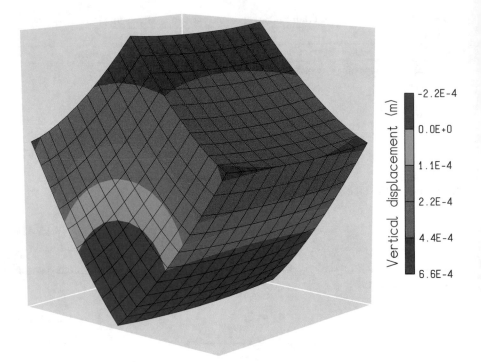

Fig. 3.17 Deformations scaled up, vertical displacements

$$P(x, y, H - x) = \frac{1}{\sqrt{2}} \rho g H \begin{pmatrix} 1 \\ 0 \\ 1 \end{pmatrix}, \qquad (3.16.1)$$

and along the top face $x \leq 0, z \geq 0$

$$P(x, y, H + x) = \frac{1}{\sqrt{2}} \rho g H \begin{pmatrix} -1 \\ 0 \\ 1 \end{pmatrix}. \qquad (3.16.2)$$

The simulation (Table 3.17) comprises one timestep to evaluate the stresses, the strains, and the displacement vector (u_x, u_y, u_z) (Fig. 3.17).

The stress tensor

$$\boldsymbol{\sigma} = \begin{pmatrix} \sigma_{11} & \sigma_{12} & \sigma_{13} \\ \sigma_{12} & \sigma_{22} & \sigma_{23} \\ \sigma_{13} & \sigma_{23} & \sigma_{33} \end{pmatrix} = \frac{\rho g}{2} \begin{pmatrix} z + H & 0 & x \\ 0 & 0 & 0 \\ x & 0 & z + H \end{pmatrix} \qquad (3.16.3)$$

satisfies the equation of mechanical equilibrium

$$\operatorname{div} \boldsymbol{\sigma} = (0, 0, \rho g) \tag{3.16.4}$$

as well as the specified surface loads because along the face $y = H$

$$\boldsymbol{\sigma} \begin{pmatrix} 0 \\ 1 \\ 0 \end{pmatrix} = \frac{\rho g}{2} \begin{pmatrix} z+H & 0 & x \\ 0 & 0 & 0 \\ x & 0 & z+H \end{pmatrix} \begin{pmatrix} 0 \\ 1 \\ 0 \end{pmatrix} = \begin{pmatrix} 0 \\ 0 \\ 0 \end{pmatrix}, \tag{3.16.5}$$

along the face $y = -H$

$$\boldsymbol{\sigma} \begin{pmatrix} 0 \\ -1 \\ 0 \end{pmatrix} = \frac{\rho g}{2} \begin{pmatrix} z+H & 0 & x \\ 0 & 0 & 0 \\ x & 0 & z+H \end{pmatrix} \begin{pmatrix} 0 \\ -1 \\ 0 \end{pmatrix} = \begin{pmatrix} 0 \\ 0 \\ 0 \end{pmatrix}, \tag{3.16.6}$$

along the top face $x \geq 0$, $z \geq 0$ where $z = H - x$

$$\boldsymbol{\sigma} \frac{1}{\sqrt{2}} \begin{pmatrix} 1 \\ 0 \\ 1 \end{pmatrix} = \frac{\rho g}{2} \begin{pmatrix} H-x+H & 0 & x \\ 0 & 0 & 0 \\ x & 0 & H-x+H \end{pmatrix} \frac{1}{\sqrt{2}} \begin{pmatrix} 1 \\ 0 \\ 1 \end{pmatrix}$$
$$= \frac{1}{\sqrt{2}} \rho g H \begin{pmatrix} 1 \\ 0 \\ 1 \end{pmatrix}, \tag{3.16.7}$$

along the top face $x \leq 0$, $z \geq 0$ where $z = H + x$

$$\boldsymbol{\sigma} \frac{1}{\sqrt{2}} \begin{pmatrix} -1 \\ 0 \\ 1 \end{pmatrix} = \frac{\rho g}{2} \begin{pmatrix} H+x+H & 0 & x \\ 0 & 0 & 0 \\ x & 0 & H+x+H \end{pmatrix} \frac{1}{\sqrt{2}} \begin{pmatrix} -1 \\ 0 \\ 1 \end{pmatrix}$$
$$= \frac{1}{\sqrt{2}} \rho g H \begin{pmatrix} -1 \\ 0 \\ 1 \end{pmatrix}, \tag{3.16.8}$$

along the bottom face $x \leq 0$, $z \leq 0$ where $z = -H - x$

$$\boldsymbol{\sigma} \frac{1}{\sqrt{2}} \begin{pmatrix} -1 \\ 0 \\ -1 \end{pmatrix} = \frac{\rho g}{2} \begin{pmatrix} -H-x+H & 0 & x \\ 0 & 0 & 0 \\ x & 0 & -H-x+H \end{pmatrix} \frac{1}{\sqrt{2}} \begin{pmatrix} -1 \\ 0 \\ -1 \end{pmatrix}$$
$$= \begin{pmatrix} 0 \\ 0 \\ 0 \end{pmatrix}, \tag{3.16.9}$$

and along the bottom face $x \geq 0$, $z \leq 0$ where $z = -H + x$

$$\sigma \frac{1}{\sqrt{2}} \begin{pmatrix} 1 \\ 0 \\ -1 \end{pmatrix} = \frac{\rho g}{2} \begin{pmatrix} -H+x+H & 0 & x \\ 0 & 0 & 0 \\ x & 0 & -H+x+H \end{pmatrix} \frac{1}{\sqrt{2}} \begin{pmatrix} 1 \\ 0 \\ -1 \end{pmatrix}$$

$$= \begin{pmatrix} 0 \\ 0 \\ 0 \end{pmatrix}. \tag{3.16.10}$$

The constitutive equations governing the present situation have already been outlined in the context of the previous example. The stress-strain relationships become

$$\epsilon_{11} = \frac{\partial u_x}{\partial x} = \frac{1}{E_i}\sigma_{11} - \frac{\nu}{E_i}\sigma_{22} - \frac{\nu_{ia}}{E_i}\sigma_{33}$$

$$= \frac{1-\nu_{ia}}{2E_i}\rho g(z+H), \tag{3.16.11}$$

$$\epsilon_{22} = \frac{\partial u_y}{\partial y} = -\frac{\nu}{E_i}\sigma_{11} + \frac{1}{E_i}\sigma_{22} - \frac{\nu_{ia}}{E_i}\sigma_{33}$$

$$= -\frac{\nu+\nu_{ia}}{2E_i}\rho g(z+H), \tag{3.16.12}$$

$$\epsilon_{33} = \frac{\partial u_z}{\partial z} = -\frac{\nu_{ia}}{E_i}\sigma_{11} - \frac{\nu_{ia}}{E_i}\sigma_{22} + \frac{1}{E_a}\sigma_{33}$$

$$= \left(\frac{1}{E_a} - \frac{\nu_{ia}}{E_i}\right)\frac{\rho g}{2}(z+H), \tag{3.16.13}$$

$$\epsilon_{12} = \frac{\partial u_x}{\partial y} + \frac{\partial u_y}{\partial x} = \frac{2(1+\nu)}{E_i}\sigma_{12} = 0, \tag{3.16.14}$$

$$\epsilon_{13} = \frac{\partial u_x}{\partial z} + \frac{\partial u_z}{\partial x} = \frac{1}{G_a}\sigma_{13} = \frac{\rho g}{2G_a}x, \tag{3.16.15}$$

$$\epsilon_{23} = \frac{\partial u_y}{\partial z} + \frac{\partial u_z}{\partial y} = \frac{1}{G_a}\sigma_{23} = 0, \tag{3.16.16}$$

and constitute a set of equations for the partial derivatives of the displacements. Integrating the strains ϵ_{11}, ϵ_{22}, and ϵ_{33} with respect to the prescribed x- and y-fixities gives the displacement vector (u_x, u_y, u_z)

$$u_x(x,z) = \frac{1-\nu_{ia}}{2E_i}\rho g(z+H)x, \tag{3.16.17}$$

$$u_y(y,z) = -\frac{\nu+\nu_{ia}}{2E_i}\rho g(z+H)y, \tag{3.16.18}$$

$$u_z(x,y,z) = \left(\frac{1}{E_a} - \frac{\nu_{ia}}{E_i}\right)\frac{\rho g}{2}\left(\frac{1}{2}z^2 + Hz\right) + h(x,y), \tag{3.16.19}$$

Table 3.17 Benchmark deposit (https://docs.opengeosys.org/books/bmb-5)

BM code	Author	Code	Files	CTest
BMB5-3.16	Peter Vogel	OGS-5	Available	ToDo

(https://oc.ufz.de/index.php/s/nlph7bhfDkj6tC7)

where $h(x, y)$ has to be determined from the shear strains and the specified z-fixity. Then

$$\epsilon_{12} = \frac{\partial u_x}{\partial y} + \frac{\partial u_y}{\partial x} = 0, \tag{3.16.20}$$

$$\epsilon_{13} = \frac{\partial u_x}{\partial z} + \frac{\partial u_z}{\partial x} = \frac{1 - \nu_{ia}}{2E_i}\rho g x + \frac{\partial h}{\partial x} = \frac{\rho g}{2G_a}x, \tag{3.16.21}$$

$$\epsilon_{23} = \frac{\partial u_y}{\partial z} + \frac{\partial u_z}{\partial y} = -\frac{\nu + \nu_{ia}}{2E_i}\rho g y + \frac{\partial h}{\partial y} = 0, \tag{3.16.22}$$

hence,

$$\frac{\partial h}{\partial x} = \frac{\rho g}{2}\left(\frac{1}{G_a} - \frac{1 - \nu_{ia}}{E_i}\right)x, \tag{3.16.23}$$

$$\frac{\partial h}{\partial y} = \frac{\rho g}{2}\frac{\nu + \nu_{ia}}{E_i}y, \tag{3.16.24}$$

and therefore,

$$h(x, y) = \frac{\rho g}{4}\left(\frac{1}{G_a} - \frac{1 - \nu_{ia}}{E_i}\right)x^2 + \frac{\rho g}{4}\frac{\nu + \nu_{ia}}{E_i}y^2 + C. \tag{3.16.25}$$

The z-fixity at the origin yields the free constant C. The displacement vector (u_x, u_y, u_z) becomes

$$u_x(x, z) = \frac{1 - \nu_{ia}}{2E_i}\rho g(z + H)x, \tag{3.16.26}$$

$$u_y(y, z) = -\frac{\nu + \nu_{ia}}{2E_i}\rho g(z + H)y, \tag{3.16.27}$$

$$u_z(x, y, z) = \frac{\rho g}{4}\left(\frac{1}{G_a} - \frac{1 - \nu_{ia}}{E_i}\right)x^2 + \frac{\rho g}{4}\frac{\nu + \nu_{ia}}{E_i}y^2$$
$$+ \frac{\rho g}{2}\left(\frac{1}{E_a} - \frac{\nu_{ia}}{E_i}\right)\left(\frac{1}{2}z^2 + Hz\right). \tag{3.16.28}$$

Fig. 3.18 Rheological analogue of the LUBBY2 model

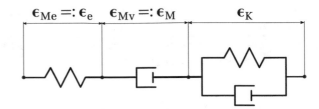

$$\epsilon_{Me} =: \epsilon_e \quad \epsilon_{Mv} =: \epsilon_M \qquad \qquad \epsilon_K$$

3.17 A Viscoelastic (LUBBY2) Material in Simple-Shear Creep

Xing-Yuan Miao, Dmitri Naumov, Thomas Nagel

This benchmark tests the OGS-6 implementation of the LUBBY2 model based on analogous tests which have been performed in OGS-5 (Kolditz et al. 2016a; Nagel et al. 2017b). The LUBBY2 constitutive model has been used extensively to model both primary and secondary creep of rock salt. It is based on the generalised Burgers model and is described by the following set of equations (Nagel et al. (2017b) (for the definition of the various strain measures, see Fig. 3.18):

$$\dot{\sigma} = \mathcal{C}_M : \left[\dot{\epsilon} - \mathcal{V}_M^{-1} : \sigma^D - \mathcal{V}_K^{-1} : (\sigma - \mathcal{C}_K : \epsilon_K) \right] \qquad (3.17.1)$$

$$\dot{\epsilon}_K = \mathcal{V}_K^{-1} : (\sigma - \mathcal{C}_K : \epsilon_K) \qquad (3.17.2)$$

$$\dot{\epsilon}_M = \mathcal{V}_M^{-1} : \sigma \qquad (3.17.3)$$

where $\mathcal{V}_M$ and $\mathcal{V}_K$ represent the viscosity tensors of the Maxwell and Kelvin elements (cf. Fig. 3.18), respectively, and $\mathcal{C}_M$ as well as $\mathcal{C}_K$ are their stiffness tensors. Making the usual assumptions of isotropy and isochoric deformations in all rheological elements except for the spring in the Maxwell element which explicitly allows elastic compressibility, the LUBBY2 model is completed by the following constitutive dependencies which define the Kelvin shear modulus and the viscosities as functions of the current stress state:

$$G_K = G_{K0} e^{m_K \sigma_{eff}} \qquad (3.17.4)$$

$$\eta_M = \eta_{M0} e^{m_1 \sigma_{eff}} \qquad (3.17.5)$$

$$\eta_K = \eta_{K0} e^{m_2 \sigma_{eff}} \qquad (3.17.6)$$

with the effective stress σ_{eff}

$$\sigma_{eff} = \sqrt{\frac{3}{2} \sigma^D : \sigma^D} \qquad (3.17.7)$$

calculated from the deviatoric stress tensor σ^D, where the m_a are material parameters characterising the stress dependency of the respective moduli.

Fig. 3.19 Loading and
boundary conditions

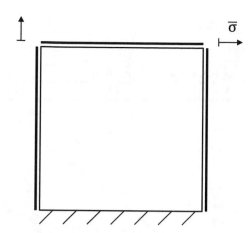

A simple-shear test was chosen as a reference solution for which the analytical
solution for the shear strain developing in a Burgers-type model following a shear-
stress jump reads (Nagel et al. 2017b):

$$\epsilon_{xy}(t) = \frac{1}{2}\left[\left(\frac{1}{G_M} + \frac{1}{\eta_M}t\right)\sigma_{xy} + \frac{1}{G_K}\left(1 - e^{-\frac{G_K}{\eta_K}t}\right)\sigma_{xy}\right] \qquad (3.17.8)$$

We here provide numerical examples to verify the implementation of the material
model in OGS-65 and OGS-6. Both 2D and 3D cases were considered. The mechanical model is either a square plate (2D) or a cube (3D) with a positive shear stress of
0.01 MPa applied along the top side or surface in the x direction; also see Fig. 3.19
for the remaining boundary conditions. For the 2D case, vertical displacements of
the all four sides are constrained, the horizontal displacement of the bottom is also
constrained to maintain the simple-shear setting. For the 3D case, the bottom surface is constrained in vertical and its normal directions, and the top, left, and right
surfaces are constrained in the vertical direction. Front and back surfaces are constrained in their normal directions. Both 2D and 3D examples were tested in OGS-5,
while only the 3D example was tested in OGS-6. A discretisation of 100 quadrilateral
elements was applied for the 2D plane strain model, while a discretisation of 1000
hexahedral elements was applied for the 3D cube model. In OGS-5, bi-/tri-quadratic
shape functions were used, while in OGS-6 hexahedral linear shape functions were
employed.

The material property set for this benchmark is listed in Table 3.18. The conjugate
gradient method (CG) was used as the linear solver with a tolerance of 10^{-16} in OGS-
6, while the direct solver (PARDISO) was used with a tolerance of 10^{-15} in OGS-5.
The Newton–Raphson method was implemented as the nonlinear solver with an
absolute displacement tolerance of 10^{-15} in OGS-5 and 10^{-12} in OGS-6. The load
was applied within one time step of 0.0001 s and subsequently held for 1 s divided

Table 3.18 Material properties used in LUBBY2 model

G_M/MPa	K_M/MPa	η_{M0}/MPa d	G_{K0}/MPa
0.8	0.8	0.5	0.8
η_{K0}/MPa d	m_1/MPa^{-1}	m_2/MPa^{-1}	m_G/MPa^{-1}
0.5	−0.3	−0.2	−0.2

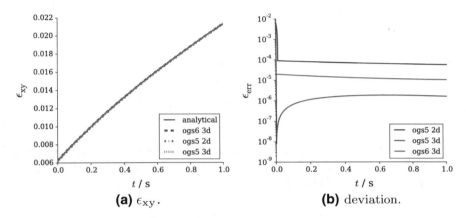

(a) ϵ_{xy}.   **(b)** deviation.

Fig. 3.20 Variation of the shear strain with time (**a**) and the deviation between analytical solution and numerical simulations (**b**)

Table 3.19 Benchmark deposit (https://docs.opengeosys.org/books/bmb-5)

BM code	Author	Code	Files	CTest
BMB5-3.17	Xing-Yuan Miao	OGS-6	See below	OK

(https://oc.ufz.de/index.php/s/nlph7bhfDkj6tC7)

into 100 equidistant time steps. The local Newton tolerance is set to 10^{-10} in both OGS-5 and OGS-6.

Comparisons of the analytical solution and the numerical simulations show the satisfactory match between these solutions, see Fig. 3.20a. The maximum deviation between the analytical solution and the 2D solution from OGS-5 is 10^{-4}, then decreases to approximately 10^{-5} by the 3D solution from OGS-6. The maximum deviation between the analytical solution and the 3D solution from OGS-5 is the minimum, approximately 10^{-6}, see Fig. 3.20b.

The OGS-6 version used for this benchmark was the master branch available at (Table 3.19)

- https://github.com/ufz/ogs

with the input files available as CTests under

- https://github.com/ufz/ogs-data/tree/master/Mechanics/Burgers.

3.18 Triaxial Compression of an Elasto-Plastic Material with Hardening Based on Ehlers' Yield Surface

Xing-Yuan Miao, Dmitri Naumov, Thomas Nagel

3.18.1 Introduction

The elasto-plastic behaviour of geomaterials is complex and usually non-linear. To capture a wide range of basic phenomena, a generic yield function for geomaterials was implemented in OGS-6 as the first elasto-plastic model. It is given by a seven-parametric ($\alpha, \beta, \gamma, \delta, \epsilon, k, m$) yield function presented by Ehlers (1995) and expressed as:

$$F = \Phi^{\frac{1}{2}} + \beta I_1 + \varepsilon I_1^2 - k(\varepsilon_{p\,eff}) \tag{3.18.1}$$

$$\Phi = J_2 (1 + \gamma \vartheta)^m + \frac{1}{2} \alpha I_1^2 + \delta^2 I_1^4 \tag{3.18.2}$$

$$\vartheta = \frac{J_3}{J_2^{\frac{3}{2}}} \tag{3.18.3}$$

where I_1 is the first principal invariant of the stress tensor, while J_2 and J_3 are the second and third principal invariants of the deviatoric stress tensor. This yield surface allows the modelling of a variable stress-dependence of the yield locus both on its position in the deviatoric plane and in the Meridian plane. Similarly, the transition between dilatant and contractile flow can be modelled when using this surface as the plastic potential. Additionally, the surface is both smooth and smoothly differentiable which is a favourable property for both numerical and theoretical analyses.

To account for isotropic hardening, the yield-surface expansion is modelled by assuming the following dependence of the strength parameter k on the plastic effective strain:

$$k(\varepsilon_{p\,eff}) = \kappa(1 + h\varepsilon_{p\,eff}) \tag{3.18.4}$$

where h represents the hardening parameter.

A different parameterization is usually chosen for the plastic potential G_F in order to model non-associative flow ($\alpha_p, \beta_p, \gamma_p, \delta_p, \epsilon_p, m_p$). Proceeding from the flow rule

$$\dot{\epsilon}_p = \lambda \frac{\partial G_F}{\partial \sigma} \tag{3.18.5}$$

the following derivatives are required:

$$\frac{\partial G_F}{\partial \sigma} = \frac{1}{2\Phi_p^{\frac{1}{2}}} \frac{\partial \Phi_p}{\partial \sigma} + \beta_p \frac{\partial I_1}{\partial \sigma} + 2\varepsilon_p I_1 \frac{\partial I_1}{\partial \sigma} \tag{3.18.6}$$

$$= \frac{1}{2\Phi_p^{\frac{1}{2}}} \frac{\partial \Phi_p}{\partial \sigma} + \beta_p \mathbf{I} + 2\varepsilon_p I_1 \mathbf{I} \tag{3.18.7}$$

In the above,

$$\frac{\partial \Phi_p}{\partial \sigma} = \left(1 + \gamma_p \vartheta\right)^{m_p} \frac{\partial J_2}{\partial \sigma} + m_p \gamma_p J_2 \left(1 + \gamma_p \vartheta\right)^{m_p - 1} \frac{\partial \vartheta}{\partial \sigma} + \alpha_p I_1 \frac{\partial I_1}{\partial \sigma} +$$
$$+ 4\delta_p^2 I_1^3 \frac{\partial I_1}{\partial \sigma} \tag{3.18.8}$$

$$= \left(1 + \gamma_p \vartheta\right)^{m_p} \sigma^D + m_p \gamma_p J_2 \left(1 + \gamma_p \vartheta\right)^{m_p - 1} \frac{\partial \vartheta}{\partial \sigma} +$$
$$+ \left(\alpha_p I_1 + 4\delta_p^2 I_1^3\right)\mathbf{I} \tag{3.18.9}$$

are required where in turn the material-independent relations

$$\frac{\partial \vartheta}{\partial \sigma} = \frac{\partial \vartheta}{\partial J_3} \frac{\partial J_3}{\partial \sigma} + \frac{\partial \vartheta}{\partial J_2} \frac{\partial J_2}{\partial \sigma} = \vartheta \left(\sigma^{D-1}\right)^D - \frac{3}{2} \frac{\vartheta}{J_2} \sigma^D \tag{3.18.10}$$

$$\frac{\partial J_3}{\partial \sigma} = J_3 \left(\sigma^{D-1}\right)^D \tag{3.18.11}$$

$$\frac{\partial J_2}{\partial \sigma} = \sigma^D \tag{3.18.12}$$

$$\frac{\partial I_1}{\partial \sigma} = \mathbf{I} \tag{3.18.13}$$

have been used. For details on the implicit integration scheme, see Nagel et al. (2017b).

3.18.2 Benchmark

For benchmarking purposes, a parameter set inspired by the one used in the work of Ehlers and Avci (2013) has been used and is given in Table 3.20. Even though analytical solutions are not available, principal features of the model can be tested. The objective of this benchmark was two-fold: (i) to test the basic elasto-plastic behaviour, namely the onset of yielding and the maintenance of the constraint $F = 0$ during plastic flow, and (ii) to test the non-associated flow for plausibility, in particular its volumetric component.

Table 3.20 Values of the parameters associated with the elastic (G, K) and plastic model components

G MPa	K MPa	h	κ MPa^{-1}	β	γ	α	δ MPa^{-1}
150	200	10	0.1	0.095	1	0.01	0.0078
ϵ	m	β_p	γ_p	α_p	δ_p MPa^{-1}	ϵ_p MPa^{-1}	m_p
0.1	0.54	0.0608	1	0.01	0.0078	0.1	0.54

For that purpose, a conventional triaxial compression test was performed under two different conditions, **load cases** 1 and 2:

1. Following the application of an isotropic confining pressure, the yield-point is approached with a constant hydrostatic stress in the elastic range. This case is referred to as $I_1 = $ const.. This is achieved by lowering the confining pressure as the axial stress is increased. Compare load paths indicated in Figs. 3.21 and 3.22.
2. Following the application of a radial confining stress at constant axial displacement, an axial compression ramp was applied. This resulted in a monotonically increasing hydrostatic stress state until the onset of yielding while transitioning through a non-deviatoric elastic loading state during the compression process. Compare load paths indicated in Fig. 3.23.

During the compression test, a continuous displacement-controlled loading process was applied axially with a displacement-rate of 1 mm/s.

A discretisation of 1000 hexahedral elements was applied to a three-dimensional cube model. The biconjugate gradient stabilized method (BiCGSTAB) was used as the linear solver and the tolerance set to 10^{-16}. The Newton–Raphson method was employed as the nonlinear solver with an absolute tolerance of 10^{-14}. The time-step size was set to 0.025 s.

The OGS-6 version used for this benchmark was the official master branch available at

- https://github.com/ufz/ogs

with the input files available as CTests under

- https://github.com/ufz/ogs-data/tree/master/Mechanics/Ehlers.

In order to test the non-associated flow feature as well as the transition between contractile and dilatant flow, the confining pressure at the onset of yielding in load case 1 is controlled such that yielding occurs either at a confining pressure corresponding to the maximum of the plastic potential in the hydrostatic plane (i.e. where plastic flow is momentarily isochoric; r_s^{max}), or to the dilatant/contractant sides of that point (r_s^{before} and r_s^{after}, respectively). This can be most clearly seen in Fig. 3.21.

To arrive at an analytical expression for this point, the consolidation pressure p_c and the parameter π_c which defines the ratio between the pressures corresponding to the curve's maximum and length (Ehlers and Avci 2013) can be expressed as functions of β_p, δ_p and ϵ_p:

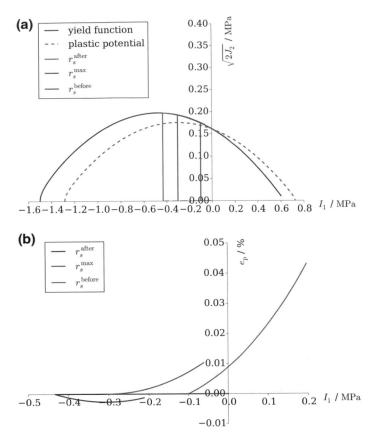

Fig. 3.21 **a** Characteristic shape of the yield function and the plastic potential in the hydrostatic plane and **b** the corresponding plastic volumetric strain when $I_1 = \text{const.}$, $\alpha = 0$. The curves labelled r_s indicate the different stress paths as described in the text

$$p_c = \frac{\beta_p}{3\left(\epsilon_p + \delta_p\right)} \tag{3.18.14}$$

$$\pi_c = \frac{1}{4\left(\epsilon_p - \delta_p\right)}\left(3\epsilon_p - \sqrt{8\delta_p^2 + \epsilon_p^2}\right) \tag{3.18.15}$$

Note that, only in the case of $\alpha = 0$, the curve's maximum is exclusively described by p_c and π_c, whereas in the general case these relations serve as an approximation. This can be seen in Figs. 3.21 and 3.22: in the case of $\alpha = 0$ (Fig. 3.21), the curve r_s^{max} hits the peak of the plastic potential and the corresponding plastic flow is isochoric at its onset, i.e. $\dot{e}_p = 0$. This is only approximately the case for $\alpha \neq 0$, see Fig. 3.22. In both cases, the load-paths r_s^{before} and r_s^{after} clearly show dilatant ($\dot{e}_p > 0$) and contractant ($\dot{e}_p < 0$) flow, respectively. Note also, how this behaviour changes as the hydrostatic

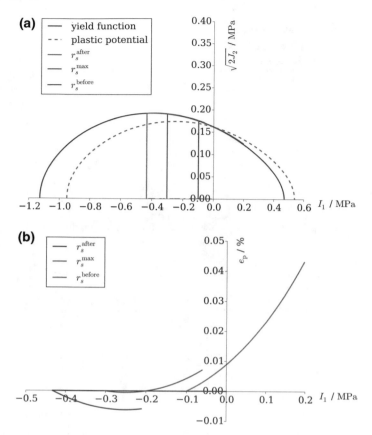

Fig. 3.22 **a** Characteristic shape of the yield function and the plastic potential in the hydrostatic plane and **b** the corresponding plastic volumetric strain when $I_1 = $ const., $\alpha = 0.01$. The curves labelled r_s indicate the different stress paths as described in the text

stress state evolves and shifts the stress state such that flow becomes more dilatant. This evolution of the stress state is correctly constrained by the yield surface ($F = 0$).

Figure 3.23 shows the triaxial compression test under the second set of boundary conditions, i.e. load case 2.

The plastic volumetric strain exhibits a strongly nonlinear progression throughout the loading process in that flow varies rapidly from contractile to dilatant as the stress state passes the peak of the plastic potential.

Note further that, (i) J_2 vanishes briefly in the elastic range once the constant confining pressure and displacement loading cause an isotropic stress state; (ii) The plastic volumetric strain achieves a minimum after which dilatant flow causes the sample volume to increase again.

In all cases, the stress state during plastic loading remains true to the yield-surface constraint $F = 0$ (Figs. 3.21, 3.22 and 3.23).

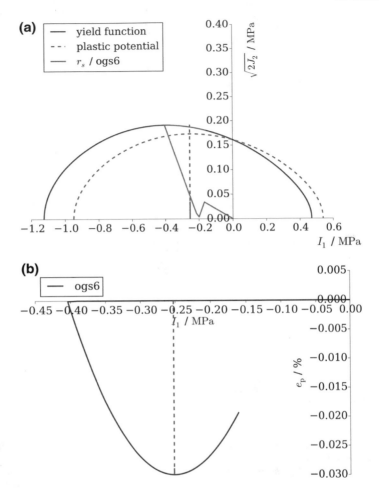

Fig. 3.23 Load case 2. **a** Stress path in the hydrostatic plane and **b** the plastic volumetric strain as it evolves with the changing stress state. The dashed line indicates I_1 at the peak of the plastic potential. The curve labelled r_s indicates the stress paths as described in the text

3.18.3 Drucker–Prager as a Special Case

Finally, an important special case was tested. By prescribing specific material parameters of the Ehlers surface to be zero, the seven-parametric yield function can be reduced to the two well-known yield conditions *Drucker–Prager* ($\alpha = \delta = \epsilon = 0$) and *von Mises* ($\alpha = \beta = \delta = \epsilon = 0$). Here, we provide a Drucker–Prager model test as an additional verification of the Ehlers material model. The boundary conditions of simulation set (2) were chosen. The typical linear shape in the hydrostatic plane and the corresponding constant-dilatancy behaviour can be seen in Fig. 3.24. Note in passing that this reduction of the Ehlers material model to Drucker–Prager cor-

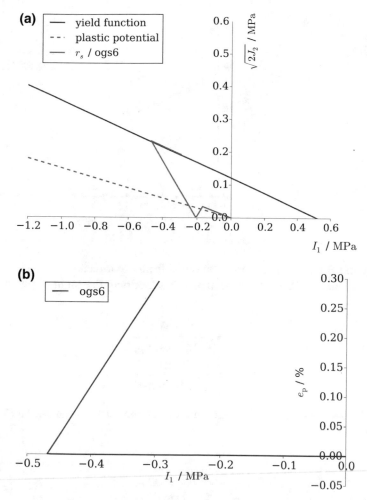

Fig. 3.24 Variations of the stress states (**a**) and the plastic volumetric strain (**b**) with the monotonic loading process following load case 2 using a Drucker–Prager yield criterion based on suitably chosen parameters in the Ehlers yield surface. The curve labelled r_s indicates the stress paths as described in the text

responds to the case where the Drucker–Prager yield surface middle circumscribes the Mohr–Coulomb yield surface, see Fig. 3.25 for a comparison of different parameterisations. The green solid line represents the results calculated from the current model, the orange, purple and pink solid lines are the Drucker–Prager yield surface (Eq. 3.18.16) circumscribes (Eq. 3.18.17), middle circumscribes (Eq. 3.18.18) and inscribes (Eq. 3.18.19) the Mohr–Coulomb yield surface, respectively.

$$\sqrt{J_2} = A + BI_1 \qquad (3.18.16)$$

Fig. 3.25 Reduction of the
Ehlers material model to
Drucker–Prager model

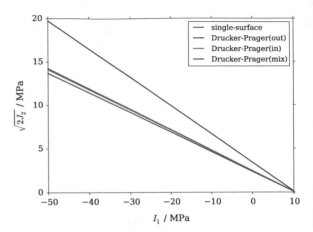

where A and B are two constants determined from experiments.

For the Drucker–Prager yield surface circumscribing the Mohr–Coulomb yield
surface

$$A = \frac{6c \cos \phi}{\sqrt{3}(3 - \sin \phi)}$$
$$B = \frac{2 \sin \phi}{\sqrt{3}(3 - \sin \phi)}$$

(3.18.17)

for the Drucker–Prager yield surface middle circumscribing the Mohr–Coulomb
yield surface

$$A = \frac{6c \cos \phi}{\sqrt{3}(3 + \sin \phi)}$$
$$B = \frac{2 \sin \phi}{\sqrt{3}(3 + \sin \phi)}$$

(3.18.18)

for the Drucker–Prager yield surface inscribing the Mohr–Coulomb yield surface

$$A = \frac{3c \cos \phi}{\sqrt{9 + 3 \sin^2 \phi}}$$
$$B = \frac{\sin \phi}{\sqrt{9 + 3 \sin^2 \phi}}$$

(3.18.19)

where c is the cohesion parameter. ϕ the angle of internal friction (Table 3.21).

Table 3.21 Benchmark deposit (https://docs.opengeosys.org/books/bmb-5)

BM code	Author	Code	Status	CTest
BMB5-3.18	Xing-Yuan Miao	OGS-6	See below	OK

(https://oc.ufz.de/index.php/s/nlph7bhfDkj6tC7)

The OGS-6 version used for this benchmark was the master branch available at

- https://github.com/ufz/ogs

with the input files available as CTests under

- https://github.com/ufz/ogs-data/tree/master/Mechanics/Ehlers.

3.19   Material Forces

Francesco Parisio, Dmitri Naumov and Thomas Nagel

Material forces, which are also referred to as configurational forces, can be seen as forces that arise as reactions to a change in configuration of the material points of a continuum. One of the most illustrative examples is the one of the zipper-force on a crack[1]: we can imagine the crack to be like a zipper which is open up to the crack tip, while the intact material is represented by the closed part of the zipper. If the forces acting on the solid reach a value higher than the crack propagation criterion (critical energy-release rate; stress-intensity factors), the crack propagates and our "zipper" opens up. Material forces at the crack tip are reactions that tend to keep the zipper in place, i.e., they point against the direction of the crack propagation as if they were a reaction to the change of configuration that is caused by the crack propagating. This is on complete analogy to the physical nodal reaction forces at external boundaries that act to keep the current configuration in its deformed shape (Fig. 3.26). The above is a simple and intuitive explanation, while a more rigorous mathematical formulation lies at the heart of material forces theory (Maugin 1995). Generally, configurational forces can be derived in different ways, such as from the Lagrangian function $\mathcal{L}$ of the continuum by investigating translational invariance (Gross et al. 2003). Their application includes error estimation in FE analyses, fracture mechanics, modelling inclusions, inhomogeneities and phase transformations (Mueller et al. 2002; Gross et al. 2003). More generally, configurational forces can be successfully employed to treat topological defects of different geometrical dimensions in continua: points (atomic displacements), lines (dislocations), planes (fractures, phase boundaries) and volumes (inclusions and voids) (Gross et al. 2003). In fracture mechanics, material forces are closely related to the J-integral of fracture (Maugin 1995).

In a small strain framework, the classical momentum equilibrium equation of the continuum reads

---

[1]White Paper–Material Forces: A Novel Approach to Fracture Mechanics in ANSYS, available at http://www.ansys-blog.com/material-forces-for-fracture/.

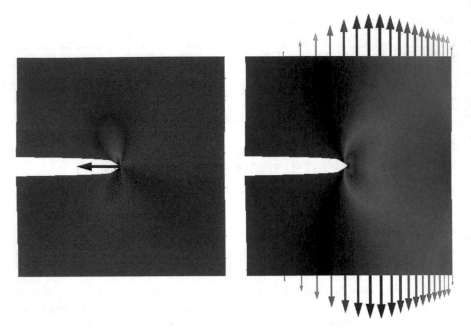

Fig. 3.26 Tensile test of a pre-cracked sample. Material force acting on the crack tip (left) and physical reaction forces at nodes associated with the vertical displacement boundary conditions

$$\text{div } \boldsymbol{\sigma} + \rho \mathbf{b} = \mathbf{0}, \tag{3.19.1}$$

where $\boldsymbol{\sigma}$ is Cauchy's stress tensor, ρ is the mass density and $\mathbf{b}$ is a body-force vector field (e.g., gravity). The gradient of the Helmholtz free energy density $\psi(\varepsilon; x)$ of a heterogeneous medium can be expanded to

$$\text{grad } \psi = \boldsymbol{\sigma} : \text{grad } \varepsilon + \left. \frac{\partial \psi}{\partial \mathbf{x}} \right|_{\text{expl}}, \tag{3.19.2}$$

and the symmetry properties of the stress and strain tensor yield

$$\sigma_{ij} \frac{\partial \varepsilon_{ij}}{\partial x_k} = \sigma_{ij} \frac{\partial^2 u_i}{\partial x_j \partial x_k} = \sigma_{ij} \frac{\partial^2 u_i}{\partial x_k \partial x_j} = \frac{\partial}{\partial x_j} \left(\sigma_{ij} \frac{\partial u_i}{\partial x_k} \right) - \frac{\partial \sigma_{ij}}{\partial x_j} \frac{\partial u_i}{\partial x_k}. \tag{3.19.3}$$

Substitution into Eq. (3.19.2) yields

$$\text{grad } \psi - \text{div} \left(\text{grad}^{\mathsf{T}} \mathbf{u} \boldsymbol{\sigma} \right) - \rho \, \text{grad}^{\mathsf{T}} \mathbf{u} \mathbf{b} - \frac{\partial \psi}{\partial \mathbf{x}} = \mathbf{0}, \tag{3.19.4}$$

which leads to the balance equation of material forces

$$\text{div } \boldsymbol{\Sigma} + \mathbf{g} = \mathbf{0}, \tag{3.19.5}$$

where $\boldsymbol{\Sigma}$ is Eshelby's stress tensor, which is in general non-symmetric and has therefore nine independent components in a three-dimensional setting, and is defined as

$$\boldsymbol{\Sigma} = \psi \mathbf{I} - \text{grad}^T \mathbf{u} \boldsymbol{\sigma}. \tag{3.19.6}$$

Furthermore, $\mathbf{g}$ is the vector field of material forces per unit volume, defined as

$$\mathbf{g} = -\rho \,\text{grad}^T \mathbf{ub} - \left.\frac{\partial \psi}{\partial \mathbf{x}}\right|_{\text{expl}} = \mathbf{g}_{\text{vol}} + \mathbf{g}_{\text{inh}} \tag{3.19.7}$$

with a volumetric component due to body forces

$$\mathbf{g}_{\text{vol}} = -\rho \,\text{grad}^T \mathbf{ub} \tag{3.19.8}$$

and a component related to inhomogeneities

$$\mathbf{g}_{\text{inh}} = - \left.\frac{\partial \psi}{\partial \mathbf{x}}\right|_{\text{expl}}. \tag{3.19.9}$$

Concerning the finite element implementation of material forces (Mueller et al. 2002) into OpenGeoSys, the set of test functions $\mathbf{v} \in \mathbf{V}_0$

$$\mathbf{V}_0 = \left\{ \mathbf{v} \in H^1(\Omega) : \mathbf{v} = \mathbf{0} \,\forall\, \mathbf{x} \in \partial\Omega \right\} \tag{3.19.10}$$

is introduced to arrive at the weak form of the material force balance

$$\int_{\Omega} (\boldsymbol{\Sigma} : \text{grad } \mathbf{v} - \mathbf{g} \cdot \mathbf{v}) \, d\Omega = 0, \tag{3.19.11}$$

which, in analogy with Cauchy's stress integration leading to nodal reactions, leads to the definition of a nodal material force vector as

$$\underline{F}_{\text{conf}} = \int_{\Omega} \underline{\underline{N}}^T \underline{g} \, d\Omega = \int_{\Omega} \underline{\underline{G}}^T \underline{\Sigma} \, d\Omega, \tag{3.19.12}$$

where $\underline{\underline{N}}$ and $\underline{\underline{G}}$ are shape-function matrices and gradient matrices, and $\underline{\bullet}$ is a vectorial representation of the tensorial quantity $\bullet$. The interested reader is referred to the work of Nagel et al. (2016b) on the general definitions and details of the involved FE-matrices as implemented in OGS-6. The described calculation of material forces is performed as a post-processing step following the solution of the physical equilibrium equations.

3.19.1 Benchmark on Material Forces: A Non-uniform Bar in Tension

To verify the numerical implementation of material forces in OGS-6 and to guide the reader towards a better understanding of the concepts illustrated in the previous section, we have designed a simple yet meaningful test: an elastic bar under uniaxial tension σ_0 with a linear distribution of Young's modulus following

$$E\,(x) = E_0 \left[1 + \alpha \left(\frac{x}{l} - \frac{1}{2} \right) \right]. \tag{3.19.13}$$

From Eq. (3.19.13), we can compute the displacement gradient as

$$\operatorname{grad} u = \epsilon = \frac{\sigma_0}{E_0} \left[1 + \alpha \left(\frac{x}{l} - \frac{1}{2} \right) \right]^{-1}, \tag{3.19.14}$$

and the free energy density as

$$\psi = \frac{1}{2}\sigma\epsilon = \frac{\sigma_0^2}{2E_0} \left[1 + \alpha \left(\frac{x}{l} - \frac{1}{2} \right) \right]^{-1}. \tag{3.19.15}$$

Based on these results, $\Sigma = -\psi$ follows. Based on $g_{\text{inh}} = -\operatorname{div} \Sigma$, the material body force density associated with the inhomogeneous stiffness distribution reads

$$g_{\text{inh}} = -\operatorname{div} \Sigma = -\frac{\alpha\sigma_0^2}{2E_0 l} \left[1 + \alpha \left(\frac{x}{l} - \frac{1}{2} \right) \right]^{-2}. \tag{3.19.16}$$

Additionally, the displacement field can be integrated to the following expression:

$$u(x) = \frac{\sigma_0 l}{\alpha E_0} \ln \frac{1 + \alpha \left(\frac{x}{l} - \frac{1}{2} \right)}{1 - \frac{\alpha}{2}}. \tag{3.19.17}$$

In the specific case presented here, the length of the bar is 1 m and its width in the y-direction is 0.05 m. The problem is solved in a plane strain two-dimensional setting and, to avoid lateral stresses developing during deformation, Poisson's ratio is set to be 0, thus maintaining uni-axial conditions. The discretization applied is 160×8 bi-linear quadratic elements of edge length 0.00625 m and the tensile stress is $\sigma_0 = 0.01$ GPa on the right hand side while the left hand side is fixed in the x-direction. The bottom right and left nodes are constrained in their vertical displacement. The distribution of elasticity follows equation Eq. 3.19.13 with $E_0 = 1$ GPa. Figure 3.27 shows the finite element mesh along with the linear distribution of Young's modulus. The material being linearly elastic, a single time step is sufficient to reach the equilibrium configuration.

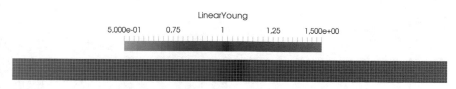

Fig. 3.27 Linear distribution of stiffness and finite element discretization of the problem

The OGS-6 version employed for this tutorial can be found in the official master branch available at

- https://github.com/fparisio/ogs/tree/SVOSDN_MF_BB.

with the input files named bar available as CTests under

- https://github.com/ufz/ogs-data/tree/master/Mechanics/Linear/MaterialForces.

The output from OGS-6 is the vectorial field of nodal material forces as in Eq. (3.19.12). For a direct comparison with the analytical solution, we need to compute the sum of material forces at every transversal section at a given coordinate x, which equals the total material force at that section. For a volume integral, we need to integrate over the transversal area A and sum in a discretized manner along the x direction the analytical solution of Eq. (3.19.16). The size of the discretization for this numerical integration is chosen as equivalent to the FEM discretization, i.e., 0.00625 m. The analytical expression of the vector of material forces at a given section x_i reads therefore

$$\underline{G}(x_i) = -\frac{\alpha \sigma_0^2}{2E_0 l} \left[1 + \alpha \left(\frac{x_i}{l} - \frac{1}{2} \right) \right]^{-2} A \Delta x_i. \tag{3.19.18}$$

where approximate linearity of $\mathbf{g}_{\text{inh}}$ in the proximity of x_i has been assumed. Figure 3.28 shows the results of the numerical analyses with OGS-6 in terms of the vectorial field of nodal material forces. It can be seen how the material forces

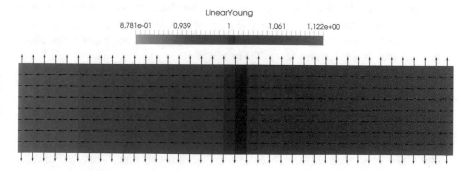

Fig. 3.28 The material forces vector field pointing toward regions of lower stiffness in the material (background color). The plot is a close-up of the central part of the bar

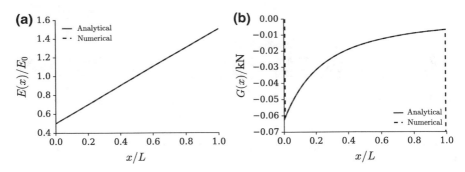

Fig. 3.29 Example for material forces acting in an inhomogeneous bar under tension: the continuous analytical distribution of Young modulus and the discretized distribution in OGS (**a**); comparison between the analytical solution and the result from the analysis with OGS (**b**)

Table 3.22 Benchmark deposit (https://docs.opengeosys.org/books/bmb-5)

BM code	Author	Code	Files	CTest
BMB5-3.19.1	Francesco Parisio	OGS-6	Available	OK

(https://oc.ufz.de/index.php/s/nlph7bhfDkj6tC7)

point toward the direction of lower stiffness in an effort to restore a lower energetic configuration. The distribution of Young's modulus in the model is shown in Fig. 3.29a for both the discrete numerical setup and the analytical continuous distribution. Figure 3.29b shows the comparison between the material forces at the nodes of the finite element discretization computed in OGS-6 following Eq. 3.19.12 (and summed over the cross-section) and the analytical solution obtained from Eq. (3.19.18). Neglecting boundary effects, there is excellent agreement between analytical and numerical solutions.

The goal of this benchmark was to introduce the users of OGS-6 to material or configurational forces as a new useful feature for the interpretation of evolving material domains. We have shown a simple example, working out explicitly all mathematical steps in a uniaxial setup and comparing the obtained analytical solution to the numerical one from OGS-6. Though far from being exhaustive, this application hints the attentive reader toward the meaningful physics that lies behind material forces: a driving force for interfaces and discontinuities (dislocations, cracks, defects, phase boundaries etc., just to name a few) (Table 3.22).

The OGS-6 version employed for this tutorial can be found on github at

- https://github.com/ufz/ogs.

The input files (bar.prj, bar.vtu and bar.gml) can be found in the source code in the folder

- https://github.com/ufz/ogs-data/tree/master/Mechanics/Linear/MaterialForces.

Chapter 4
T Processes

Vinay Kumar, Jobst Maßmann and Rainer Helmig

4.1 Effective Thermal Conductivity

4.1.1 Introduction and Theory

Thermal processes in porous media involve both conduction and convection. In low conducting media, or under very low Peclet numbers, the contribution of thermal conduction to the total heat transport might be just as or even more important than the contribution of thermal convection. For a purely diffusive system, the equation for pure conduction gives the total heat flux q_T through a given area of the domain

$$q_T = -\lambda_{pm} \nabla T \qquad (4.1.1)$$

The diffusion coefficient of the system λ_{pm} is the effective thermal conductivity of the porous media. In its simplest form, it is a function of the thermal conductivities of the constitutive phases and the volumetric distribution of the phases (given by the porosity ϕ) in a considered Representative Elementary Volume (REV). For a porous medium fully saturated with a solid and a fluid phase each, the effective thermal conductivity of the medium is given by

$$\lambda_{eff} = f\left(\lambda_f, \lambda_s, \phi\right) \qquad (4.1.2)$$

V. Kumar (✉) · J. Maßmann
BGR, Federal Institute for Geosciences and Natural Resources,
Hanover, Germany
e-mail: Vinay.Kumar@bgr.de

R. Helmig
Department of Hydromechanics and Modelling of Hydrosystems,
University of Stuttgart, Stuttgart, Germany

© Springer International Publishing AG 2018
O. Kolditz et al. (eds.), *Thermo-Hydro-Mechanical-Chemical Processes in Fractured Porous Media: Modelling and Benchmarking*,
Terrestrial Environmental Sciences, https://doi.org/10.1007/978-3-319-68225-9_4

In the following examples three simple averaging laws (cf. Aplin et al. 1999; Dong et al. 2015; Woodside and Messmer 1961) to determine effective thermal conductivity are compared with each other in numerical simulations of heat conduction in a low-permeability, fully-saturated porous medium. They are

- Arithmetic averaging,

$$\lambda_{\mathrm{arith}} = \phi\lambda_{\mathrm{f}} + (1 - \phi)\lambda_{\mathrm{s}} \tag{4.1.3}$$

- Geometric averaging,

$$\lambda_{\mathrm{geom}} = \lambda_{\mathrm{s}}^{1-\phi} \cdot \lambda_{\mathrm{f}}^{\phi} \tag{4.1.4}$$

- Harmonic averaging

$$\lambda_{\mathrm{harm}} = \frac{1}{\frac{\phi}{\lambda_{\mathrm{f}}} + \frac{1-\phi}{\lambda_{\mathrm{s}}}} \tag{4.1.5}$$

These averaging laws are derived from a simplified model concept which considers heat flow in a medium analogously to the flow of electrical current in a resistor network. This model concept explains heat conduction in a porous medium whose micro-structure shows a greater degree of pore-network connectivity in one plane compared to the direction perpendicular to it. In such media the conduction of heat can occur either simultaneously through both the liquid and solid phases, similar to resistors in parallel or sequentially with heat flowing first through one phase and then trough the other, similar to resistors in series. In the former case the effective thermal conductivity is given by the arithmetic average and in the latter case, the harmonic average. For media without a distinctly oriented pore structure, the conduction of heat can neither be approximated to a strict series nor to a parallel arrangement. In this case the geometric mean serves as a good approximation for the effective thermal conductivity of the system.

Considering constant values for λ_{s}, λ_{f} and the temperature gradient, the effective thermal conductivity of a system changes with the porosity and with the change in arrangement of the phases relative to the direction of heat flow. The variation of effective thermal conductivity is plotted as a function of porosity in the Fig. 4.1a, b below. Figure (4.1a) shows the case where the solid-phase thermal conductivity is around 6 times larger than the liquid-phase thermal conductivity. Figure (4.1b) shows the case where the solid-phase thermal conductivities is only around 1.5 times larger than the liquid-phase thermal conductivity. The graphs indicate that the influence of the averaging law is greater when the thermal conductivity ratio is greater.

If the model concept allows for different conducting pathways in different directions, then the averaging laws can be chosen differently in different principal directions. When anisotropy axes are not aligned with the coordinate system, the Eigenvalue problem

$$|\lambda_{\mathrm{s}} - \beta\mathbf{I}| = 0 \tag{4.1.6}$$

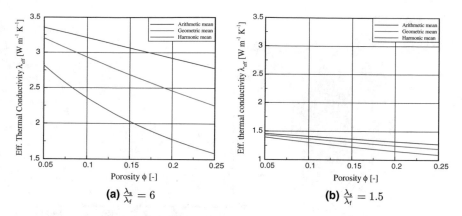

Fig. 4.1 Variation of thermal conductivity with porosity for two different solid-to-fluid thermal conductivity ratios

has to be solved, where $\mathbf{I}$ is the identity tensor and β the vector of Eigen-values which give the thermal conductivity in the principal directions. The solution yields the effective thermal conductivity tensor $\boldsymbol{\lambda'}_{\text{eff}}$, given by

$$\boldsymbol{\lambda'}_{\text{eff}} = \begin{bmatrix} \lambda'_{\text{eff,i}} & 0 & 0 \\ 0 & \lambda'_{\text{eff,j}} & 0 \\ 0 & 0 & \lambda'_{\text{eff,k}} \end{bmatrix} = \lambda_{ii}. \tag{4.1.7}$$

The principal directions are given by the Eigen-vectors, which form the components of the transformation matrix $\mathbf{Q}$. $\boldsymbol{\lambda'}_{\text{eff}}$ is then transformed back to the original direction using $\mathbf{Q}$ and its transpose $\mathbf{Q}^{\mathsf{T}}$ by

$$\boldsymbol{\lambda}_{\text{eff}} = \mathbf{Q}^{\mathsf{T}} \cdot \boldsymbol{\lambda'}_{\text{eff}} \cdot \mathbf{Q} \tag{4.1.8}$$

whereby $\boldsymbol{\lambda}_{\text{eff}}$ is used in the global coordinate system for further calculations.

4.1.2   Model Setup

The previously discussed concepts are illustrated in a numerical benchmark of a heater test in homogenous, anisotropic, saturated claystone. According to it, λ_s is the thermal conductivity of the grains(λ_{grains}) and λ_f is the thermal conductivity of water(λ_{water}). An axisymmetric domain of $100\,\text{m} \times 100\,\text{m}$ consisting of a centrally placed cylindrical heater of 0.5 m radius and 2 m height producing a time dependent thermal power output is chosen. For the sake of simplicity, the technical and geotechnical barriers are neglected and the anisotropy (bedding) in the claystone is assumed

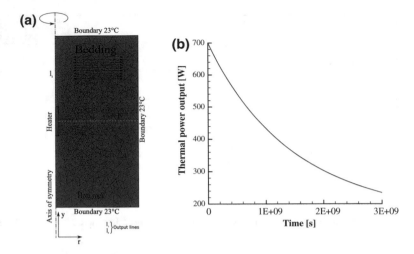

Fig. 4.2 Model geometry (**a**) and thermal power output (**b**) used in simulations

Table 4.1 Parameter set used in simulations

Parameter	Value	Unit
Porosity(ϕ)	0.12	$[-]$
sp. heat capacity of grains	626	$J\ kg^{-1}\ K^{-1}$
sp. heat capacity of water	4180	$J\ kg^{-1}\ K^{-1}$
Grain density	2454	$kg\ m^{-3}$
Water density	1000	$kg\ m^{-3}$
Grain thermal cond. ($\lambda_{grain,parallel}$)	3.5	$W\ m^{-1}\ K^{-1}$
Grain thermal cond. ($\lambda_{grain,perpendicular}$)	1.5	$W\ m^{-1}\ K^{-1}$
Water thermal cond. (λ_{water})	0.6	$W\ m^{-1}\ K^{-1}$

to be oriented along the coordinate axis in the horizontal (r) direction. The schematic of the experiment and the thermal power of the heater is given in Fig. 4.2a, b.

The parameters used are given in Table 4.1.

The following four models are compared:

(a) Arithmetic averaging in all directions.
(b) Geometric averaging in all directions.
(c) Harmonic averaging in all directions.
(d) Arithmetic averaging in r and harmonic averaging in y direction.

A comparison of the resulting temperature distribution, exemplary at one specific time step ($t = 1.52 \times 10^8$ s) is shown for the cases a–c in the Fig. 4.3.

A comparison between the temperature distribution of the models a–c and the model d with direction-dependent averaging for the effective thermal conductivity is

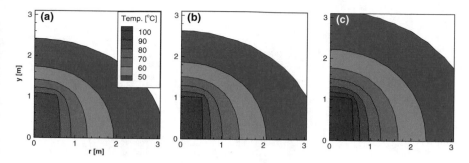

Fig. 4.3 Temperature distribution considering effective thermal conductivities by **a** arithmetic, **b** geometric and **c** harmonic averaging between the solid and liquid phase thermal conductivities

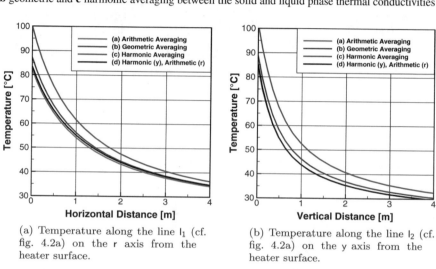

(a) Temperature along the line l_1 (cf. fig. 4.2a) on the r axis from the heater surface.

(b) Temperature along the line l_2 (cf. fig. 4.2a) on the y axis from the heater surface.

Fig. 4.4 Comparison of temperature distribution between different averaging laws for effective thermal conductivity of saturated claystone

Table 4.2 Benchmark deposit (https://docs.opengeosys.org/books/bmb-5)

BM code	Author	Code	Files	CTest
BMB5-4.1	Vinay Kumar	OGS-5.7.1 https://github.com /VinayGK/ogs5/tree /thermal-cond-averaging	Available	ToDo

shown graphically for the r direction in Fig. 4.4a and for the y direction in Fig. 4.4b and Table 4.2.

Part III
Coupled Processes

Chapter 5
HH Processes

Aaron Peche, Thomas Graf, Lothar Fuchs, Insa Neuweiler,
Jobst Maßmann, Markus Huber, Sara Vassolo, Leonard Stoeckl,
Falk Lindenmaier, Christoph Neukum, Miao Jing and Sabine Attinger

Urban Water Resources Management (UWRM)

5.1 Coupling OpenGeoSys and HYSTEM-EXTRAN for the Simulation of Pipe Leakage

Aaron Peche, Thomas Graf, Lothar Fuchs and Insa Neuweiler

OpenGeoSys was coupled to the pipe flow model HYSTEM-EXTRAN [HE, version 7.7 or newer] (itw 2010) in order to simulate pipe leakage in a variably saturated subsurface. The newly developed weak coupling scheme is applicable for the Richards flow process in a modified version of OGS 5 that can be downloaded from the custom branch available at https://github.com/APeche/OGS-HYSTEM-EXTRAN.git. The shared-memory-based coupling was implemented using the inter-process communication method Named Pipes, which is considered a cost-effective and easy-to-implement push-migration solution (Laszewski and Nauduri 2012). The

A. Peche (✉) · T. Graf · I. Neuweiler
Institute of Fluid Mechanics and Environmental Physics in Civil Engineering,
Leibniz Universität Hanover, Hanover, Germany
e-mail: peche@hydromech.uni-hannover.de

L. Fuchs
Institute for Technical and Scientific Hydrology (itwh GmbH), Hanover, Germany

M. Huber
geo:tools, Mittenwald, Germany

J. Maßmann · S. Vassolo · L. Stoeckl · F. Lindenmaier · C. Neukum
BGR, Federal Institute for Geosciences and Natural Resources, Hanover, Germany

M. Jing · S. Attinger
UFZ, Helmholtz Centre for Environmental Research, Hanover, Germany

© Springer International Publishing AG 2018
O. Kolditz et al. (eds.), *Thermo-Hydro-Mechanical-Chemical Processes in Fractured Porous Media: Modelling and Benchmarking*,
Terrestrial Environmental Sciences, https://doi.org/10.1007/978-3-319-68225-9_5

125

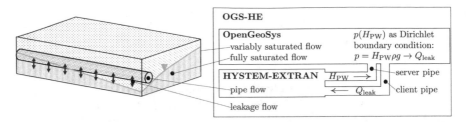

Fig. 5.1 Conceptual model of the coupled OpenGeoSys-HYSTEM-EXTRAN model (OGS-HE [Modified from Peche et al. (2017)])

implementation of the coupling scheme is based on a timestep-wise update of boundary conditions and source terms. Bidirectional interprocess data transfer is realized using in total two Named Pipes. First pipe is the server pipe, mainly used to send the HE-calculated pipe water level H_{PW} [L] to OGS. In OGS, H_{PW} is used as a Dirichlet-type boundary condition assigned to the pipe defect surface. It is converted to a hydrostatic pressure p [M L^{-1} T^{-2}] using fluid density ρ [M L^{-3}] and gravitational acceleration g [L T^{-2}] in the form of $p = H_{PW} \cdot \rho \cdot g$. The second Named Pipe is the client pipe used to send the OGS-calculated leakage flow Q_{leak} [L^3 T^{-1}] to HE, where it is used as a source term. The coupled model will be referred to as OGS-HE. The conceptual model is visualized in Fig. 5.1.

An extensive validation for the Richards flow process in OGS is given in Kolditz et al. (2012d). HE is widely used in practice. The calculation of pipe flow in HE is based on SWMM (Gironás et al. 2009), which undergoes a quality assurance program to ensure numerical result accuracy (Rossman 2006). The present chapter serves the purpose to show that the implemented interprocess data transfer between OGS and HE is correct. It further demonstrates the correctness of OGS-internal conversion of H_{PW} to p and the update of boundary conditions. In this first set of benchmarks, this is delimited for the transferred water levels H_{PW}. A detailed description of the model including all benchmark examples is given in Peche et al. (2017).

Governing Equations

HE describes pipe flow in the form of open channel flow using the complete one-dimensional Saint-Venant equations (de Saint-Venant 1871), where pressurized pipe flow is approximated using a Preissmann slot (Preissmann 1961) or an iterative approach. The Saint-Venant equations consist of continuity equation (5.1.1) and equation of motion (5.1.2) (itw 2010; Chaudhry 2008).

$$\frac{\partial A}{\partial t} + \frac{\partial (v \cdot A)}{\partial x} = 0 \qquad (5.1.1)$$

$$\frac{1}{g}\frac{\partial v}{\partial t} + \frac{v}{g}\frac{\partial v}{\partial x} + \frac{\partial H_{\text{PW}}}{\partial x} = I_s - I_r \tag{5.1.2}$$

where A [L^2] is cross-sectional area, t [T] is time, v [L T^{-1}] is flow velocity, x [L] is pipe length, g [L T^{-2}] is gravitational acceleration, H_{PW} [L] is pipe water level, I_s [-] is bottom slope, and I_r [-] is friction slope. I_r can be parametrized using the Darcy-Weißbach equation in combination with the Prandtl–Colebrook friction law or the approach of Gauckler–Manning–Strickler (itwh:2010). In HE, the Saint-Venant equations are solved with a finite volume scheme.

In OGS, flow in the variably saturated porous medium is calculated using Richards' equation (Richards 1931a; Kolditz et al. 2012d).

$$\phi\rho\frac{\partial S}{\partial p}\frac{\partial p}{\partial t} - \nabla \cdot \left[\rho\frac{K K_{\text{rel}}}{\mu}(\nabla p - \rho g)\right] = \frac{\rho Q_{\text{leak}}}{V_u} \tag{5.1.3}$$

where ϕ [-] is porosity, ρ [M L^{-3}] is fluid density, S [-] is water saturation, p [M L^{-1} T^{-2}] is water pressure, ∇ [L^{-1}] is nabla operator, K [L^2] is intrinsic permeability, K_{rel} [-] is relative permeability, μ [M L^{-1} T^{-1}] is fluid dynamic viscosity, and V_u [L^3] is the cell volume. The saturated hydraulic conductivity k_s [L T^{-1}] is calculated with $k_s = K\rho g/\mu$. Equation (5.1.3) is solved using a retention function $S(p)$ and relative hydraulic conductivity function $K_{\text{rel}}(S)$ after e.g. van Genuchten (1980).

Verification and Validation of OGS-HE

The following sections describe the verification and validation examples. Examples include one analytical solution and measurement results from two physical experiments (Skaggs et al. 1970; Siriwardene et al. 2007). We chose examples that are relevant for pipe leakage.

5.1.1 Analytical Solution of Stationary Constant Water Level-Driven Infiltration into a Horizontally Layered Soil Column

Brunner et al. (2009) proposed a one-dimensional analytical model for surface water-groundwater interaction. That analytical model can be modified such that it is a broad approximation of pipe leakage into a horizontally layered soil, where a pipe defect is located above a colmation layer overlaying the aquifer strata. This modified setup enables to derive an analytical solution used for the validation of OGS-HE. The setup of the analytical model is visualized in Fig. 5.2a. Flow is driven by a Dirichlet boundary condition of constant pressure at the domain bottom and a constant pipe water level at the domain top. Assuming a constant cross-sectional area and

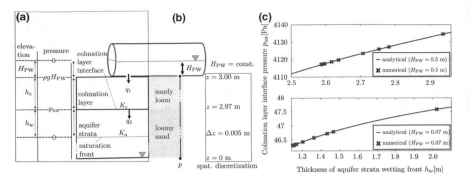

Fig. 5.2 Comparison of OGS-HE against an analytical solution describing constant pipe water level-driven infiltration into a layered soil column reprinted from Peche et al. (2017). **a** Analytical model modified from Brunner et al. (2009), **b** conceptual model of OGS-HE, **c** verification results

volume conservation results in the same specific discharge through the fully saturated colmation layer q_1 [L T^{-1}] and the fully saturated part of the aquifer strata q_2.

$$q_1 = q_2 \qquad (5.1.4)$$

Assuming linear pressure profiles, reformulation using Darcy's law results in

$$-k_{s,c}\frac{(H_{\text{PW}} + h_c - p_{\text{int}} \cdot (\rho g)^{-1} + 0)}{h_c} = -k_{s,a}\frac{(p_{\text{int}} \cdot (\rho g)^{-1} + h_w - 0 + 0)}{h_w} \qquad (5.1.5)$$

where $k_{s,c}$ and $k_{s,a}$ [L T^{-1}] are saturated hydraulic conductivities of colmation layer and aquifer strata at full saturation, respectively, h_c [L] is colmation layer thickness, p_{int} [M L^{-1} T^{-2}] is pressure at the colmation layer interface, and h_w [L] is the thickness of the saturation front in the aquifer strata. Equation (5.1.5) can be reformulated to a form that describes p_{int} as a function of H_{PW} and h_w:

$$p_{\text{int}} = \left(\left[\left(\frac{k_{s,c}h_w}{h_c k_{s,a}}(H_{\text{PW}} + h_c)\right) - h_w\right]\left(1 + \frac{k_{s,c}h_w}{h_c k_{s,a}}\right)^{-1}\right) \cdot \rho g \qquad (5.1.6)$$

The setup of the numerical model is visualized in Fig. 5.2b. The pipe flow model represents a pipe of 30 m length. It was controlled by an upstream Dirichlet-type boundary condition, where H_{PW} was either 0.07 m or 0.5 m. Downstream boundary condition of the pipe flow model was set as free outflow boundary condition. It should be noted that the details of the pipe flow setup are irrelevant as long as they lead to a fixed and known water level above the pipe defect. This was the case in our examples. Geometry and spatial discretization of the porous subsurface model are given in Fig. 5.2b. The steady-state was calculated using different bottom boundary conditions regulated by p_{bot} (band width $-6000\,\text{Pa} \le p_{\text{bot}} \le -50\,\text{Pa}$). The initial condition was pre-calculated, representing the steady-state pressure distribution from a similar

Table 5.1 Fluid properties and material parameters

Soil properties

	$\alpha\ [m^{-1}]$	m [-]	$K\ [m^2]$	S_r [-]	S_{max} [-]	ϕ [-]
Aquifer strata	3	0.56	$4.1297 \cdot 10^{-12}$	0.14	1	0.41
Colmation layer	3	0.47	$1.3052 \cdot 10^{-12}$	0.16	1	0.41
Fluid properties						
$\rho\,[\mathrm{kg\ m^{-3}}]$	1000					
$\mu\,[\mathrm{kg\ m^{-1}\ s^{-1}}]$	0.001					

same setup based on a constant global pressure of $-100\,\mathrm{Pa}$. Material properties of aquifer strata and colmation layer represent sandy loam and loamy sand, respectively. Parameters originate from Carsel and Parrish (1988) and Roth (2006) and are given in Table 5.1, where α and m are parameters of the van Genuchten parametrization and S_r and S_{max} [-] are residual and maximal saturation, respectively. Fluid properties are also given in Table 5.1. Varying H_{PW} and p_{bot} enabled a variation of h_w and thus lead to a different result of p_{int}, which was then compared with the analytical solution. Clearly, results agree very well, and are given in Fig. 5.2c.

5.1.2 Physical Experiment of Transient, Constant Water Level-Driven Infiltration into a Homogeneous Soil Column

Skaggs et al. (1970) presented a physical experiment, which describes one-dimensional infiltration into an initially dry soil column. In that experiment, infiltration was driven by keeping a constant water table on top of the soil column. Skaggs et al. (1970) measured the infiltration rate and depth of the wetting front over a time interval of 90 min. That experiment was numerically reproduced using OGS-HE, where infiltration is driven by a time-constant pipe water level. The domain of the pipe flow model is similar to the one described in Sect. 5.1.1. The upstream boundary was set to the Dirichlet-type boundary condition $H_{\mathrm{PW}} = 0.0075\,\mathrm{m}$. The downstream boundary of the pipe flow model was assigned as free outflow boundary condition. Initial condition, boundary conditions, temporal and spatial discretization as well as material properties of the porous subsurface model are given in Fig. 5.3a. Values for retention curve and relative conductivity curve are taken from Tian and Liu (2011) and Vogel (1987) and given in Table 5.2. Fluid properties are $\rho = 999.7\,\mathrm{kg\ m^{-3}}$ and $\mu = 0.00125\,\mathrm{kg\ m^{-1}\ s^{-1}}$. A detailed description of a numerical model setup for the reproduction of the physical experiment is given in Vogel (1987). Again, results agree very well, and are given in Fig. 5.3b.

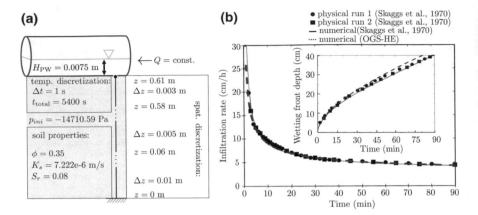

Fig. 5.3 Comparison of OGS-HE against the physical experiment by Skaggs et al. (1970) reprinted from Peche et al. (2017): **a** conceptual model of OGS-HE, **b** validation results

Table 5.2 Retention curve and relative conductivity function data

$S(P_c)$		$K_{rel}(S)$	
S [-]	P_c [Pa]	S [-]	K_{rel} [-]
0.080	19614.1	0.080	0.001
0.177	9807.1	0.143	0.008
0.243	7845.7	0.286	0.015
0.331	5884.2	0.429	0.030
0.509	3922.8	0.500	0.050
0.757	1961.4	0.571	0.082
0.874	980.7	0.643	0.250
1	0	0.714	0.550
		0.786	0.886
		0.821	0.963
		0.857	0.992
		0.874	0.997
		1	1

5.1.3 Physical Experiment of Transient, Variable Water Level-Driven Infiltration into a Homogeneous Soil Column

In the physical experiment conducted by Siriwardene et al. (2007), a column was filled with a sand layer and a gravel layer on top. Water levels were varied in the gravel layer. The gravel layer was filled such that a water column of 0.75 m was on top of the sand-gravel-interface. Then, the column was drained such that water levels above the sand-gravel-interface continually decreased to a minimum of 0.05 m. Siriwardene et al. (2007) conducted a total of 14 refill-drainage cycles over a time interval of approximately 24 h, and continuously measured the outflow at the bottom of the column.

In the one-dimensional OGS-HE numerical model, the sand-gravel-interface represents the pipe defect. This defect is located in a pipe flow model domain similar to the model domain described in Sect. 5.1.1. Refill-drainage cycles were reproduced using a time-variable Dirichlet-type boundary condition located at the upstream boundary of the pipe flow model. This way, H_{PW} was regulated according to Siriwardene et al. (2007), representing water levels between 0.05 and 0.75 m. The downstream boundary condition of the pipe flow model was set as free outflow boundary condition. The setup of the porous subsurface model in form of initial condition, boundary conditions, spatial and temporal discretization as well as material properties is given in Fig. 5.4a. Fluid properties are $\rho = 998.2 \, \text{kg m}^{-3}$ and $\mu = 0.001 \, \text{kg}$ $\text{m}^{-1} \, \text{s}^{-1}$. Simulation results were compared to results from Siriwardene et al. (2007) from experiment runtime of 11–24 h and to results of a numerical model from Browne et al. (2008) as shown in Fig. 5.4b. Clearly, simulation results of OGS-HE agree well with measured data at late times. At early times, model results show a deviation compared with physical experiment results, which might be because the warm-up phase of the physical experiment has an effect on the solution (Browne et al. 2008).

Integrated Water Resources Management (IWRM)

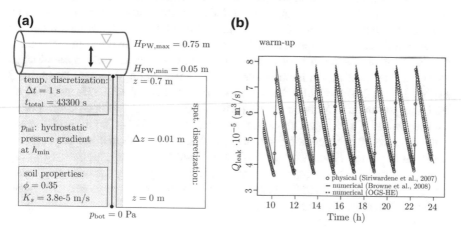

Fig. 5.4 Comparison of OGS-HE against the physical experiment by Siriwardene et al. (2007) reprinted from Peche et al. (2017): **a** conceptual model of OGS-HE, **b** validation results

5.2 Coupling WEAP and OpenGeoSys for a Decision Support System for Integrated Water Resources Management

Jobst Maßmann, Markus Huber, Sara Vassolo, Leonard Stoeckl, Falk Lindenmaier, Christoph Neukum

5.2.1 Background

With regard to limited resources in semi-arid regions, the sustainable use of groundwater is of crucial importance. Prerequisites for a sustainable use are, among others, a realistic estimation of the groundwater volume stored in the underground, the groundwater recharge or renewal as well as the losses through natural discharge and pumping activities. Therefore, knowledge about the current state and further development of the whole social, economic, cultural and environmental framework is needed. To understand this complex interacting system and the outcomes of any changes, a Decision Support System (DSS), based on computational models, renders assistance.

The presented DSS is based on the water evaluation and planning system WEAP,[1] developed by the SEI (Yates et al. 2005). It balances groundwater, soil water and surface water on catchment, subcatchment and landuse class scale for simulating future changes in demand and supply. In order to get detailed information about the local development of the aquifers, as water storage, flow and water table, a coupling to a groundwater flow model is obligatory. So far, the coupling between WEAP and a groundwater flow model has been limited to MODFLOW 2000 (Harbaugh et al. 2000), as presented in Maßmann et al. (2012). This coupling has been developed in the framework of a technical cooperation project (BGR 2017). The finite difference method, however bears certain limitations, and in many cases, the use of programs based on the Finite Element Method (FEM) would be more appropriate. Not only because of their far better representation of complex aquifer geometries and structures, but also because model refinements, as generally required in areas of strong gradients, like at wells, are easily prepared.

5.2.2 General Concept

The general linkage concept is presented in Fig. 5.5. Since the WEAP elements do not have any spatial reference, the spatial allocation is done by *Mask Layers*, which are

---

[1] http://www.weap21.org.

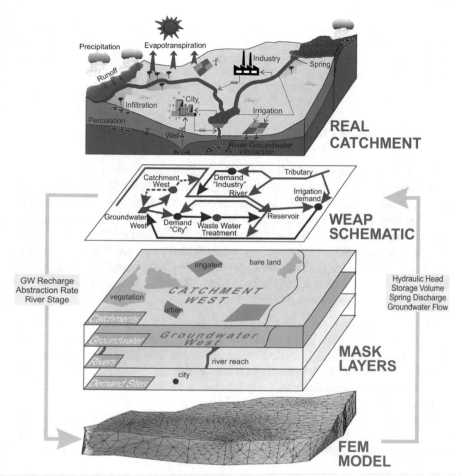

Fig. 5.5 Schematic representation of the WEAP-FEM coupling

mesh independent ESRI shapefiles (ESRI 1998). They are created within the BGR software *LinkKitchen FEM* (developed by geo:tools), and contain all groundwater relevant WEAP Elements, which are: Demand Sites (points), Rivers (polylines), Groundwater (polygons), Catchments (polygons) and flow boundaries (polygons). In LinkKitchen FEM, the groundwater withdrawals and renewals, as calculated by WEAP, are in a first step, projected on the corresponding Mask Layer and in a second step on the FE mesh. Based on this projection, the FEM boundary conditions and source terms are written in the corresponding input files and the FE program is executed. Results from the FE model are read by LinkKitchen and provided to WEAP, as far as needed.

Considering the initial setup of a WEAP-FEM DSS it has to be mentioned that the model construction is carried out independently in WEAP and FEM software with the customary tools. The coupling of WEAP and FEM is carried out with LinkKitchen

FEM in a second step. For this purpose, in addition to the WEAP and FEM models, further information must be available so that the WEAP elements can be located. This can be done by ESRI shapefiles containing the location of wells, catchments, rivers or landuse classes.

5.2.3 Software Concept

WEAP is considered as the basis of the DSS. Groundwater extraction, losses due to discharge as well as groundwater recharge or return flows from irrigated areas are calculated by WEAP as time- and scenario-dependent variables. To demonstrate the transferability of the approach, three FE codes have been chosen, covering commercial and open source software: OpenGeoSys, Spring and FEFLOW. The implementation of the WEAP-FEM coupling is subdivided into three phases, as depicted in Fig. 5.6. Phase one (unidirectional coupling), has been completed successfully, phase two (bidirectional coupling) is in process. In the first phase, the unidirectional coupling is carried out in three steps:

1. WEAP calculation: All withdrawals and renewals for the various WEAP elements are calculated within the software WEAP for each scenario over the entire calculation period (e.g. 2014–2020) subdivided in time steps (e.g. months). This includes, for example, the withdrawals to meet the drinking water demand of cities or the irrigation requirements, the groundwater recharge or inflows and outflows from reservoirs, rivers, etc.
2. LinkKitchen FEM: Reading WEAP results, processing, output of input data for FEM-Software: Once the WEAP calculation is finished, the withdrawals/renewals as well as river infiltration rates calculated for WEAP elements are read for each

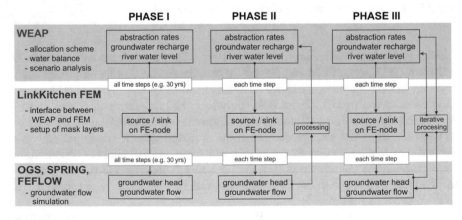

Fig. 5.6 Software concept and phases of implementation for WEAP-FEM coupling

time step and scenario with the help of LinkKitchen FEM, distributed to the corresponding FEM nodes and transferred as input data to the FEM program.

3. FEM calculation: Following the WEAP calculation, the FEM program computes the groundwater flow and groundwater levels for each time step at each node.

By incorporating a groundwater flow model to the DSS, it is possible to assess the effects of different water management scenarios on groundwater levels and flows. A new development of the program LinkKitchen was essential to provide input data for the FEM model based on the WEAP results.

In the second phase, a bidirectional coupling is targeted. At each time step, WEAP calculates the withdrawals, the groundwater recharge and the river stages at the corresponding WEAP element. LinkKitchen FEM reads these variables every time step, calculates the distribution to the FEM nodes and creates the input data for the FEM model. The FEM model is executed for each WEAP time step (e.g. month). In the second phase, a bidirectional coupling is targeted. In each time step WEAP calculates the withdrawals, the groundwater recharge, and the river stages at the corresponding WEAP element. LinkKitchen FEM reads these variables every time step, calculates the distribution to the FEM nodes and creates the input data for the FEM model. The FEM model is executed for each WEAP time step (e.g. month). After the flow calculation by the FEM programs, the groundwater levels, the groundwater storage, the spring discharges, and the groundwater flows are transferred to WEAP via LinkKitchen FEM, and serve as the calculation base for the next time step. A feedback from the FEM software to WEAP is therefore possible. In such a way, decisions based on groundwater levels like reduction of pumping rates or calculation of irrigation volumes can be considered in the DSS.

The bidirectional coupling could lead to slight inconsistencies in the water balance, since WEAP receives the actual water levels only at the end of the time step (e.g. 1 month). For example, a drying out of a well in WEAP will only be taken into account in the following time step. A bidirectional coupling is not suitable for fast-reacting aquifers such as karst. A strong change in groundwater level within one time step generally leads to numerical instability, especially if WEAP modifies the water allocation based on groundwater levels. An iterative coupling scheme, as proposed in phase three, should be able to cope with these shortcomings.

5.2.4 Benchmark Problem

In order to verify and demonstrate the implementation of the WEAP-OGS coupling a benchmark problem is defined (Fig. 5.7). Even though it is a generic setup, it represents a typical hydrogeological system featuring catchments with different infiltration characteristics, well fields, aquifers and aquitards.

An overview of the model setup is given in Fig. 5.8. A representation of all water relevant processes is given within the WEAP model. The WEAP schematic represents the water allocation scheme. WEAP allows many possibilities to define the

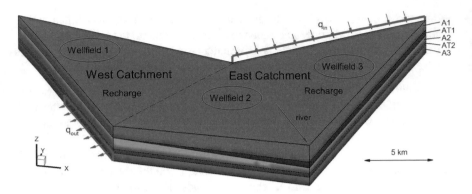

Fig. 5.7 WEAP-OGS coupling benchmark: hydrogeological system

demand, including complex algorithms considering socio-economic changes. Here, the demand is controlled by the number of inhabitants and their annual water consumption. Groundwater recharge is computed by rainfall, evaporation and a simple rainfall runoff model on catchment scale. Depending on the water level in the river and the groundwater head, the groundwater-river interaction is calculated. Lateral in- and outflows are covered by external source/sink.

In LinkKitchen FEM, all groundwater-relevant WEAP elements are spatially related to GIS-features, read from shapefiles. The user friendly GUI analyses in the WEAP model and displays the WEAP elements in an interactive tree view in order to facilitate the assignment. As a result, Mask Layers are produced, which combine WEAP elements with shape files. Additional data, as river bed conductance, flow-stage-width-relations, depth of wells and alike will be stored together with the graphical data.

The OGS reference model represents the static geohydraulic system. Since transient data is handled by WEAP, no time dependent boundary conditions must defined here. The setup consists of the mesh including the definition of material groups, permeabilities, storage coefficients and fluid properties. Furthermore, the initial head needs to be defined at all nodes.

As depicted in Fig. 5.9 the presentation of the modeling results is twofold: whereas WEAP provides diversified opportunities to evaluate, display and export water balances with different emphases, the standard output of OGS enables convenient postprocessing of heads and the three dimensional flow field by Tecplot or Paraview. Although OGS is run for each WEAP time step separately (usually one month), LinkKitchen FEM provides a comprehensive *.pvd export, which combines the OGS results of all time steps for each scenario.

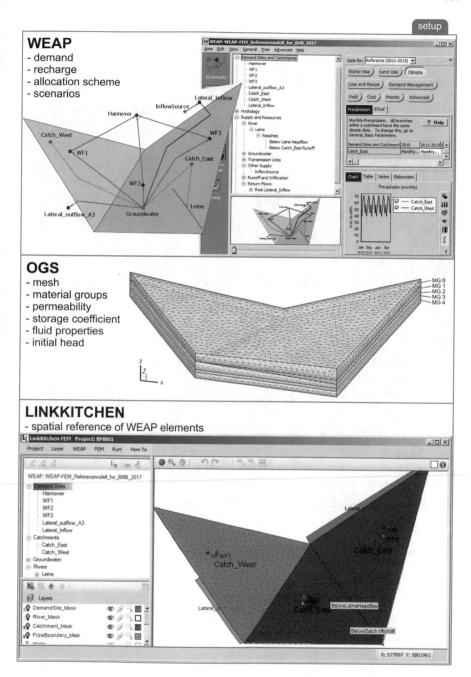

Fig. 5.8 Setup of a coupled WEAP-OGS model

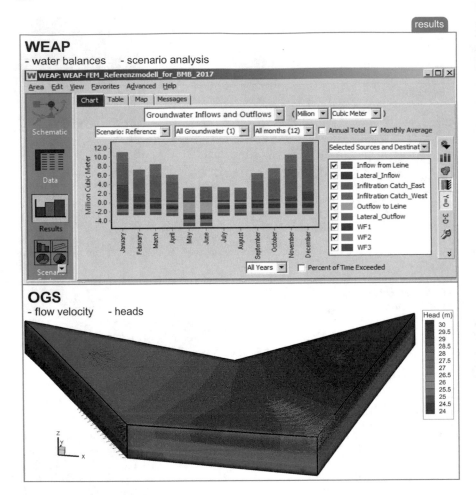

Fig. 5.9 Exemplary results of a coupled WEAP-OGS model

Catchment Scale Hydrological Modeling (CSHM)

5.3 Coupling mHM with OGS for Catchment Scale Hydrological Modeling

Miao Jing, Falk Heße, Wenqing Wang, Thomas Fischer, Marc Walther, Sabine Attinger

Most of the current hydrological models do not contain a physically-based groundwater flow component. The motivation of developing the coupled model mHM#OGS

Fig. 5.10 The concept of the mesoscale hydrologic model mHM (Samaniego et al. 2010)

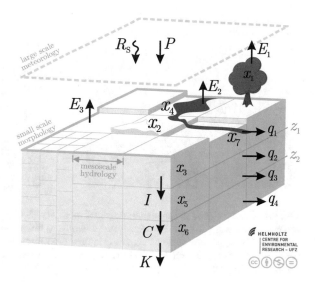

is to bridge the gap between catchment hydrology and groundwater hydrology. The coupled model mHM#OGS mechanistically accounts for surface flow, soil-zone flow and subsurface flow using an off-line coupling method. We applied the coupled model mHM#OGS in a case study of a catchment in central Germany to simulate the hydrologic processes, especially focusing on the long-term groundwater flow dynamics. The model is calibrated in two steps using catchment discharge and groundwater heads, respectively. Furthermore, the time series of groundwater heads are used to verify the model (Jing et al. 2017).

5.3.1 mHM

The mesoscale Hydrologic Model mHM (Samaniego et al. 2010) is a spatially explicit distributed hydrologic model that uses grid cells as a primary modeling unit, and accounts for the following processes: canopy interception, snow accumulation and melting, soil moisture dynamics, infiltration and surface runoff, evapotranspiration, subsurface storage and discharge generation, deep percolation and baseflow and discharge attenuation and flood routing (Fig. 5.10). The total runoff generated at each grid cell is routed to the neighbouring downstream grid cell following the river network using the Muskingum–Cunge routing algorithm. The model is driven by hourly or daily meteorological forcings (e.g., precipitation, temperature), and it utilizes observable basin physical characteristics (e.g., soil textural, vegetation, and geological properties) to infer the spatial variability of the required parameters.

**(a) Conceptual catchment
representation**

(b) Schematic of mHM#OGS

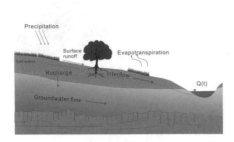

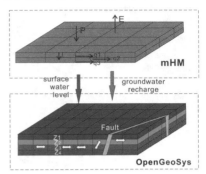

Fig. 5.11 The concept and modeling target of mHM#OGS (Jing et al. 2017). **a** The conceptual representation of hydrological processes in a catchment; **b** the schematic used to couple the mHM and OGS. The upper box depicts the canopy interception, atmospheric forcing, and the land surface processes represented by mHM. The lower box shows the saturated zone represented by the OGS groundwater model

5.3.2 Structure of the Coupled Model mHM#OGS

The coupled mHM#OGS model was developed to simulate coupled surface-water and groundwater (SW/GW) flow in one or more catchments by simultaneously calculating flow across the land surface and within subsurface materials. mHM#OGS simulates flow within three hydrological regions. The first region is limited by the upper bound of plant canopy and the lower bound of the soil zone bottom. The second region includes surface streams. The third region is the saturated aquifer beneath the soil zone. mHM is used to simulate the processes in the first and second region, while OGS is used to simulate the hydrological processes in the third region (Fig. 5.11) (Jing et al. 2017).

The two models are coupled in a sequential manner by fed fluxes and variables from one model to another at every time step. Technically, we use a self-developed data communication code GIS2FEM to convert data format and exchange variables and fluxes. The coupling interface converts time-series of variables and fluxes to Neumann boundary conditions, which can be directly read by OGS. The modified OGS source code can produce raster files containing the time-series of flow-dependent variables and volumetric flow rates with the same resolution of mHM grid cells, which can be directly read by mHM. The detailed workflow of the coupling technique is shown in Table 5.3 (Jing et al. 2017).

5.3.3 Model Setup in a Catchment

We use a mesoscale catchment upstream of Naegelstedt catchment in central Germany with a drainage area of 845 km^2 to verify our model. The Naegelstedt catchment comprises the headwaters of the Unstrut river basin. It was selected in this

Table 5.3 Description of computational sequence for mHM#OGS using a sequential coupling scheme (Jing et al. 2017)

Sequence no.	Computation
1	**Initialize, assign, and read** — Run mHM and OGS initialize procedures, OGS assign, read and prepare parameters and subroutines for later simulation
2	**Compute land-surface and soil-zone hydrologic processes in mHM** — Distribute precipitation, compute canopy interception and evapotranspiration, determine snowpack accumulation and snowmelt, infiltration and groundwater recharge, and surface runoff for each grid cell
3	**Compute the long-term mean of land-surface and soil-zone hydrologic processes** — Compute the long-term mean of the entire simulation period and write them as a set of raster files
4	**Transfer primary variables and volumetric flow rates to OGS** — Transfer flow-dependent variables and volumetric flow rates needed for computing saturated flow as Neumann boundary conditions or Dirichlet boundary conditions in OGS
5	**Steady-state calibration** — Run OGS-only steady-state simulation using boundary conditions given by mHM in step 4. The calibrated hydrogeological parameters are fed to the transient model. The steady-state groundwater head serves as an initial condition of the transient mHM#OGS modeling
6	**Start transient simulation of mHM#OGS** — Sequence through coupled mHM and OGS components
7	**Compute stepwise soil-zone flow and storage in mHM** — Compute evapotranspiration, interflow, surface runoff and groundwater recharge
8	**Transfer primary variables and volumetric flow rates to OGS** — The same as step 4, except this step generates time-dependent raster files of flow rates
9	**Solve the groundwater flow equation** — Calculate ground-water heads and head-dependent flows in study region
10	**Compute budgets** — Run water budget package to check overall water balance as well as time-dependent water budgets in each storages
11	**Write results** — Output the simulations results
12	**Check for the end of simulation** — Repeat time loop (steps 7–12) until end of simulation period
15	**End of simulation** — Close files and clear computer memory

study because there are many groundwater monitoring wells operated by Thuringian State office for the Environment and Geology (TLUG). We use a spatially distributed aquifer model to explicitly present groundwater storage. The thickness of the layers follows values provided by well logs and geophysical data, and was further linearly interpolated as a 3D geological model. To convert data format, we use the workflow developed by Fischer et al. (2015) to convert the complex 3D geological model into open-source VTU format file that can be used by OGS. Model elements of OGS were set to a 250×250 m horizontal resolution and a 10 m vertical resolution over the whole model domain (Fig. 5.12).

We start the modeling by performing the daily simulation of mHM to calculate near-surface and soil-zone hydrological processes. Several resolutions ranging from

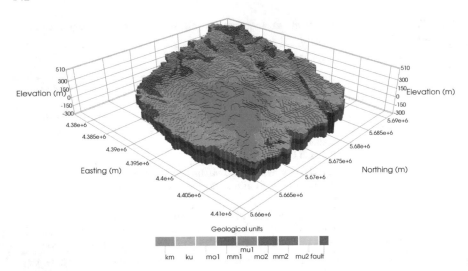

Fig. 5.12 Three dimensional and cross section characterization of the hydrogeologic framework model illustrating vertical and horizontal model discretization and hydrogeologic zones (Jing et al. 2017)

200 m to 2 km are applied in mHM to account different scale of spatial heterogeneity. The mHM-alone model is calibrated independent of OGS for the period 1970–2005 by matching the observed streamflows.

For the second step, OGS is run independent of mHM using a steady state stress period. Long-term average recharge rates estimated by the mHM simulations are used for the steady state boundary condition. The long-term average baseflow rates estimated by mHM simulations are used as boundary condition at stream beds. The groundwater levels obtained from a couple of monitoring wells are averaged over the whole simulation period. The steady state groundwater model is calibrated by adjusting aquifer hydraulic conductivity values by matching groundwater levels in the catchment (Jing et al. 2017).

To further test the effectiveness of the coupled model mHM#OGS, the results are assessed using the historical groundwater head time series. Calibrated hydraulic conductivities from steady state groundwater model are used as the parameter set for transient groundwater model. Instead of calibrating parameters of storages, we use a homogeneous storage for all groundwater aquifers. The specific storage are set to 1.0×10^{-5} for all groundwater aquifers. The groundwater head distribution in steady state model are used as initial condition of the transient model (Jing et al. 2017).

5.3.4   Model Verification

First, we show the calibration result of mHM in Fig. 5.13. The calibration result is good with a Pearson correlation coefficient $R_{cor} > 0.9$ for the monthly discharge simulation at the Naegelstedt station (see Fig. 5.13).

Simulated water-table depth over the whole catchment is shown in Fig. 5.14. The simulation provides a reasonable distribution of the hydraulic head. Groundwater-table depth is as large as 40 m in the higher southwestern and northern mountainous areas, whereas less than 5 m in the central lowlands. This simulation result is coincident with regionalized observations of groundwater heads (Wechsung 2005).

The model performances in the study area were verified by the comparison of simulated groundwater head time series to observations at 19 monitoring wells. Instead of plotting the absolute values of heads, namely h_{mod} and h_{obs}, we plotted the model results and corresponding observed data in their anomalies by subtracting their mean values $\overline{h}_{mod}$ and $\overline{h}_{obs}$. Figure 5.15 presents observed and simulated groundwater heads for the period 1975–2005. The Pearson correlation coefficient R_{cor} is used to assess the correlation between observation and simulation. Both of those two wells in Fig. 5.15 show promising simulation results with the R_{cor} of 0.745 and 0.786.

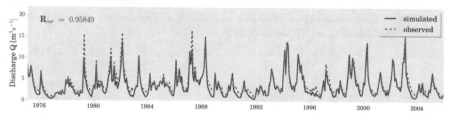

Fig. 5.13 Observed and modeled monthly streamflow at the outlet of Naegelstedt catchment (Jing et al. 2017)

Fig. 5.14 Simulated water-table depth over Naegelstedt catchment (Jing et al. 2017)

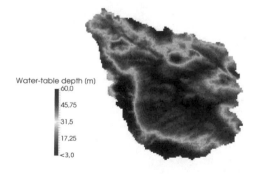

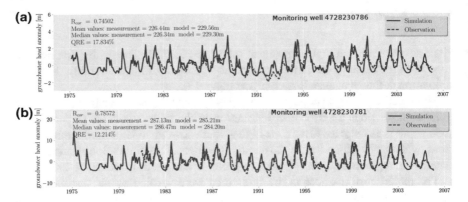

Fig. 5.15 The comparison between measurements data of groundwater heads (green dashed line) and model outputs (orange solid line). **a** Monitoring well 4728230786 located at upland near stream. **b** Monitoring well 4828230754 located at mountainous area

Reference:

M. Jing, F. Heße, W. Wang, T. Fischer, M. Walther, M. Zink, A. Zech, R. Kumar, L. Samaniego, O. Kolditz, S. Attinger, Improved representation of groundwater at a regional scale – coupling of mesocale hydrologic model (mhm) with opengeosys (ogs). Geosci. Model Dev. Discuss. (2017)

Chapter 6
H² Processes

Yonghui Huang and Haibing Shao

Richards Flow

The Richards flow model has been implemented in the newly developed OGS-6 version. In this section, we will use the single continuum model benchmark to demonstrate the correctness of the implementation.

The Richards equation is often used to mathematically describe water movement in the unsaturated zone. It has been introduced by Richards (1931b), who suggested that Darcy's law under consideration of the mass conservation principle, is also appropriate for unsaturated flow conditions in porous media. The Richards flow is also regarded as a simplification of a two phase flow system (water and air) with the ignorance of the air transport.

The pressure based formulation of this governing equation (6.0.1), which selects the unknown primary variable as p, can be written as:

$$\phi \rho_w \frac{\partial S}{\partial p_c} \frac{\partial p_c}{\partial t} + \nabla \cdot \left(\rho_w \frac{k_{rel} \mathbf{k}}{\mu_w} (\nabla p_w - \rho_w \mathbf{g}) \right) = Q_w \qquad (6.0.1)$$

Y. Huang (✉)
UFZ, Helmholtz Centre for Environmental Research, Leipzig, Germany
e-mail: Yonghui.Huang@ufz.de

Y. Huang
TU Dresden, Dresden, Germany

H. Shao
UFZ, Helmholtz Centre for Environmental Research, Leipzig, Germany

H. Shao
TU Freiberg, Freiberg, Germany

© Springer International Publishing AG 2018
O. Kolditz et al. (eds.), *Thermo-Hydro-Mechanical-Chemical Processes in Fractured Porous Media: Modelling and Benchmarking*,
Terrestrial Environmental Sciences, https://doi.org/10.1007/978-3-319-68225-9_6

6.1 Infiltration in Homogeneous Soil

6.1.1 Definition

This infiltration problem refers to a classical field experiment described by Warrick et al. (1971), who examined simultaneous solute and water transfer in unsaturated soil within the Panoche clay loam, an alluvial soil of the Central Valley of California. A quadratic 6.10 m plot, which had an average initial saturation of 0.455, was wetted for 2.8 h with 0.076 m of 0.2 N CaCl₂, followed by 14.7 h infiltration of 0.229 m solute-free water. The soil-water pressure was monitored by duplicate tensiometer installations at 0.3, 0.6, 0.9, 1.2, 1.5 and 1.8 m below surface.

6.1.2 Model Configuration

Two fixed pressure boundary conditions are used in the flow equation with a uniform initial saturation in the whole domain of 45.5%. At the top, the 2 m high soil column is open to the atmosphere, i.e. the capillary pressure is 0 Pa. The bottom of the column has a capillary pressure of 21,500 Pa. Homogeneous material properties are assumed within the whole domain. The average saturated moisture content, which is equal to the porosity of the soil, is 0.38. The saturated permeability is $9.35\,e{-}12$ m². The relative permeability and capillary pressure versus saturation data are fitted by the soil characteristic functions respectively. A detailed description can be found in Kolditz et al. (2012c). The parameters used for this benchmark are summarized in Table 6.1.

The geometry domain is uniformly discretized into 100 quad elements. A fixed time stepping control is applied over the whole simulation with $\Delta t = 1$ s. The whole simulation runs for a time span of 2×10^4 s. The simulation results are compared against the results generated by OGS-5.

Table 6.1 Material parameters for the Infiltration problem

Property	Symbol	Unit	Value (formulation)
Porosity	ϕ	–	0.38
Permeability	k	m^2	$4.46e - 13$
Liquid dynamic viscosity	μ_w	$Pa.s$	1.0000×10^{-3}
Liquid density	ρ_w	$kg.m^{-3}$	1.0000×10^3
Capillary pressure	P_c	Pa	Curve
Relative permeability	$k_{rel\,w}$	–	Curve

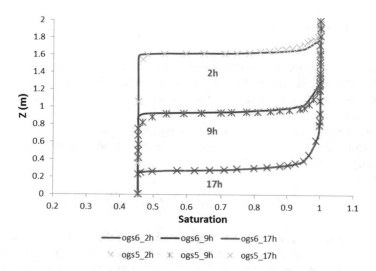

2
1.8
1.6
1.4
1.2
1
0.8
0.6
0.4
0.2
0

Fig. 6.1 Comparison of OGS-6(line) and OGS-5(scatter) simulation results at different time for the infiltration problem

Table 6.2 Benchmark deposit (https://docs.opengeosys.org/books/bmb-5)

BM code	Author	Code	Files	CTest
BMB5-6.1	Yonghui Huang	OGS-6	See below	OK

(https://oc.ufz.de/index.php/s/nlph7bhfDkj6tC7)

6.1.3   Results

The simulated and experimental saturation data at various time steps are plotted in Fig. 6.1. The OGS-6 simulated infiltration front propagates through the soil column and resembles well the saturation results of OGS-5. Figure 6.1 shows the saturation contours after 2, 9 and 17 h for structured meshes (Table 6.2).

The source code of the Richards flow process can be found in the following link:

• https://github.com/ufz/ogs/blob/master/ProcessLib/RichardsFlow

The input files for the infiltration benchmark can be found in the following link:

• https://github.com/ufz/ogs-data/tree/master/Parabolic/Richards.

Isothermal Two-Phase Flow
The isothermal two-phase two-component flow model has been implemented in the newly developed OGS-6 version. In this section, we will first introduce the mathematical framework of the isothermal two-phase flow model. Then a few benchmarks are selected and presented in order for the verification of the correctness of these implementation. The comparisons are made against the previous OGS-5 version (Kolditz et al. 2012b) or analytical solutions as well.

Mathematical Framework

Mass Balance Equation
We consider isothermal two-phase flow process. Thermal effects are neglected. The partially saturated sample is treated as an immiscible two-phase system, which indicates the mass transfer between different phases are ignored as well. Furthermore, both fluids are assumed imcompressible.

Consider two-phase flow in porous media, typically liquid phase (denoted by l) and gas phase(denoted by g). For each phase in two-phase fluid flow, mass conservation is given by the following equation,

$$\frac{\partial}{\partial t}(\phi S^g \rho_k^g + \phi S^l \rho_k^l) + \nabla \cdot (\mathbf{J}_k^g + \mathbf{J}_k^l) = Q_k \qquad (6.1.1)$$

where S is saturation, ρ stands for phase density, ϕ is the porosity, $\mathbf{J}$ is total flux. Q_k represents the source and sink term. The subscript k in Eq. (6.1.1) denotes the component, e.g. air ($k = a$) or water ($k = w$), within each phase, $\gamma \in \{g, l\}$. For any phase $\gamma \in \{g, l\}$, an advection vector $\mathbf{J}_{Ak}^\gamma$ and a diffusion vector $\mathbf{J}_{Dk}^\gamma$ comprise the total flux, i.e.

$$\mathbf{J}_k^\gamma = \mathbf{J}_{Ak}^\gamma + \mathbf{J}_{Dk}^\gamma \qquad (6.1.2)$$

According to generalized Darcy's law, the advective flux can be written as:

$$\mathbf{J}_{Ak}^\gamma = -\rho_k^\gamma \frac{\mathbf{k}k_{rel}^\gamma}{\mu^\gamma} (\nabla p^\gamma - \rho^\gamma \mathbf{g}), \qquad (6.1.3)$$

where $\mathbf{k}$ is the intrinsic permeability, k_{rel}^γ is the relative permeability of the phase, and μ^γ is the viscosity.

The diffusive part of the total flux is given by Fick's law

$$\mathbf{J}_{Dk}^\gamma = -\phi S^\gamma \rho^\gamma \mathbb{D}_k^\gamma \nabla \left(\frac{\rho_k^\gamma}{\rho^\gamma}\right), \qquad (6.1.4)$$

where $\mathbb{D}$ is the diffusion coefficient tensor. Since $\rho^\gamma = \rho_a^\gamma + \rho_w^\gamma$, we have

$$\mathbf{J}_{Dw}^\gamma + \mathbf{J}_{Da}^\gamma = \mathbf{0} \qquad (6.1.5)$$

under the assumption $\mathbb{D}_a^\gamma = \mathbb{D}_w^\gamma$.

For simplicity, we consider a water-air mixture system. We expand the mass balance equation (6.1.1) with the flux defined based upon the above Eqs. (6.1.2, 6.1.3 and 6.1.4). For the water component, the diffusive part of the total flux takes the form

$$\mathbf{J}_{Dw}^l = -\phi S^l \rho^l \mathbb{D}_w^l \nabla \left(\frac{\rho_w^l}{\rho^l}\right), \quad \mathbf{J}_{Dw}^g = -\phi S^g \rho^g \mathbb{D}_w^g \nabla \left(\frac{\rho_w^g}{\rho^g}\right). \qquad (6.1.6)$$

Obviously, $\mathbb{D}_w^l = \mathbf{0}$. Therefore, the mass balance equation for water component can be written as follows

$$\frac{\partial}{\partial t}\left(\phi S^g \rho_w^g + \phi S^l \rho_w^l\right) - \nabla \cdot \left[\rho_w^l \frac{kk_{rel}^l}{\mu^l}\left(\nabla p^l - \rho^l \mathbf{g}\right)\right]$$

$$-\nabla \cdot \left[\rho_w^g \frac{kk_{rel}^g}{\mu^g}\left(\nabla p^g - \rho^g \mathbf{g}\right)\right] - \nabla \cdot \left[\phi S^g \rho^g \mathbb{D}_w^g \nabla \left(\frac{\rho_w^g}{\rho^g}\right)\right] = Q_w. \qquad (6.1.7)$$

While the mass balance equation for the air component is analogous to Eq. 6.1.7.

Pressure-Pressure (pp) Scheme

In this process, the gas pressure and capillary pressure are selected as the combination of the primary variables.

Based on the description of isothermal two-phase flow above, mass balance equation (6.1.1) can be modified in order to obtain governing equations for isothermal two-phase flow in a porous medium. In this formulation primary variables are gas pressure p^g, and capillary pressure p^c.

The basic equations of the isothermal two-phase flow system are:

$$\phi \rho_w \frac{\partial S_w}{\partial p_c}\dot{p}_c + \nabla \cdot \left[\rho_w \frac{kk_{relw}}{\mu_w}\left(-\nabla p^g + \nabla p^c + \rho_w \mathbf{g}\right)\right] = Q_w \qquad (6.1.8)$$

$$-\phi \rho^a \frac{\partial S_w}{\partial p_c}\dot{p}_c + \phi(1 - S_w)\left(\frac{\partial \rho_a}{\partial p^g}\dot{p}^g + \frac{\partial \rho_a}{\partial p_c}\dot{p}_c\right) +$$

$$\nabla \cdot \left[\rho_a \frac{kk_{rela}}{\mu_a}\left(-\nabla p^g + \rho_a \mathbf{g}\right)\right] = Q_a \qquad (6.1.9)$$

Numerical Solution

In this implementation, the standard Galerkin finite element method is employed for spacial discretization, with a backward Euler fully implicit scheme for the time integration. The monolithic strategy is applied for solving the differential equation system, which indicates that both gas pressure and capillary pressure are obtained in the same time within one nonlinear iteration loop.

Standard Newton scheme is applied for the linearization, with an absolute tolerance set to be 1e-7 and the maximum iteration number set to be 50. After the linearization of the global governing equations, a sparse and asymmetric linear system is assembled and needs to be solved. The BiCGStab solver from the LIS library (Nishida 2010) or Eigen library (Guennebaud et al. 2010) is employed with an ILU preconditioner to obtain the solution.

6.2 Liakopoulos Experiment

6.2.1 Definition

This benchmark is based on an experiment by Liakopoulos (1965). The experiment mainly represents the natural drainage of a sand column which is fully saturated with water in the beginning. The sand column undergoes a desaturation process with the only force of gravity. Various numerical work have been made to simulate this experiment, here we refer to Lewis and Schrefler (1987)(pp 167–174).

6.2.2 Model Configuration

In the model, the geometry is represented by a 2-D 0.1×1 m column, and it is uniformly discritized into 72 quad elements as Fig. 6.2 shown.

The initial pressure is assumed to be atmospheric pressure. The initial capillary pressure is zero since no gas exists in the column in the beginning. The top and bottom of the column is open to the atmosphere. The initial and boundary conditions are summarized in Fig. 6.2.

The material and fluid properties adopted in this simulation are listed in Table 6.3. The simulation runs over a time span of 7200 s. A fixed time step size control is applied for the simulation with $\Delta t = 1$ s.

For verification, the model is compared with the results generated by OGS-5 Kolditz et al. (2012c). The same model geometry and model parameters are adopted between two models.

Fig. 6.2 Schematic of the benchmark formulated to simulate the Liakopoulos experiment

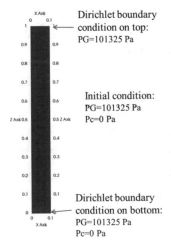

Dirichlet boundary condition on top:
PG=101325 Pa

Initial condition:
PG=101325 Pa
Pc=0 Pa

Dirichlet boundary condition on bottom:
PG=101325 Pa
Pc=0 Pa

Table 6.3 Material parameters for the Liakopoulos problem

Property	Symbol	Unit	Value(formulation)
Porosity	n	–	2.975×10^{-1}
Permeability	κ	m^2	4.5000×10^{-13}
Liquid dynamic viscosity	μ_w	$Pa.s$	1.0000×10^{-3}
Gas dynamic viscosity	μ_a	$Pa.s$	1.8×10^{-5}
Liquid density	ρ_w	$kg.m^{-3}$	1.0000×10^{3}
Gas density	ρ_a	$kg.m^{-3}$	Ideal Gas Law's
Capillary pressure	P_c	Pa	Experimental curve
Relative permeability	$k_{rel\,w}$	–	Experimental curve
Relative permeability	$k_{rel\,a}$	–	Brook-Corey functions

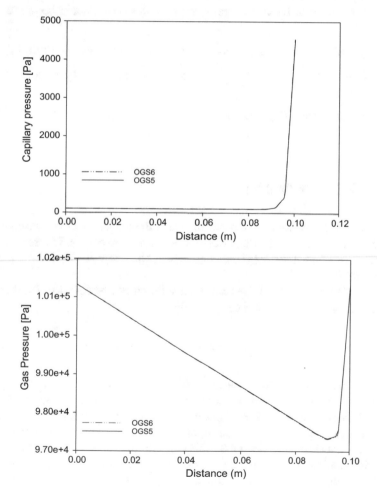

Fig. 6.3 Vertical profiles of capillary (top) and gas pressures (bottom), comparison against OGS5 results

Table 6.4 Benchmark deposit (https://docs.opengeosys.org/books/bmb-5)

BM code	Author	Code	Files	CTest
BMB5-6.2	Yonghui Huang	OGS-6	See below	OK

(https://oc.ufz.de/index.php/s/nlph7bhfDkj6tC7)

6.2.3 Results

The vertical profiles results of primary variables (capillary and gas pressure) at $t = 20\,$s are given in Fig. 6.3. The red dash line represents the OGS-6 results, while the black line is the OGS-5 results. It can be observed that the results agree well with the results generated by OGS-5 (Table 6.4).

The source code of the isothermal two phase flow process can be found in the following link:

- https://github.com/ufz/ogs/tree/master/ProcessLib/TwoPhaseFlowWithPP.

The input files for the Liakopoulos experiment benchmark can be found in the following link:

- https://github.com/ufz/ogs-data/tree/master/Parabolic/TwoPhaseFlowPP/Liakopoulos

6.3 McWhorter Problem

In order to further investigate the performance of the implementation, we consider a diffusion dominated oil-water displacement transport problem: McWhorter problem, for which a quasi-analytic solution exists which was proposed by McWhorter and Sunada (1990).

The detailed derivation of the analytical solution can be found in Kolditz et al. (2012c). The analytical solution can be written as:

$$x(S_w, t) = \frac{Q_t}{A\phi}\frac{dF}{dS_w}t, \tag{6.3.1}$$

where F is the fractional flow function *with* capillary pressure, and normally must be determined numerically. The solution looks very similar to the Buckley–Leverett solution. However, there is a very important difference: On contrary to the Buckley–Leverett problem, F is *always* concave, and hence there *never* is shock front. This, of course, is not surprising since the PDE with capillary effects is a parabolic PDE, and therefore all solutions must be smooth. Like this, it can be used to verify the performance of a numerical simulator and judge the exactness of a discretization scheme.

6.3.1   Definition

The test benchmark problem for capillary effects is formulated as if the instantaneous displacement occurs in one-dimensional horizontal reservoir initially occupied by oil. A special attention has been paid on full consideration of the capillary effects and for arbitrary capillary-hydraulic properties. The numerical solution has been obtained through solving the governing equation (6.1.9) by pressure-pressure scheme described in Sect. 6.1.3. Initially the porous media is fully saturated with nonwetting phase oil, and the wetting phase water flows in, gradually displace the oil by the capillary drive.

6.3.2   Model Configuration

The porous media is represented by a horizontal 1D domain with the length of 2.6 m. Line elements has been used with space discretization $\Delta x = 0.01$ m with overall 260 elements.

Here the flow is governed by capillary force when water saturation at the left end of the horizontal column is kept to be one, while the right end is kept to be no flux at all. No source term is accounted. The schematic of this benchmark is shown in Fig. 6.4. The initial and boundary conditions formulated in terms of gas pressure and capillary pressure are given in the figure as well.

The simulation runs for a time span of 1×10^4 s. A fixed time step size control is applied for the simulation. For the first 100 time step, $\Delta t = 1$ s, then the time step size increases to 10 s.

The material and fluid properties used in this benchmark are listed in Table 6.5.

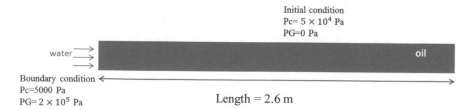

Fig. 6.4 Schematic of the benchmark formulated to test McWhorter problem in terms of the P-P scheme

Table 6.5 Material and fluid properties for the McWhorter problem

Property	Symbol	Unit	Value (formulation)
Column length	L	m	2.6
Wetting dynamic viscosity	μ_w	$Pa.s$	1.0×10^{-3}
Non-wetting dynamic viscosity	μ_{nw}	$Pa.s$	1.0×10^{-3}
Wetting phase density	ρ_w	$kg.m^{-3}$	1.0×10^3
Non-wetting phase density	ρ_{nw}	$kg.m^{-3}$	1.0×10^3
Permeability	$\mathbf{k}$	m^2	1.0×10^{-10}
Porosity	ϕ	–	3.0×10^{-1}
Residual saturation of water	S_{rw}	–	0
Residual saturation of oil	S_{nrw}	–	0
Entry pressure	p_d	Pa	5.0×10^3
Soil distribution index	λ	–	2.0
Capillary pressure	$p^c(S_{\text{eff}})$	Pa	Brooks-Corey model
Relative permeability	$k_{rel}(S_{\text{eff}})$	–	Brooks-Corey model

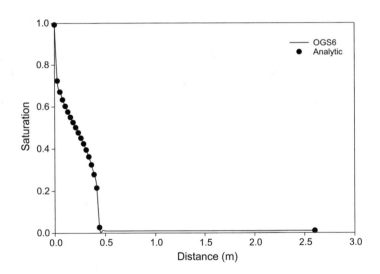

Fig. 6.5 Water saturation, S_w profile of the present result along with analytical solution based on one by McWhorter

Table 6.6 Benchmark deposit (https://docs.opengeosys.org/books/bmb-5)

BM code	Author	Code	Files	CTest
BMB5-6.3	Yonghui Huang	OGS-6	See below	OK

(https://oc.ufz.de/index.php/s/nlph7bhfDkj6tC7)

6.3.3 Results

Based on the above discussion, OGS-6 implementation produces agreeable solution compared against the analytical solution. Figure 6.5 shows water saturation profile, S_w with a fine grid along with 2.6 m long horizontal column at $t = 1000$ s. The red curve represents the OGS-6 results while the scatter points are the analytical solution (Table 6.6).

The source code of the isothermal two phase flow process can be found in the following link:

- https://github.com/ufz/ogs/tree/master/ProcessLib/TwoPhaseFlowWithPP.

The input files for the McWhorter problem benchmark can be found in the following link:

- https://github.com/ufz/ogs-data/tree/master/Parabolic/TwoPhaseFlowPP/Mc Worts.

Chapter 7
HT (Convection) Processes

Fabien Magri, Mauro Cacace, Thomas Fischer, Dmitri Naumov,
Wenqing Wang, Norihiro Watanabe, Tianyuan Zheng, Xing-Yuan Miao,
Thomas Nagel and Marc Walther

7.1 3D Benchmark of Free Thermal Convection in a Faulted System

Fabien Magri, Mauro Cacace, Thomas Fischer, Dmitri Naumov, Wenqing Wang, Norihiro Watanabe, Marc Walther

7.1.1 Introduction

In a geothermal system, unstable fluid density profiles due to temperature variations can trigger the onset and development of free thermal convective processes (Elder 1967). Early studies on the problem showed that the development of free thermal

F. Magri (✉) · T. Fischer · D. Naumov · W. Wang ·
T. Zheng · X.-Y. Miao · T. Nagel · M. Walther
UFZ, Helmholtz Centre for Environmental Research, Leipzig, Germany
e-mail: Fabien.Magri@ufz.de

F. Magri
FU Berlin, Berlin, Germany

T. Zheng · X.-Y. Miao · M. Walther
TU Dresden, Dresden, Germany

T. Nagel
Trinity College Dublin, Dublin 2, Ireland

M. Cacace
GFZ, Helmholtz Centre Potsdam, German Research Centre for Geosciences,
Potsdam, Germany

N. Watanabe
AIST, National Institute of Advanced Industrial Science and Technology,
Renewable Energy Research Center, Koriyama, Fukushima, Japan

© Springer International Publishing AG 2018
O. Kolditz et al. (eds.), *Thermo-Hydro-Mechanical-Chemical Processes
in Fractured Porous Media: Modelling and Benchmarking*,
Terrestrial Environmental Sciences, https://doi.org/10.1007/978-3-319-68225-9_7

157

convection in the Earth's crust require a relatively high permeability of the porous rocks (Lapwood 1948). Since the permeability inside the damaged area of major fault zones can far exceed the permeability of the enclosing rocks (Wallace and Morris 1986), one can expect the development of free thermal convective instabilities to occur in such tectonically perturbed rocks. The onset of thermal convection of a single-phase fluid in a vertical fault enclosed in impermeable rocks was considered in a full 3D approximation by Wang et al. (1987). A fundamental result of those investigations was that highly permeable faults allow for onset of free thermal convection even under a normal (e.g. $30\,°C \cdot km^{-1}$) geothermal gradient. In contrast to simple homogenous 1D and 2D systems, no appropriate analytical solutions can be derived to test numerical models for more complex 3D systems that account for variable fluid density and viscosity as well as permeability heterogeneity (e.g. presence of faults). Owing to the efficacy of thermal convection for the transport of thermal energy and dissolved minerals in the moving fluid, a benchmark case study for density/viscosity driven flow is crucial to ensure that the applied numerical model accurately simulates the physical processes.

The presented chapter proposes a 3D benchmark test for the simulation of thermal convection in a faulted system that accounts for temperature dependent fluid viscosity. The linear stability analysis recently developed by Malkovsky and Magri (2016) is used to estimate the critical Rayleigh number within the fault: by definition, thermal convection occurs or is absent if a small perturbation develops or decays with time, i.e. at Rayleigh numbers higher or lower than the critical value, respectively. OGS-6 results are compared to those obtained using the commercial software FEFLOW (Diersch 2014) to test the ability of the open source code in matching both the critical Rayleigh number at which convection occurs and the dynamical features of convective processes. Additionally, the results derived from an application relying on the GOLEm simulator (Cacace and Jacquey 2017) are also presented. Despite some internal differences between the three software in handling the nonlinear coupling of thermal and hydraulic processes, all models are qualitatively comparable.

7.1.2   Problem Formulation HT

Let us consider a faulted geological system, with height H, as shown in Fig. 7.1. It is assumed that temperatures T at lower ($z = 0$) and upper ($z = H$) boundaries of the rock layer are fixed and equal to T_h and T_c, respectively, and $T_c < T_h$ (subscripts h and c for hot and cold, respectively). Rock properties such as density, specific heat capacity and thermal conductivity of the whole system (i.e. fault zone and enclosing rocks) are homogeneous and temperature-invariant. The fault zone permeability is k, whereas that of the enclosing rocks is negligibly small.

Fig. 7.1 Schematic representation of a single fault embedded in conductive rocks. The fault width is 2δ, and $\Delta = \frac{\delta}{H}$ is half of the aspect ratio of the fault. T_c and T_h are top (cold) and bottom (hot) temperature boundary conditions

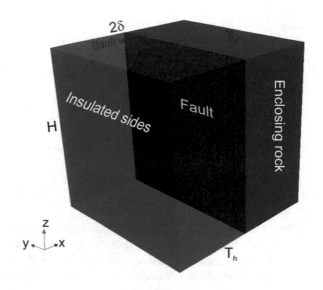

Governing Equations

Fluid velocity components v_x, v_y, and v_z satisfy Darcy's law

$$v_x = -\frac{k}{\mu}\frac{\partial p}{\partial x}, \quad v_y = -\frac{k}{\mu}\frac{\partial p}{\partial y}, \quad v_z = -\frac{k}{\mu}\left(\frac{\partial p}{\partial z} + \rho g\right)$$

where x, y, and z are cartesian coordinates, p is pressure, k is the permeability of the rocks, μ is dynamic viscosity of the fluid, ρ is density of the fluid. The permeability distribution in the considered rock layer can be represented as:

$$k = \begin{cases} k, & |y| < \delta, \ 0 < z < H, \\ 0, & |y| \geq \delta, \ 0 < z < H \end{cases}$$

where 2δ is width of the fault zone.

We assume that over given temperature ranges at hydrostatic pressure the dependence of fluid density on temperature can be approximated by a linear function (Bear 1972)

$$\rho = \rho_0 \left[1 - \beta (T - T_c)\right]$$

where ρ_0 is fluid density at $T = T_c$, β is fluid thermal expansion coefficient.

Temperature dependence of fluid viscosity can be approximated by the function

$$\mu(T) = \mu_0 \exp\left(-\frac{T - T_c}{T_v}\right)$$

where $\mu_0 = \mu(T_c)$ and T_v are approximation constants. This equation of state (EOS) provides good fit over temperature variations of $150°$ C as shown later in the numerical examples.

The continuity equation in the Boussinesq's approximation takes the form

$$\frac{\partial v_x}{\partial x} + \frac{\partial v_y}{\partial y} + \frac{\partial v_z}{\partial z} = 0$$

Temperature distribution in the fault zone satisfies the equation

$$\rho_r c_r \frac{\partial T}{\partial t} + \rho c \left(v_x \frac{\partial T}{\partial x} + v_y \frac{\partial T}{\partial y} + v_z \frac{\partial T}{\partial z} \right) = \lambda \left(\frac{\partial^2 T}{\partial x^2} + \frac{\partial^2 T}{\partial y^2} + \frac{\partial^2 T}{\partial z^2} \right)$$

where t is the time, ρ_r and c_r are density and specific heat capacity of the rock, $\lambda = n\lambda_l + (1 - n)\lambda_s$ is averaged thermal conductivity of the fluid (subscript l for liquid) and the saturated rocks (subscript s for solid), and n is rock porosity.

The fluid velocity is zero outside the fault. Therefore, the temperature distribution outside the fault satisfies the equation

$$\rho_r c_r \frac{\partial \vartheta}{\partial t} = \lambda \left(\frac{\partial^2 \vartheta}{\partial x^2} + \frac{\partial^2 \vartheta}{\partial y^2} + \frac{\partial^2 \vartheta}{\partial z^2} \right)$$

where ϑ is the temperature in the considered layer outside the fault ($0 < z < H$; $|y| > \delta$).

Boundary Conditions

The upper and lower boundaries of the layer are impermeable (i.e. the normal component of the velocity at the top and bottom sides of the model is equal to zero). Therefore, conditions for the pressure at these boundaries take the form:

$$z = 0, \quad \frac{\partial p}{\partial z} + \rho_0 g [1 - \beta (T_h - T_c)] = 0;$$
$$z = H, \quad \frac{\partial p}{\partial z} + \rho_0 g = 0.$$

The lower and upper boundaries of the layer are isothermal, i.e.:

$$z = 0, \ T = \vartheta = T_h; \quad z = H, \ T = \vartheta = T_c,$$

$$T_h > T_c$$

Sides are insulated.

Rayleigh Theory

The OGS implementation given above allows to directly calculating the Rayleigh number, by example, at the top cold temperature T_c, using the following formula:

Fig. 7.2 Modeled faulted system displaying the finite element mesh and temperature boundary conditions. 107'000 hexahedral elements discretize the faulted block. A mesh refinement is performed within the 40 m wide fault (zoom). The top and bottom temperature boundary conditions are 10 and 160 °C respectively, corresponding to an initial geothermal gradient of approximately 27 °C · km^{-1}

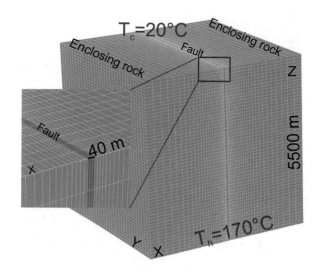

$$\text{Ra} = \frac{k\rho_0^2 cg\beta\,(T_h - T_c)\,H}{\lambda\,\mu_0} \tag{7.1.1}$$

Malkovsky and Magri (2016) provide a simple polynomial fit to estimate the critical Rayleigh number Ra_{crit} for temperature-dependent viscous fluids

$$\text{Ra}_{\text{crit}} = \left[\left(\tfrac{8.19}{\Delta}\right)^{5/4} + \left(4\pi^2\right)^{5/4}\right]^{4/5} \left(1 + 0.493\gamma + 0.12\gamma^2\right) e^{-\gamma},$$
$$0.003 \le \Delta \le 0.05 \text{ and } 0 < \gamma \le 2.5 \tag{7.1.2}$$

where $\gamma = \frac{T_h - T_c}{T_v}$ characterizes the viscous property of the saturating fluid and $\Delta = \frac{\delta}{H}$ is half of the fault aspect ratio. Thermal convection develops for $\text{Ra} > \text{Ra}_{\text{crit}}$.

7.1.3 Numerical Benchmark

All models refer to an embedded 3D fault with aspect ratio $\frac{2\delta}{H} = 7.3 \cdot 10^{-3}$ (i.e. $\Delta = 3.65 \cdot 10^{-3}$). Specifically, the dimensions of the modeled fault are $H = 5000$ m and $\delta = 40$ m discretized with prismatic elements (Fig. 7.2).

Material and Solid Properties

The physical properties of both fault and enclosing rocks are homogenous and isotropic, as given in Table 7.1.

Table 7.1 Medium and solid properties. For simplicity, storage and heat capacity are set to 0

Properties	$n[-]$	$k[m^2]$	$\lambda_s[W \cdot m^{-1} \cdot K^{-1}]$
Fault	10^{-3}	$1.7 \cdot 10^{-13}$	3
Enclosing rock	10^{-3}	10^{-18}	3

Table 7.2 Fluid properties. $\mu_0 \rho_0$ at $T_0 = 20\,^{\circ}C$ are $10^{-3}Pa \cdot s$ and $1000\,kg \cdot m^{-3}$, respectively

Properties	$\gamma = \frac{T_h - T_c}{T_v}[-]$	$\beta[K^{-1}]$	$c_f[J \cdot kg^{-1}]$	$\lambda_l[W \cdot m^{-1} \cdot K^{-1}]$
Fluid	2	$4.3 \cdot 10^{-4}$	4200	0.65

Fluid Properties

At the given pressure, the EOS for fluid viscosity $\mu(T) = \mu_0 \exp\left(-\frac{T-T_c}{T_v}\right)$ and fluid density $\rho = \rho_0\left[1 - \beta\left(T - T_c\right)\right]$ are sufficient to describe fluid properties at liquid phase over temperature ranges $T_h - T_c \approx 150\,^{\circ}C$ or smaller. Both EOS have been implemented into the applied software. Fluid properties used in the benchmark are summarized in Table 7.2.

Initial and Boundary Conditions

This benchmark illustrates a low enthalpy geothermal system, where $20 \leq T \leq 170\,^{\circ}C$. An example of a high enthalpy system can be found in Malkovsky and Magri (2016).

In all models, the initial pressure is hydrostatic and the initial temperature T^{init} increases linearly with depth, from 20 to $170\,^{\circ}C$ (top and bottom Dirichlet boundary conditions, Fig. 7.2).

Within the fault, a perturbation of the form $\varepsilon(x, z) = \sin(\pi z)\cos(\pi x)$ is added to T^{init} in order to trigger a circular convective cell within the fault plane (x-z). The amplitude of the perturbation is $\pm 1\,^{\circ}C$.

Rayleigh Number Setup

Ra is calculated with Eq. (7.1.1), i.e. at the reference top temperature $T_c = 20\,^{\circ}C$, with $T_h - T_c = 150\,^{\circ}C$. Using the values given in the tables, this benchmark refers to an estimated Ra = 829.

According to Eq. (7.1.2), at $\Delta = 3.65 \cdot 10^{-3}$ and $\gamma = 2$, the onset of convection is triggered at the critical Rayleigh number $Ra^{crit} \approx 753$.

Simulation Time, Time Scheme

The simulated period is $2 \cdot 10^{11}s$. Simulations were run using a fixed-time stepping (user defined).

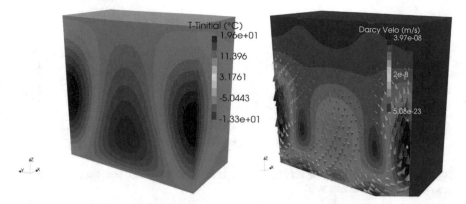

Fig. 7.3 OGS-6 results for the final time at the fault plane; Ra = 829; left: temperature perturbation $\varepsilon = T - T^{\text{init}}$, right: Darcy velocity field

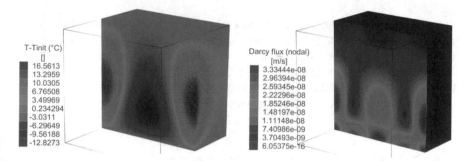

Fig. 7.4 FEFLOW results for the final time at the fault plane; Ra = 829; left: temperature perturbation, right: Darcy velocity field

7.1.4 Results

The calculated Darcy flow field and temperature anomaly $\varepsilon = T - T^{\text{init}}$ are illustrated in Fig. 7.3 at the end of the simulated period for the case $Ra = 829$.

OGS results compare well against those obtained with FEFLOW (Fig. 7.4). It can be seen that the patterns as well as the temperature and velocity ranges are in excellent agreement with OGS results. Minor differences can only be observed in the peak values.

Remarks to the Users

Here it is worth recalling that in density-driven flow problems different type of numerical implementation (e.g. weak coupling vs. monolithic) or applied solver induces numerical errors which propagation modifies the developing fingers. As a result, the temporal evolution of the temperature perturbation vary, depending on the

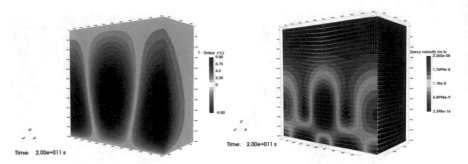

Fig. 7.5 MOOSE results for the final time at the fault plane; Ra $= 829$; left: temperature perturbation, right: Darcy velocity field

software applied, as observed in the different peak values calculated with OGS-6 and
FEFLOW.

As an additional example, Fig. 7.5 shows the results obtained with GOLEm which
uses a full monolithic formulation while dealing with the nonlinearities between the
two processes without making use of the Boussinesq approximation.

It can be seen that patterns, though qualitatively very similar to those derived from
OGS (Fig. 7.3) and FEFLOW (Fig. 7.4), show some deviations in the magnitudes and
geometry of computed thermal anomalies in the fault plane.

7.1.5 OGS-6 and OGS-5 Computing Time

Scaling tests were performed on two discretized meshes, obtained from sequential
refinements of the original grid with 107'000 elements illustrated in Fig. 7.2. The
first and second refinements (Ref1, Ref2) led to 859'000 and 6'871'000 hexahedral
elements, respectively. The OGS simulations were performed on the EVE cluster
system at the UFZ using computing nodes with 20 cores (Intel® Xeon® CPU E5-
2670 v2, 20 MB cache per core). For each simulation we reserved a 20 core compute
node with 64 GB main memory such that a simulation run fits a compute node. Full
nodes were allocated to avoid internode communication.

Table 7.3 summarizes the computational time needed to solve the described 3D
benchmark: (1) OGS-6 runtimes decrease for single core and parallel appliances in
comparison to OGS-5; (2) it can be seen that on refined meshes (e.g. Ref1, 2) with
a high degree of parallelization (e.g. >60 cores) OGS-6 is up to a factor 20 faster
than OGS-5; (3) for Ref1 in both applications, both OGS-5 and OGS-6, show a clear
decrease of the efficiency for 60 or more cores.

The better performance and scaling of OGS-6 versus OGS-5 relies on: (1) a fully
coupled HT approach, i.e. monolithic scheme in OGS-6 using a Newton–Raphson
method versus a staggered scheme in OGS-5 using fixed-point iterations, and (2)
computation of shape functions and derivatives. In OGS-6, this is done only once at

Table 7.3 Computing time (in [s]) OGS-6 versus OGS-5 for different refinements

Cores	Original mesh		Ref1	
	OGS-5	OGS-6	OGS-5	OGS-6
1	20'235	14'636	–	–
20	8'300	815	83'915	10'380
40	–	–	58'568	5'766
60	–	–	63'242	3'741
80	–	–	62'645	2'846
100	–	–	75'640	2'548

the beginning of the simulation run, while in OGS-5, computation of shape functions is performed at each integration point for each local assembly.

7.1.6 Summary

A numerical benchmark for the simulations of thermal convection onset of a single-phase fluid with temperature dependent viscosity in a vertical permeable fault is proposed.

OGS-6 successfully simulates the onset of thermal convection at critical Rayleigh numbers inferred from recently developed linear stability analysis (Malkovsky and Magri 2016). The parallelization in OGS-6 for this type of problem shows great computational and scaling improvements with respect to the previous version. OGS results are consistent with those obtained with FEFLOW and GOLEm. Owing to strong coupling of the partial differential equations, the numerical solutions are highly sensitive to numerical errors that can be due to different applied methods and/or solver settings. Consequently, the calculated patterns are all qualitatively similar while differences in the calculated values exist. Therefore, results of case-studies must be interpreted carefully and require additional constraints (field data).

The given example provides a useful benchmark that can be applied to any numerical code of thermally-driven flow processes at basin-scale (Table 7.4).

Table 7.4 Benchmark deposit (https://docs.opengeosys.org/books/bmb-5)

BM code	Author	Code	Files	CTest
BMB5-7.1	Fabien Magri	OGS-5	Available	OK
	Fabien Magri	FEFLOW	Available	-
	Mauro Cacace	MOOSE	Available	-

7.2 2D Benchmark of Large-Scale Free Thermal Convection

Tianyuan Zheng, Fabien Magri, Dmitri Naumov, Thomas Nagel

This is a summary of a 2D benchmark test for the simulation of thermal convection in a km-scale porous media that accounts for temperature-dependent fluid viscosity and density. This problem is analogous to the 3D case presented in Sect. 7.1. By definition, thermal convection occurs or is absent if a small perturbation develops or decays with time.

OGS-6 results (monolithic approach) are compared to those obtained with the commercial software FEFLOW (Diersch 2013) to test the ability of the open-source code in matching the dynamical features of convective processes.

7.2.1 Problem Formulation

Let us consider a 2D vertical square as illustrated in Fig. 7.6, i.e. the fault plane as previously illustrated in the 3D case (Fig. 7.2). It is assumed that temperatures at lower ($z = 0$) and upper ($z = H$) boundaries of the rock layer are fixed and equal to T_h and T_c, respectively, and $T_c < T_h$. Rock properties such as density, specific heat capacity and thermal conductivity of the whole system are considered homogeneous and temperature-invariant.

Fig. 7.6 2D domain and FE mesh. Square 5.5×5.5 km; 32×32 elements

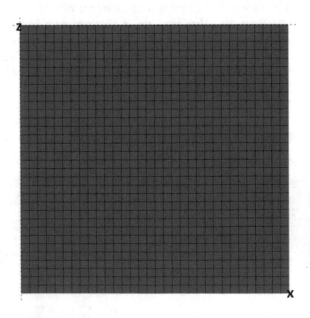

Governing Equations

Fluid velocity components v_x, v_z satisfy Darcy's law:

$$v_x = -\frac{k}{\mu}\frac{\partial p}{\partial x}, \quad v_z = -\frac{k}{\mu}\left(\frac{\partial p}{\partial z} + \rho g\right)$$

(7.2.1)

where p is the pressure, k is the intrinsic permeability of the rocks, μ is the dynamic viscosity of the fluid and ρ is the density of the fluid.

We assume that over given temperature ranges at hydrostatic pressure the dependence of fluid density on temperature can be approximated by a linear function (Bear 1972).

$$\rho = \rho_0 \left[1 - \beta \left(T - T_c\right)\right]$$

(7.2.2)

where ρ_0 is the fluid density at $T = T_c$ and β is the fluid thermal expansion coefficient.

Temperature dependence of fluid viscosity can be approximated by the function:

$$\mu(T) = \mu_0 \exp\left(-\frac{T - T_c}{T_v}\right)$$

(7.2.3)

where $\mu_0 = \mu(T_c)$ and T_v are approximation constants.

The continuity equation in the Boussinesq's approximation takes the form:

$$\frac{\partial v_x}{\partial x} + \frac{\partial v_z}{\partial z} = 0$$

(7.2.4)

Temperature distribution in the porous media satisfies:

$$\rho_r c_r \frac{\partial T}{\partial t} + \rho c \left(v_x \frac{\partial T}{\partial x} + v_z \frac{\partial T}{\partial z}\right) = \lambda \left(\frac{\partial^2 T}{\partial x^2} + \frac{\partial^2 T}{\partial z^2}\right)$$

(7.2.5)

where t is the time, ρ_r and c_r are density and specific heat capacity of the fluid saturated rocks, $\lambda = n\lambda_f + (1 - n)\lambda_s$ is thermal conductivity of the fluid saturated rocks and n is the rock porosity.

Boundary Conditions and Initial Conditions

Boundary and initial conditions are identical to those of the 3D case, as given in Sect. 7.1.

7.2.2 Numerical Benchmark

Here, the values of the different physical properties of the numerical benchmark are given in Tables 7.6 and 7.5.

Table 7.5 Medium and solid properties. For simplicity storage and heat capacity are set to 0

Porosity	n	0.001	–
Permeability	**k**	10^{-14}	m^2
Thermal conductivity	λ_f	3	W/m/K

Table 7.6 Fluid properties μ_0 and p_0 at $T_0 = 20\,^\circ C$ are $10^{-3}\,Pa\,s$ and $1000\,kg/m^3$, respectively

–	$\gamma = \frac{T_h - T_c}{T_v}$	2	–
Thermal expansion coefficient	β	4.3×10^{-4}	1/K
Heat capacity	C_p	4200	$J/m^3/K$
Thermal conductivity	λ_s	0.65	W/m/K

Material Properties

For simplicity, these values are constant and allowed for thermal convection.

At a given pressure, the EOS for fluid viscosity (Eq. 7.2.3) and fluid density (Eq. 7.2.2 are sufficient to describe fluid properties at liquid phase over temperature ranges $T_h - T_c \approx 150\,^\circ C$ or smaller. Both EOS have been implemented into OGS. Fluid properties used in the benchmark are summarized in Table 7.6.

Additionally, the case with constant viscosity ($\mu = 10^{-3} Pa\,s$) will also be illustrated.

Numerical Setup

Due to the high non-linearity in this numerical model, the linear solver and time stepping were elaborately chosen. The direct solver SparseLU (Guennebaud et al. 2010) were used so solve the linear system and Picard iterations were employed for the non-linear iterations until convergence with a relative tolerance of 10^{-5} was reached.

7.2.3 Results

Temperature-Dependent Viscosity

The calculated Darcy-flow field and temperature anomaly $\varepsilon = T - T^{init}$ are illustrated respectively in Fig. 7.7a and b at the end of the simulated period (Fig. 7.8).

A single convective cell covering the entire domain forms. A thermally-driven plume flows along the bottom side of the PM at peak velocity of $6.7 \cdot 10^{-9}$ m/s. The resulting upwelling/downwelling increases/decreases the initial temperature profile. The solution displays periodical oscillations.

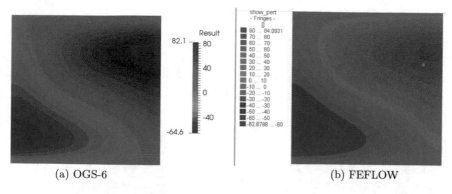

(a) OGS-6   (b) FEFLOW

Fig. 7.7 Temperature perturbation (°C) at t = $5 \cdot 10^{10}$ s

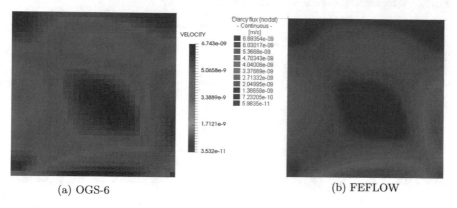

(a) OGS-6   (b) FEFLOW

Fig. 7.8 Darcy velocity field (m/s) at t = $5 \cdot 10^{10}$ s

Constant Viscosity

Here the fluid viscosity is set to its reference value, ($\mu = 1 \cdot 10^{-3}$ Pa s). Compared to the previous case, the temperature anomaly and the velocity field are less vigorous which highlights the strong destabilizing effects of temperature-dependent viscosity. These solutions are steady state (Figs. 7.9 and 7.10).

7.2.4 Summary

- A numerical benchmark for the simulations of 2D km-scale thermal convection onset of a single-phase fluid with temperature dependent viscosity and density is proposed.
- OGS-6 successfully simulates the onset of thermal convection. OGS-6 results are consistent with those obtained with the previous version and FEFLOW (Table 7.7).

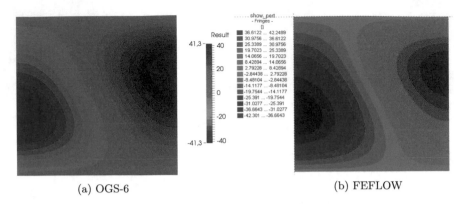

(a) OGS-6   (b) FEFLOW

Fig. 7.9 Temperature perturbation (°C) at $t = 5 \cdot 10^{10}$ s

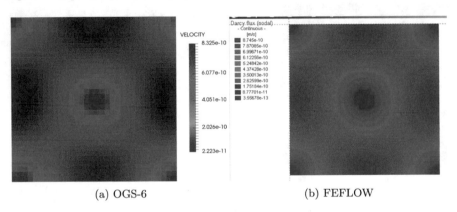

(a) OGS-6   (b) FEFLOW

Fig. 7.10 Darcy velocity field (m/s) at $t = 5 \cdot 10^{10}$ s

Table 7.7 Benchmark deposit (https://docs.opengeosys.org/books/bmb-5)

BM code	Author	Code	Files	CTest
BMB5-7.3	Tianyuan Zheng	OGS-6	See below	ToDo

(https://oc.ufz.de/index.php/s/nlph7bhfDkj6tC7)

The OGS-6 version performed for this benchmark can be found in the master branch under:

• https://github.com/ufz/ogs.

The input files named square_5500x5500.prj are available at:

• https://github.com/ufz/ogs-data/tree/master/Parabolic/HT/ConstViscosity.

7.3   Two-Dimensional Transient Thermal Advection

Tianyuan Zheng, Xing-Yuan Miao, Dmitri Naumov, Thomas Nagel

Heat transport in a moving liquid is discussed in this section, and a series of bench-marks is presented comprising the transient two-dimensional temperature distribution in a moving liquid.

The following assumptions are made to simplify the problem:

- Constant material properties
- Neglecting viscous dissipation effects
- Local thermal equilibrium
- No phase change process

Hence, flow is not affected by temperature, but couples into heat transport via an advection term:

$$(\varrho c_p)\frac{\partial T}{\partial t} - \nabla \cdot \boldsymbol{\lambda}\nabla T + \varrho_f c_f \mathbf{q}\nabla T = 0 \tag{7.3.1}$$

$$\phi\kappa\frac{\partial p}{\partial t} = \nabla \cdot \left(\frac{\mathbf{k}}{\mu}\nabla p - \varrho_f\mathbf{g}\right) \tag{7.3.2}$$

where ϱc_p represents the heat capacity of the porous media $\phi\varrho_f c_f + (1 - \phi)\varrho_s c_s$, λ represents the heat conductivity $\phi\lambda_f + (1 - \phi)\lambda_s$ and $\mathbf{q}$ is the Darcy velocity. The different parameters and their values are given in Table 7.8.

Table 7.8 Physical parameters

Symbol	Parameter	Value	Unit
ϱ	Solid density	2000	kg· m^{-3}
ϱ	Fluid density	1000	kg· m^{-3}
c_s	Solid heat capacity	250	J· kg^{-1}· K^{-1}
c_f	Fluid heat capacity	1100	J· kg^{-1}· K^{-1}
λ_s	Solid thermal conductivity	50	W·m^{-1}·K^{-1}
λ_f	Fluid thermal conductivity	10	W·m^{-1}·K^{-1}
$\phi\kappa$	Liquid storage	2.5e^{-10}	Pa^{-1}
κ	Compressibility	0	Pa^{-1}
ϕ	Porosity	0.1	–
$\mathbf{k}$	Permeability	10e^{-11}	m^2
μ	Viscosity	0.001	Pa· s
β	Thermal expansion coefficient	2.07e^{-4}	K^{-1}

Within the finite element framework, the above equations were cast into the following matrix form:

$$
\begin{pmatrix} \mathbf{M}_{\mathrm{TT}} & \mathbf{0} \\ \mathbf{0} & \mathbf{M}_{\mathrm{pp}} \end{pmatrix} \begin{pmatrix} \dot{T} \\ \dot{p} \end{pmatrix} + \begin{pmatrix} \mathbf{K}_{\mathrm{TT}} & \mathbf{K}_{\mathrm{Tp}} \\ \mathbf{0} & \mathbf{K}_{\mathrm{pp}} \end{pmatrix} \begin{pmatrix} T \\ p \end{pmatrix} = \begin{pmatrix} f_{\mathrm{T}} \\ f_{\mathrm{p}} \end{pmatrix}
\tag{7.3.3}
$$

in which $\mathbf{M}_{TT} = (\mathbf{N})^{\mathrm{T}} \varrho c_{\mathrm{p}} \mathbf{N}$ indicates the thermal storage term, $\mathbf{M}_{pp} = (\mathbf{N})^{\mathrm{T}} \phi \kappa \mathbf{N}$ indicates the pressure-associated storage term for liquid flow, $\mathbf{K}_{TT} = (\nabla \mathbf{N})^{\mathrm{T}} \lambda \nabla \mathbf{N} + \mathbf{N}^{\mathrm{T}} \varrho_{\mathrm{f}} c_{\mathrm{f}} \mathbf{q} \nabla \mathbf{N}$ indicates the advection-diffusion matrix for heat transport, $\mathbf{K}_{pp} = (\nabla \mathbf{N})^{\mathrm{T}} \frac{k}{\mu} \nabla \mathbf{N}$ is the Laplace matrix for liquid flow and $\mathbf{K}_{Tp} = (-\mathbf{N})^{\mathrm{T}} \beta_{\mathrm{f}} \mathbf{N}$ is the thermal expansion term which is only considered in the density-driven heat convection process described in the previous 2d thermal convection benchmark.

7.3.1 Transient 2D Heat Transport with Moving Liquid

In this benchmark, the storage term of the liquid flow equation is neglected, thus considering the flow process to be in steady state. A transient heat transport process is coupled with the steady-state flow.

The domain is a 2D rectangular domain extending 2 m by 1.5 m along the positive x and y axes. The liquid and solid are both incompressible and the gravity has been explicitly neglected. The storage term of the liquid flow equation is assumed to be zero. Pressure $p = 200000$ Pa at the left boundary and zero at the right boundary drive with the steady-state flow along the x axis. A constant temperature $T_{\mathrm{b}} = 274.15$ K is given in the centre of the left boundary (from 0.5 to 1 m), whereas in the remaining domain the initial temperature is set to $T_0 = 273.15$ K. The simulation evaluates the transient temperature distribution in the entire domain after 4000 s and the result is validated with an analytical solution (Fig. 7.11).

7.3.1.1 Analytical Solution

A similar analytical solution was described in Kolditz et al. (2016c), the rectangle with 2×1.5 m located in the x-y plane and divided into 200×150 square elements. The permeable porous media is considered, with isotropic permeability and porosity. The specified inlet temperature reads,

$$
g(y) = \begin{cases} 0 & \text{for } y \leq a \\ T_{\mathrm{b}} & \text{for } a \leq y \leq b \\ 0 & \text{for } y \geq b \end{cases}
\tag{7.3.4}
$$

The solution of pressure distribution and specific discharge is same with Kolditz et al. (2016c). For 1D flow along the x-axis, the pressure comes

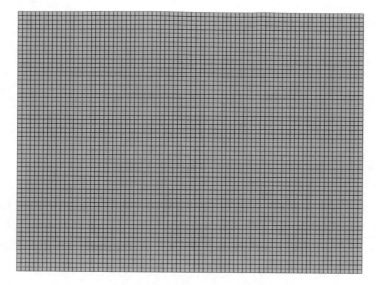

Fig. 7.11 Spatial discretization

$$p(x) = p_0 \left(1 - \frac{x}{L_x} \right) \tag{7.3.5}$$

and the specific discharge q is obtained by Darcy's law

$$q = \frac{k}{\mu} \frac{P_0}{L_x} \tag{7.3.6}$$

Then the heat transport equation is considered (see Eq. 7.3.1). The following notations introduced:

$$w = \frac{\phi \varrho_f c_f}{\phi \varrho_f c_f + (1 - \phi) \varrho_s c_s} \frac{q}{\phi}$$
$$\chi = \frac{\phi \lambda_f + (1 - \phi) \lambda_s}{\phi \varrho_f c_f + (1 - \phi) \varrho_s c_s} \tag{7.3.7}$$

The heat transport equations becomes

$$\frac{\partial T}{\partial t} + w \frac{\partial T}{\partial x} = \chi \left(\frac{\partial^2 T}{\partial x^2} + \frac{\partial^2 T}{\partial y^2} \right) \tag{7.3.8}$$

and is complemented by the initial conditions

$$T(x, y, 0) = T_0 \tag{7.3.9}$$

and the boundary conditions

$$
\begin{aligned}
T(0, y, t) &= g(y) \quad \text{for} \quad t > 0 \\
\lim_{x \to \infty} T(x, y, t) &= 0 \quad \text{for} \quad t > 0 \\
\lim_{y \to \infty} T(x, y, t) &= 0 \quad \text{for} \quad t > 0 \\
\lim_{y \to -\infty} T(x, y, t) &= 0 \quad \text{for} \quad t > 0
\end{aligned}
\tag{7.3.10}
$$

Performing a Laplace transformation with respect to t gives

$$
s \frac{\partial \bar{T}}{\partial t} + w \frac{\partial \bar{T}}{\partial x} = \chi \left(\frac{\partial^2 \bar{T}}{\partial x^2} + \frac{\partial^2 \bar{T}}{\partial y^2} \right)
\tag{7.3.11}
$$

where T is the Laplace transform of T and s is the transformation parameter. The boundary conditions convert to

$$
\begin{aligned}
T(0, y, s) &= \frac{g(y)}{s} \\
\lim_{x \to \infty} T(x, y, s) &= 0 \\
\lim_{y \to \infty} T(x, y, s) &= 0 \\
\lim_{y \to -\infty} T(x, y, s) &= 0
\end{aligned}
\tag{7.3.12}
$$

Then the Fourier transform is implemented

$$
\chi \bar{U}'' - w \bar{U} - (s + \chi r^2) \bar{U} = 0
\tag{7.3.13}
$$

where $\bar{U}$ is the Fourier transform of $\bar{T}$ and r is the Fourier transformation parameter. The boundary condition now becomes

$$
\begin{aligned}
\bar{U}(0, r, s) &= \frac{G(r)}{s} \\
\lim_{x \to \infty} \bar{U}(x, r, s) &= 0
\end{aligned}
\tag{7.3.14}
$$

where $G(r)$ is the Fourier transform of $g(y)$. This yields

$$
\bar{U}(x, r, s) = \frac{G(r)}{s} \exp \left[x \left(\frac{w}{2\chi} - \sqrt{ \left(\frac{w}{2\chi} \right)^2 + \frac{s}{\chi} + r^2 } \right) \right]
\tag{7.3.15}
$$

Application of the inverse Laplace transformation results in

$$U(x,r,t) = \frac{x}{(4\pi\chi)^{1/2}} \int_0^t \frac{G(r)\exp(-\chi r^2 t')}{(t')^{3/2}}\exp\left(-\frac{(x-t'w)^2}{4\chi t'}\right)dt' \quad (7.3.16)$$

Knowing the inverse Fourier transform

$$F^{-1}G(r)\exp(-\chi r^2 t') = \frac{\exp(-y^2/(4\chi t'))}{(2\chi t')^{1/2}} \quad (7.3.17)$$

and using the convolution theorem of Fourier transformation

$$F^{-1}G(r)\exp(-\chi r^2 t') = F^{-1}G(r)F^{-1}\exp(-\chi r^2 t')$$
$$= T_b \int_a^b \frac{(2\pi)^{-1/2}{}^{1/2}}{2\chi t'}\exp\left(-\frac{(y-v)^2}{4\chi t'}\right)dv \quad (7.3.18)$$

one finally arrives at the temperature solution

$$T = \frac{T_b x}{4(\pi\chi)^{1/2}} \int_0^t \left(-\frac{(x-t'w)^2}{4\chi t'}\right)\left(\mathrm{erf}\frac{y-a}{(4\chi t')^{1/2}} - \mathrm{erf}\frac{y-b}{(4\chi t')^{1/2}}\right)t'^{-3/2}dt' \quad (7.3.19)$$

The analytical solution was implemented with Python and Romberg integration scheme is applied to integral. It has to be noticed is that the analytical solution does not contain the boundary values ($x = 0$) which have to be assigned manually (Fig. 7.12).

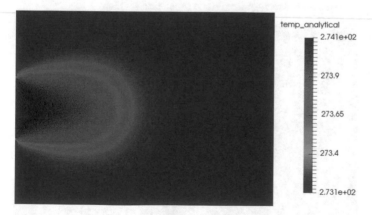

Fig. 7.12 2D hydrothermal temperature profile (analytical solution)

7.3.1.2 Numerical Solution

The governing equation (7.3.1) is treated with the finite element method in a monolithic scheme. The biconjugate gradient stabilized method (BiCGSTAB) is used as the linear solver with a tolerance of 10^{-16}. The Picard method was implemented as the nonlinear solver with the tolerance of 10^{-5}. The element size is set to $0.01 \times 0.01\,\text{m}^2$ and a time step of $100\,\text{s}$ is chosen.

In order to quantify the difference between analytical solution and numerical solution, the relative error ($\varepsilon = |T_{\text{analytical}} - T_{\text{numerical}}|/T_{\text{analytical}}$) is calculated.

From Figs. 7.13 and 7.14, it can be seen that the numerical solution fits well with the analytical solution and maximum relative error is in the order of 10^{-4}. The maximum error exists in the two corners of the left boundary. This is because

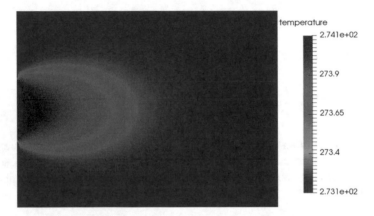

Fig. 7.13 2D hydrothermal temperature profile (numerical solution)

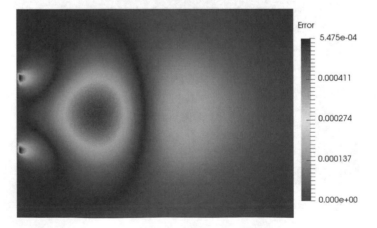

Fig. 7.14 The relative error between numerical solution and analytical solution

Table 7.9 Benchmark deposit (https://docs.opengeosys.org/books/bmb-5)

BM code	Author	Code	Files	CTest
BMB5-7.3	Tianyuan Zheng	OGS-6	See below	ToDo

(https://oc.ufz.de/index.php/s/nlph7bhfDkj6tC7)

the analytical solution tends to be infinite when the x coordinate is closed to zero (Table 7.9).

The OGS-6 version performed for this benchmark can be found in the master branch under:

- https://github.com/ufz/ogs.

The input files named quad_80x60 are available at:

- https://github.com/ufz/ogs-data/tree/TH-monolithic/Parabolic/TH.

Chapter 8
HM Processes

Gesa Ziefle, Jobst Maßmann, Norihiro Watanabe, Dmitri Naumov, Herbert Kunz and Thomas Nagel

8.1 Fluid Injection in a Fault Zone Using Interface Elements with Local Enrichment

This section focuses on coupled hydraulic–mechanical processes in a fault zone. The presented benchmark is motivated by the "Fault Slip (FS)" experiment of the Mont Terri Project.[1] In this experiment, a fluid injection into a fault zone is carried out and the resulting hydraulic and mechanical effects are monitored. The fault zone is characterized by a range of minor and major faults and the experiment comprises several steps where various locations are influenced by an injection. More information about this experiment as well as similar approaches can be found in Guglielmi et al. (2015b), and Guglielmi et al. (2015a), and Derode et al. (04/2015).

Modelling this system is a challenging task that has been selected to be part of the DECOVALEX-2019 project[2]–where Task B focuses on related modeling approaches and their comparison.

---

[1] https://www.mont-terri.ch/
[2] www.decovalex.org
[3] https://docs.opengeosys.org/docs

G. Ziefle (✉) · J. Maßmann · H. Kunz
BGR, Federal Institute for Geosciences and Natural Resources, Hanover, Germany
e-mail: Gesa.Ziefle@bgr.de

N. Watanabe
National Institute of Advanced Industrial Science and Technology, Renewable Energy Research Center (AIST), Koriyama, Fukushima, Japan

D. Naumov · T. Nagel
Helmholtz Centre for Environmental Research - UFZ, Leipzig, Germany

T. Nagel
Trinity College Dublin, Dublin, Ireland

© Springer International Publishing AG 2018
O. Kolditz et al. (eds.), *Thermo-Hydro-Mechanical-Chemical Processes in Fractured Porous Media: Modelling and Benchmarking*, Terrestrial Environmental Sciences, https://doi.org/10.1007/978-3-319-68225-9_8

Fig. 8.1 Representation of
the fracture elements in the
LIE modelling approach

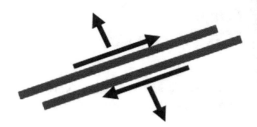

The following section presents a simplified 2-dimensional model approach using interface elements (LIE) of co-dimension one with local enrichment functions of the finite-element solution space to simulate the fracture behavior due to a fluid injection (for implementational details see Watanabe et al. 2012). The simulations are carried out with OpenGeoSys 6.[3] More information on modelling approaches can also be found in Rutqvist et al. (2015).

8.1.1 Model Approach

The detailed model approach is presented in Watanabe et al. (2012). It is based on the usage of lower-dimensional interface elements (LIE) for simulating flow through the fracture in combination with a Heaviside-enriched solution space to model the displacement discontinuity across the fracture. The discrete fracture is modeled by a pair of surfaces between which normal and shear displacements are permissible as presented in Fig. 8.1.

The effective-stress approach is used to calculate the stress field in the fracture:

$$\boldsymbol{\sigma}_{\text{tot},f} = \boldsymbol{\sigma}_{\text{eff},f} - \alpha_f \, p_f \mathbf{1} \tag{8.1}$$

with $\boldsymbol{\sigma}_{\text{tot},f}$ and $\boldsymbol{\sigma}_{\text{eff},f}$ being the total and effective stresses in the fracture linked via the Biot coefficient in the fracture α_f and the fluid pressure in the fracture p_f.

The relationship between the effective stress and the fracture relative displacement vector results from:

$$\mathrm{d}\boldsymbol{\sigma}_{\text{eff},f} = \mathbf{K} \, \mathrm{d}\mathbf{w} \tag{8.2}$$

with the fracture shear and normal displacements being part of the displacement vector $\mathbf{w}$, the vector of the effective stress increment defined analogously, and the stiffness matrix comprising the normal, shear and coupled stiffness of the fracture given by:

$$\mathbf{K} = \begin{bmatrix} k_{tt} & k_{tn} \\ k_{nt} & k_{nn} \end{bmatrix} \tag{8.3}$$

Concerning the fluid flow, the fracture is modeled based on the parallel-plate assumption where the fluid flow along the fracture $\mathbf{q}_f$ is directly related to fracture aperture b_h by the cubic law such that:

$$\mathbf{q}_f = \frac{b_h^2}{12\mu_{FR}}\mathbf{I}\left(-\nabla p_f + \rho_{FR}\mathbf{g}\right) \tag{8.4}$$

with the fluid viscosity μ_{FR}, the fluid density ρ_{FR} and the gravity vector $\mathbf{g}$. Assuming a uniform pressure across the fracture width, this yields to the mass balance equation:

$$b_h S_f \frac{\partial p_f}{\partial t} + \alpha_f \frac{\partial b_h}{\partial t} + \nabla \cdot (b_h \mathbf{q}_f) + q^+ + q^- = 0 \tag{8.5}$$

with the leakage flux from the opposing fracture surfaces to the surrounding porous media q^+ and q^- and the specific storage of the fracture S_f resulting from $S_f = \frac{1}{K_f}$ with the compressibility of the fracture K_f and the time t.

The fracture is assumed to be characterized by elasto-plastic material behavior following the Mohr–Coulomb failure criterion. Based on the mentioned elasto–plastic behaviour, the variable fracture aperture b_h is calculated from elastic and inelastic contributions:

$$b_h = b_{h,init} + \Delta b_{h,elastic} + \Delta b_{h,shear} \tag{8.6}$$

with the initial fracture aperture $b_{h,init}$, the elastic part of the aperture $b_{h,elastic}$ (change following from Eq. (8.2)) and the opening due to the shear dilation of the fracture $b_{h,shear}$.

8.1.2 Model Set-Up

The modeled two-dimensional domain has an extent of 20 m times 20 m as it is presented in Fig. 8.2.

The initial effective stresses are assumed to be 6 MPa in horizontal and 7 MPa in vertical direction. The pore pressure is assumed to be initially 0.5 MPa in the

Fig. 8.2 Geometry and initial and boundary conditions of the modeled domain

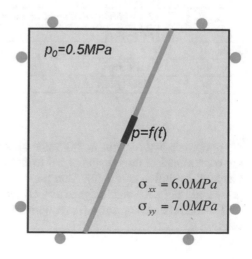

p_0=0.5MPa

p=f(t)

σ_{xx} = 6.0MPa

σ_{yy} = 7.0MPa

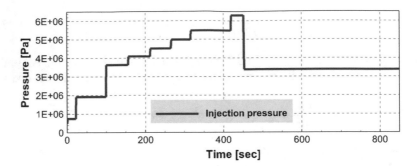

Fig. 8.3 Temporal evolution of the pressure boundary condition at the injection point

Table 8.1 Material parameters: fluid injection in a fault zone

Symbol	Parameter	Unit	Value
Host rock (elastic)			
E	Young's modulus	GPa	6.1
ν	Poisson number	–	0.3275
α_r	Biot coefficient	–	0
ρ_{SR}	Density of solid	$kg \cdot m^{-3}$	2450
ϕ	Porosity	–	0.0
$\mathbf{K}$	Intrinsic permeability	m^2	1e-17
S_r	Storage	1/Pa	1.0e-10
Fault zone (elasto-plastic)			
K_n	Normal stiffness	GPa/m	20
K_s	Shear stiffness	GPa/m	20
α_f	Biot coefficient	–	1
ψ_{MC}	Dilatancy angle	deg	10
ϕ_{MC}	Friction angle	deg	22
c_{MC}	Cohesion	–	0
$b_{h,init}$	Initial fracture hydraulic aperture	m	1e-5
S_f	Storage	1/Pa	4.4e-10
Fluid			
μ_{FR}	Viscosity	$Pa \cdot s$	1e-3
ρ_{FR}	Density of water	$kg \cdot m^{-3}$	1000

entire domain and remains at this value at the model boundary. The displacements are constrained in directions normal to the boundary as indicated in Fig. 8.2. The temporal evolution of the injection pressure in the center of the model domain is presented in Fig. 8.3 and assigned as a Dirichlet boundary condition.

The material parameters are summarized in Table 8.1 and the FEM-mesh is given in Fig. 8.4.

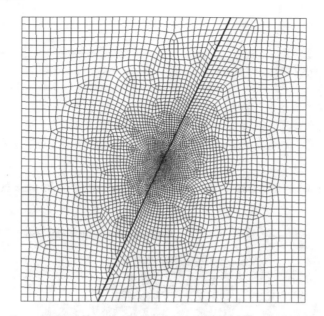

Fig. 8.4 Two-dimensional mesh and lower-dimensional representation of the fault zone

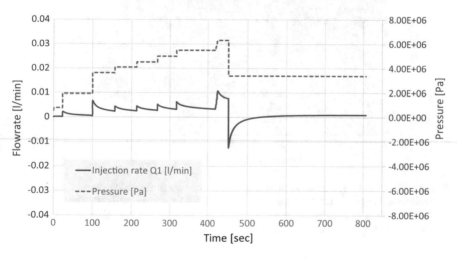

Fig. 8.5 Temporal evolution of the flowrate and the pressure in the injection point

8.1.3 Results

The injection causes significant stress redistribution in the fault zone, resulting in elasto-plastic normal as well as shear deformations due to the Mohr–Coulomb failure criterion.

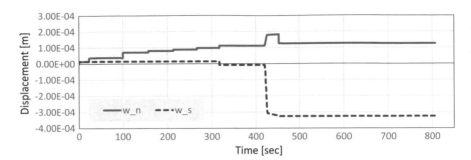

Fig. 8.6 Temporal evolutions of the normal and shear displacement of the crack in the injection point

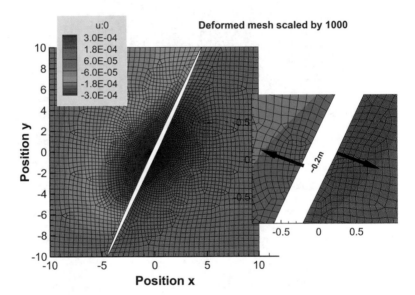

Fig. 8.7 Horizontal displacements presented as contour plot in the entire domain after 450 s at maximal injection pressure. The displaced mesh is scaled by a factor of 1000. The zoomed area indicates the shear displacements in the center of the domain. The discontinuity embodied via solution-space enrichment is clearly visible

The resulting time-dependent flowrate into the crack is presented in Fig. 8.5 in combination with the injection pressure at the injection point. The step-wise pressure increase leads to an inflow into the fault domain while the subsequently applied decrease of the injection pressure initially leads to a backflow since the pressure in the fault is higher than the injection pressure at this time step.

The temporal evolution of the normal and shear displacements of the crack in the injection point are presented in Fig. 8.6. A moderate increase of normal displacements and hardly any shear displacements can be seen during the first 420 s. After that, the

Table 8.2 Benchmark deposit (https://docs.opengeosys.org/books/bmb-5)

BM code	Author	Code	Files	CTest
BMB5-8.1	Gesa Ziefle	OGS-6	Available	ToDo

(https://oc.ufz.de/index.php/s/nlph7bhfDkj6tC7)

failure criterion is reached and a significant increase of the normal displacements due to dilatancy is accompanied by significant shear displacements. The normal displacements feature an elastic rebound due to the decrease of the injection pressure. A contour plot of the displacements at maximum injection pressure is presented in Fig. 8.7. Information about the benchmark deposit can be found in (Table 8.2).

The OGS-6 version used is available at:

- https://github.com/endJunction/ogs/tree/LIE3D_0.

Chapter 9
TM Processes

Peter Vogel, Jobst Maßmann, Xing-Yuan Miao and Thomas Nagel

Thermoelastic Beams

Peter Vogel and Jobst Maßmann

This section presents problems on thermal stresses in beams. We focus on the closed form solutions. The associated simulation exercises have been checked by OGS; they may serve as verification tests. For the underlying theory of linear thermoelasticity see Carlson (1972), for more details on thermoelastic beams see Hetnarski and Eslami (2009).

9.1 A Linear Temperature Distribution

Given length $L = 2$ m and thickness $H = 0.2$ m the domain represents the rectangular beam $[0, L] \times [0, H] \times [0, H]$. It is discretized by $40 \times 2 \times 2$ equally sized hexahedral elements. The solid material has been selected elastic with Young's modulus $E = 25,000$ MPa, Poisson's ratio $\nu = 0.25$, zero heat capacity, and thermal expansion $\alpha_T = 3 \cdot 10^{-5}$ 1/K. Gravity is neglected via zero material density. Fixities have been prescribed on the entire surface of the domain with zero x-displacement

P. Vogel (✉) · J. Maßmann
BGR, Federal Institute for Geosciences and Natural Resources,
Hanover, Germany
e-mail: Peter.Vogel@bgr.de

X.-Y. Miao · T. Nagel
UFZ, Helmholtz Centre for Environmental Research, Leipzig, Germany

X.-Y. Miao
TU Dresden, Dresden, Germany

T. Nagel
Trinity College Dublin, Dublin 2, Ireland

© Springer International Publishing AG 2018
O. Kolditz et al. (eds.), *Thermo-Hydro-Mechanical-Chemical Processes in Fractured Porous Media: Modelling and Benchmarking*, Terrestrial Environmental Sciences, https://doi.org/10.1007/978-3-319-68225-9_9

Fig. 9.1 Temperature distribution

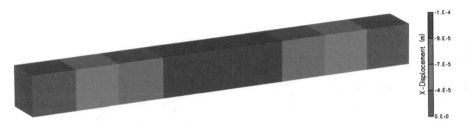

Fig. 9.2 X-Displacements

Table 9.1 Benchmark deposit (https://docs.opengeosys.org/books/bmb-5)

BM code	Author	Code	Files	CTest
BMB5-9.1	Peter Vogel	OGS-5	Available	TBD

(https://oc.ufz.de/index.php/s/nlph7bhfDkj6tC7)

along the faces $x = 0$ and $x = L$, zero y-displacement along the faces $y = 0$ and $y = H$, and zero z-displacement along the faces $z = 0$ and $z = H$. The simulation (Table 9.1) starts from zero initial temperature and comprises one timestep applying an instant temperature change with temperature $T_0 = 8\,^{\circ}\text{C}$ at $x = L$ and zero applied temperature at $x = 0$.

The formal solution proceeds in two steps, first to solve for the temperature distribution (Fig. 9.1) and then to evaluate displacements (Fig. 9.2), strains, and stresses.

Due to the setup the thermal problem depends on the x-coordinate only and the Laplace equation governing the steady-state temperature distribution $T(x)$ becomes

$$\frac{d^2 T}{dx^2} = 0. \tag{9.1.1}$$

The temperature distribution

$$T(x) = T_0 \frac{x}{L} \tag{9.1.2}$$

satisfies the Laplace equation and the imposed thermal boundary conditions, hence, this is the solution of the thermal problem.

For the solution of the mechanical problem note that the applied temperature change is identical to the above $T(x)$. Let (u_x, u_y, u_z) denote the displacement vector and

$$e = \frac{\partial u_x}{\partial x} + \frac{\partial u_y}{\partial y} + \frac{\partial u_z}{\partial z} \tag{9.1.3}$$

the volumetric strain. The displacement-temperature equations of equilibrium

$$\frac{\partial^2 u_x}{\partial x^2} + \frac{\partial^2 u_x}{\partial y^2} + \frac{\partial^2 u_x}{\partial z^2} + \frac{1}{1-2\nu}\frac{\partial e}{\partial x} = \frac{2(1+\nu)}{1-2\nu}\alpha_T\frac{\partial T}{\partial x} = \frac{2(1+\nu)}{1-2\nu}\alpha_T\frac{T_0}{L},$$

$$\frac{\partial^2 u_y}{\partial x^2} + \frac{\partial^2 u_y}{\partial y^2} + \frac{\partial^2 u_y}{\partial z^2} + \frac{1}{1-2\nu}\frac{\partial e}{\partial y} = \frac{2(1+\nu)}{1-2\nu}\alpha_T\frac{\partial T}{\partial y} = 0, \tag{9.1.4}$$

$$\frac{\partial^2 u_z}{\partial x^2} + \frac{\partial^2 u_z}{\partial y^2} + \frac{\partial^2 u_z}{\partial z^2} + \frac{1}{1-2\nu}\frac{\partial e}{\partial z} = \frac{2(1+\nu)}{1-2\nu}\alpha_T\frac{\partial T}{\partial z} = 0$$

and the imposed fixities are satisfied by

$$u_x = u_x(x), \tag{9.1.5}$$
$$u_y = 0, \tag{9.1.6}$$
$$u_z = 0, \tag{9.1.7}$$

if $u_x(x)$ satisfies the ordinary differential equation

$$\frac{d^2 u_x}{dx^2} + \frac{1}{1-2\nu}\frac{d^2 u_x}{dx^2} = \frac{2(1+\nu)}{1-2\nu}\alpha_T\frac{T_0}{L} \tag{9.1.8}$$

subject to the boundary conditions

$$u_x(0) = u_x(L) = 0. \tag{9.1.9}$$

The x-displacement becomes

$$u_x(x) = \frac{1+\nu}{1-\nu}\alpha_T\frac{T_0}{2}\left(\frac{x^2}{L} - x\right). \tag{9.1.10}$$

The strains read

$$\epsilon_{11} = \frac{\partial u_x}{\partial x} = \frac{1+\nu}{1-\nu}\alpha_T T_0\left(\frac{x}{L} - \frac{1}{2}\right), \tag{9.1.11}$$

$$\epsilon_{22} = \frac{\partial u_y}{\partial y} = 0, \tag{9.1.12}$$

$$\epsilon_{33} = \frac{\partial u_z}{\partial z} = 0, \tag{9.1.13}$$

$$\epsilon_{12} = \frac{\partial u_x}{\partial y} + \frac{\partial u_y}{\partial x} = 0, \tag{9.1.14}$$

$$\epsilon_{13} = \frac{\partial u_x}{\partial z} + \frac{\partial u_z}{\partial x} = 0, \tag{9.1.15}$$

$$\epsilon_{23} = \frac{\partial u_y}{\partial z} + \frac{\partial u_z}{\partial y} = 0. \tag{9.1.16}$$

The constitutive equations

$$\epsilon_{11} - \alpha_T T(x) = \alpha_T T_0 \left(\frac{2\nu x}{(1-\nu)L} - \frac{1+\nu}{2(1-\nu)} \right)$$
$$= \frac{1}{E}[\sigma_{11} - \nu(\sigma_{22} + \sigma_{33})],$$
$$\epsilon_{22} - \alpha_T T(x) = -\alpha_T T_0 \frac{x}{L} = \frac{1}{E}[\sigma_{22} - \nu(\sigma_{11} + \sigma_{33})],$$
$$\epsilon_{33} - \alpha_T T(x) = -\alpha_T T_0 \frac{x}{L} = \frac{1}{E}[\sigma_{33} - \nu(\sigma_{11} + \sigma_{22})], \tag{9.1.17}$$
$$\epsilon_{12} = 0 = \frac{2(1+\nu)}{E}\sigma_{12},$$
$$\epsilon_{13} = 0 = \frac{2(1+\nu)}{E}\sigma_{13},$$
$$\epsilon_{23} = 0 = \frac{2(1+\nu)}{E}\sigma_{23}$$

yield the stress tensor

$$\boldsymbol{\sigma} = \begin{pmatrix} \sigma_{11} & \sigma_{12} & \sigma_{13} \\ \sigma_{12} & \sigma_{22} & \sigma_{23} \\ \sigma_{13} & \sigma_{23} & \sigma_{33} \end{pmatrix}$$
$$= -\frac{E\alpha_T T_0}{1-\nu} \begin{pmatrix} \frac{1-\nu}{2(1-2\nu)} & 0 & 0 \\ 0 & \frac{\nu}{2(1-2\nu)} + \frac{x}{L} & 0 \\ 0 & 0 & \frac{\nu}{2(1-2\nu)} + \frac{x}{L} \end{pmatrix}, \tag{9.1.18}$$

which satisfies the equation of mechanical equilibrium

$$\text{div } \boldsymbol{\sigma} = 0. \tag{9.1.19}$$

9.2 A Steady-State Antisymmetric Temperature Distribution

Given length $L = 2\,\text{m}$ and thickness $H = 0.2\,\text{m}$ the domain represents the rectangular beam $[0, L] \times [0, H] \times [0, H]$. It is discretized by $40 \times 2 \times 2$ equally sized hexahedral elements. The solid material has been selected elastic with Young's modulus $E = 25,000\,\text{MPa}$, Poisson's ratio $\nu = 0.25$, zero heat capacity, and thermal expansion $\alpha_T = 3 \cdot 10^{-5}$ 1/K. Gravity is neglected via zero material density.

Fixities have been prescribed on the entire surface of the domain with zero x-displacement along the faces $x = 0$ and $x = L$, zero y-displacement along the faces $y = 0$ and $y = H$, and zero z-displacement along the faces $z = 0$ and $z = H$. The simulation (Table 9.2) starts from zero initial temperature and comprises one timestep applying an instant temperature change with temperature $T_0 = 4\,°C$ at $x = 0$ and temperature $-T_0 = -4\,°C$ at $x = L$.

The formal solution proceeds in two steps, first to solve for the temperature distribution (Fig. 9.3) and then to evaluate displacements (Fig. 9.4), strains, and stresses.

Due to the setup the thermal problem depends on the x-coordinate only and the Laplace equation governing the steady-state temperature distribution $T(x)$ becomes

$$\frac{d^2 T}{dx^2} = 0. \tag{9.2.1}$$

The temperature distribution

$$T(x) = T_0 \left(1 - \frac{2x}{L}\right) \tag{9.2.2}$$

satisfies the Laplace equation and the imposed thermal boundary conditions, hence, this is the solution of the thermal problem.

For the solution of the mechanical problem note that the applied temperature change is identical to the above $T(x)$. Let (u_x, u_y, u_z) denote the displacement vector and

$$e = \frac{\partial u_x}{\partial x} + \frac{\partial u_y}{\partial y} + \frac{\partial u_z}{\partial z} \tag{9.2.3}$$

the volumetric strain. The displacement-temperature equations of equilibrium

Table 9.2 Benchmark deposit (https://docs.opengeosys.org/books/bmb-5)

BM code	Author	Code	Files	CTest
BMB5-9.2	Peter Vogel	OGS-5	Available	TBD

(https://oc.ufz.de/index.php/s/nlph7bhfDkj6tC7)

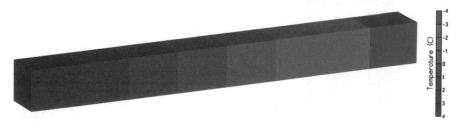

Fig. 9.3 Temperature distribution

Fig. 9.4 X-Displacements

$$\frac{\partial^2 u_x}{\partial x^2} + \frac{\partial^2 u_x}{\partial y^2} + \frac{\partial^2 u_x}{\partial z^2} + \frac{1}{1-2\nu}\frac{\partial e}{\partial x} = \frac{2(1+\nu)}{1-2\nu}\alpha_T\frac{\partial T}{\partial x} = -\frac{2(1+\nu)}{1-2\nu}\alpha_T\frac{2T_0}{L},$$

$$\frac{\partial^2 u_y}{\partial x^2} + \frac{\partial^2 u_y}{\partial y^2} + \frac{\partial^2 u_y}{\partial z^2} + \frac{1}{1-2\nu}\frac{\partial e}{\partial y} = \frac{2(1+\nu)}{1-2\nu}\alpha_T\frac{\partial T}{\partial y} = 0, \qquad (9.2.4)$$

$$\frac{\partial^2 u_z}{\partial x^2} + \frac{\partial^2 u_z}{\partial y^2} + \frac{\partial^2 u_z}{\partial z^2} + \frac{1}{1-2\nu}\frac{\partial e}{\partial z} = \frac{2(1+\nu)}{1-2\nu}\alpha_T\frac{\partial T}{\partial z} = 0$$

and the imposed fixities are satisfied by

$$u_x = u_x(x), \qquad (9.2.5)$$

$$u_y = 0, \qquad (9.2.6)$$

$$u_z = 0, \qquad (9.2.7)$$

if $u_x(x)$ satisfies the ordinary differential equation

$$\frac{d^2 u_x}{dx^2} + \frac{1}{1-2\nu}\frac{d^2 u_x}{dx^2} = -\frac{2(1+\nu)}{1-2\nu}\alpha_T\frac{2T_0}{L} \qquad (9.2.8)$$

subject to the boundary conditions

$$u_x(0) = u_x(L) = 0. \qquad (9.2.9)$$

The x-displacement becomes

$$u_x(x) = -\frac{1+\nu}{1-\nu}\alpha_T T_0\left(\frac{x^2}{L} - x\right). \qquad (9.2.10)$$

The strains read

$$\epsilon_{11} = \frac{\partial u_x}{\partial x} = \frac{1+\nu}{1-\nu}\alpha_T T(x), \qquad (9.2.11)$$

$$\epsilon_{22} = \frac{\partial u_y}{\partial y} = 0, \qquad (9.2.12)$$

$$\epsilon_{33} = \frac{\partial u_z}{\partial z} = 0, \tag{9.2.13}$$

$$\epsilon_{12} = \frac{\partial u_x}{\partial y} + \frac{\partial u_y}{\partial x} = 0, \tag{9.2.14}$$

$$\epsilon_{13} = \frac{\partial u_x}{\partial z} + \frac{\partial u_z}{\partial x} = 0, \tag{9.2.15}$$

$$\epsilon_{23} = \frac{\partial u_y}{\partial z} + \frac{\partial u_z}{\partial y} = 0. \tag{9.2.16}$$

The constitutive equations

$$\epsilon_{11} - \alpha_T T(x) = \frac{2\nu}{1-\nu}\alpha_T T(x) = \frac{1}{E}[\sigma_{11} - \nu(\sigma_{22} + \sigma_{33})], \tag{9.2.17}$$

$$\epsilon_{22} - \alpha_T T(x) = -\alpha_T T(x) = \frac{1}{E}[\sigma_{22} - \nu(\sigma_{11} + \sigma_{33})], \tag{9.2.18}$$

$$\epsilon_{33} - \alpha_T T(x) = -\alpha_T T(x) = \frac{1}{E}[\sigma_{33} - \nu(\sigma_{11} + \sigma_{22})], \tag{9.2.19}$$

$$\epsilon_{12} = 0 = \frac{2(1+\nu)}{E}\sigma_{12}, \tag{9.2.20}$$

$$\epsilon_{13} = 0 = \frac{2(1+\nu)}{E}\sigma_{13}, \tag{9.2.21}$$

$$\epsilon_{23} = 0 = \frac{2(1+\nu)}{E}\sigma_{23} \tag{9.2.22}$$

yield the stress tensor

$$\sigma = \begin{pmatrix} \sigma_{11} & \sigma_{12} & \sigma_{13} \\ \sigma_{12} & \sigma_{22} & \sigma_{23} \\ \sigma_{13} & \sigma_{23} & \sigma_{33} \end{pmatrix} = -\frac{E\alpha_T}{1-\nu}T(x)\begin{pmatrix} 0 & 0 & 0 \\ 0 & 1 & 0 \\ 0 & 0 & 1 \end{pmatrix}, \tag{9.2.23}$$

which satisfies the equation of mechanical equilibrium

$$\text{div } \sigma = 0. \tag{9.2.24}$$

9.3 A Transient Antisymmetric Temperature Distribution

Given length $L = 2\,\text{m}$ and thickness $H = 0.2\,\text{m}$ the domain represents the rectangular beam $[0, L] \times [0, H] \times [0, H]$. It is discretized by $200 \times 1 \times 1$ equally sized hexahedral elements. The solid material has been selected elastic with Young's modulus $E = 25,000\,\text{MPa}$, Poisson's ratio $\nu = 0.25$, density $\rho = 2000\,\text{kg/m}^3$, thermal conductivity $k = 2.7\,\text{W/(m·K)}$, heat capacity $c = 0.45\,\text{J/(kg·K)}$, and thermal expansion $\alpha_T = 3 \cdot 10^{-4}\,\text{1/K}$. Gravity is neglected via explicit assignment.

Fixities have been prescribed on the entire surface of the domain with zero x-displacement along the faces $x = 0$ and $x = L$, zero y-displacement along the faces $y = 0$ and $y = H$, and zero z-displacement along the faces $z = 0$ and $z = H$. Temperatures $T_0 t$ ($T_0 = 1\,°\text{C/s}$) and $-T_0 t$ are applied at the faces $x = L$ and $x = 0$, respectively, for times $t > 0$. Starting from zero initial temperature the simulation (Table 9.3) evaluates the transient temperature distribution $T(x, t)$ as well as the associated mechanical load with output after 5 and 10 s.

The formal solution proceeds in two steps, first to solve for the temperature distribution (Fig. 9.5) and then to evaluate displacements (Fig. 9.6), strains, and stresses.

The heat conduction equation

$$\rho c \frac{\partial T}{\partial t} = \text{div}(k \text{ grad } T) \tag{9.3.1}$$

is the governing equation describing the transient thermal problem. Introducing the notation

$$\chi = \frac{k}{\rho c} \tag{9.3.2}$$

the present temperature distribution $T(x, t)$ is governed by the parabolic equation

$$\frac{1}{\chi} \frac{\partial T}{\partial t} = \frac{\partial^2 T}{\partial x^2}, \tag{9.3.3}$$

the initial condition

$$T(x, 0) = 0 \quad \text{for} \quad 0 \le x \le L, \tag{9.3.4}$$

Table 9.3 Benchmark deposit (https://docs.opengeosys.org/books/bmb-5)

BM code	Author	Code	Files	CTest
BMB5-9.3	Peter Vogel	OGS-5	Available	TBD

(https://oc.ufz.de/index.php/s/nlph7bhfDkj6tC7)

Fig. 9.5 Temperature distribution after 10 s

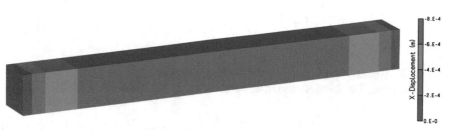

Fig. 9.6 X-Displacements after 10 s

and the boundary conditions imposed at the beams ends

$$T(0, t) = -T_0 t \quad \text{for} \quad t > 0, \tag{9.3.5}$$

$$T(L, t) = T_0 t \quad \text{for} \quad t > 0. \tag{9.3.6}$$

Applying the Laplace transform with respect to t yields the ordinary differential equation

$$\chi \bar{T}'' - s \bar{T} = 0, \tag{9.3.7}$$

where $\bar{T}$ is the transform of T, s is the transformation parameter, and the prime denotes the derivative with respect to x. This equation has to be solved with respect to the transformed boundary conditions. The Laplace transform of the temperature distribution becomes

$$\bar{T}(x, s) = T_0 \frac{\sinh[\sqrt{s/\chi}(x - L/2)]}{s^2 \sinh[\sqrt{s/\chi}L/2]}. \tag{9.3.8}$$

Then, following Churchill (1958)

$$
\begin{aligned}
\bar{T}(x, s) &= \frac{T_0}{s^2} \frac{\exp[\sqrt{s/\chi}(x - L/2)] - \exp[-\sqrt{s/\chi}(x - L/2)]}{\exp[\sqrt{s/\chi}L/2] - \exp[-\sqrt{s/\chi}L/2]} \\
&= \frac{T_0}{s^2} \frac{\exp[\sqrt{s/\chi}(x - L)] - \exp[-\sqrt{s/\chi}x]}{1 - \exp[-\sqrt{s/\chi}L]} \\
&= \frac{T_0}{s^2} \left\{ \exp[\sqrt{s/\chi}(x - L)] - \exp[-\sqrt{s/\chi}x] \right\} \sum_{n=0}^{\infty} \exp[-\sqrt{s/\chi}nL] \\
&= \frac{T_0}{s^2} \sum_{n=0}^{\infty} \left\{ \exp[-\sqrt{s/\chi}(nL + L - x)] - \exp[-\sqrt{s/\chi}(nL + x)] \right\}.
\end{aligned} \tag{9.3.9}
$$

Abramowitz and Stegun (1972) give the inverse Laplace transform

$$\mathscr{L}^{-1}\left\{\frac{1}{s^2}\exp(-K\sqrt{s})\right\} = 4t \cdot i^2\mathrm{erfc}\frac{K}{2\sqrt{t}}, \qquad (9.3.10)$$

where $K \geq 0$ and $i^2\mathrm{erfc}$ denotes the 2nd repeated integral of the complementary error function. The temperature distribution $T(x, t)$ becomes

$$T(x, t) = 4T_0 t \sum_{n=0}^{\infty} \left(i^2\mathrm{erfc}\frac{(n+1)L - x}{2\sqrt{\chi t}} - i^2\mathrm{erfc}\frac{nL + x}{2\sqrt{\chi t}}\right). \qquad (9.3.11)$$

For the solution of the mechanical problem note that the applied temperature change is identical to the above $T(x, t)$. Let (u_x, u_y, u_z) denote the displacement vector and

$$e = \frac{\partial u_x}{\partial x} + \frac{\partial u_y}{\partial y} + \frac{\partial u_z}{\partial z} \qquad (9.3.12)$$

the volumetric strain. The displacement-temperature equations of equilibrium

$$\frac{\partial^2 u_x}{\partial x^2} + \frac{\partial^2 u_x}{\partial y^2} + \frac{\partial^2 u_x}{\partial z^2} + \frac{1}{1 - 2\nu}\frac{\partial e}{\partial x} = \frac{2(1+\nu)}{1 - 2\nu}\alpha_T\frac{\partial T}{\partial x},$$

$$\frac{\partial^2 u_y}{\partial x^2} + \frac{\partial^2 u_y}{\partial y^2} + \frac{\partial^2 u_y}{\partial z^2} + \frac{1}{1 - 2\nu}\frac{\partial e}{\partial y} = \frac{2(1+\nu)}{1 - 2\nu}\alpha_T\frac{\partial T}{\partial y} = 0, \quad (9.3.13)$$

$$\frac{\partial^2 u_z}{\partial x^2} + \frac{\partial^2 u_z}{\partial y^2} + \frac{\partial^2 u_z}{\partial z^2} + \frac{1}{1 - 2\nu}\frac{\partial e}{\partial z} = \frac{2(1+\nu)}{1 - 2\nu}\alpha_T\frac{\partial T}{\partial z} = 0$$

and the imposed fixities are satisfied by

$$u_x = u_x(x, t), \qquad (9.3.14)$$
$$u_y = 0, \qquad (9.3.15)$$
$$u_z = 0, \qquad (9.3.16)$$

if $u_x(x, t)$ satisfies the differential equation

$$\frac{\partial^2 u_x}{\partial x^2} + \frac{1}{1 - 2\nu}\frac{\partial^2 u_x}{\partial x^2} = \frac{2(1+\nu)}{1 - 2\nu}\alpha_T\frac{\partial T}{\partial x} \qquad (9.3.17)$$

subject to the boundary conditions

$$u_x(0, t) = u_x(L, t) = 0 \quad \text{for} \quad t > 0. \qquad (9.3.18)$$

The x-displacement becomes

$$u_x(x,t) = 8\alpha_T \frac{1+\nu}{1-\nu} T_0 \sqrt{\chi t^3} \sum_{n=0}^{\infty} \left(i^3\text{erfc}\frac{(n+1)L-x}{2\sqrt{\chi t}} + i^3\text{erfc}\frac{nL+x}{2\sqrt{\chi t}} \right.$$

$$\left. -i^3\text{erfc}\frac{(n+1)L}{2\sqrt{\chi t}} - i^3\text{erfc}\frac{nL}{2\sqrt{\chi t}} \right) \quad (9.3.19)$$

where i^3erfc denotes the 3rd repeated integral of the complementary error function. The strains read

$$\epsilon_{11} = \frac{\partial u_x}{\partial x} = \frac{1+\nu}{1-\nu}\alpha_T T(x,t), \quad (9.3.20)$$

$$\epsilon_{22} = \frac{\partial u_y}{\partial y} = 0, \quad (9.3.21)$$

$$\epsilon_{33} = \frac{\partial u_z}{\partial z} = 0, \quad (9.3.22)$$

$$\epsilon_{12} = \frac{\partial u_x}{\partial y} + \frac{\partial u_y}{\partial x} = 0, \quad (9.3.23)$$

$$\epsilon_{13} = \frac{\partial u_x}{\partial z} + \frac{\partial u_z}{\partial x} = 0, \quad (9.3.24)$$

$$\epsilon_{23} = \frac{\partial u_y}{\partial z} + \frac{\partial u_z}{\partial y} = 0. \quad (9.3.25)$$

The constitutive equations

$$\epsilon_{11} - \alpha_T T(x,t) = \frac{2\nu}{1-\nu}\alpha_T T(x,t) = \frac{1}{E}[\sigma_{11} - \nu(\sigma_{22} + \sigma_{33})], \quad (9.3.26)$$

$$\epsilon_{22} - \alpha_T T(x,t) = -\alpha_T T(x,t) = \frac{1}{E}[\sigma_{22} - \nu(\sigma_{11} + \sigma_{33})], \quad (9.3.27)$$

$$\epsilon_{33} - \alpha_T T(x,t) = -\alpha_T T(x,t) = \frac{1}{E}[\sigma_{33} - \nu(\sigma_{11} + \sigma_{22})], \quad (9.3.28)$$

$$\epsilon_{12} = 0 = \frac{2(1+\nu)}{E}\sigma_{12}, \quad (9.3.29)$$

$$\epsilon_{13} = 0 = \frac{2(1+\nu)}{E}\sigma_{13}, \quad (9.3.30)$$

$$\epsilon_{23} = 0 = \frac{2(1+\nu)}{E}\sigma_{23} \quad (9.3.31)$$

yield the stress tensor

$$\sigma = \begin{pmatrix} \sigma_{11} & \sigma_{12} & \sigma_{13} \\ \sigma_{12} & \sigma_{22} & \sigma_{23} \\ \sigma_{13} & \sigma_{23} & \sigma_{33} \end{pmatrix} = -\frac{E\alpha_T}{1-\nu}T(x,t)\begin{pmatrix} 0 & 0 & 0 \\ 0 & 1 & 0 \\ 0 & 0 & 1 \end{pmatrix}, \quad (9.3.32)$$

which satisfies the equation of mechanical equilibrium

Table 9.4 Benchmark deposit (https://docs.opengeosys.org/books/bmb-5)

BM code	Author	Code	Files	CTest
BMB5-9.4	Peter Vogel	OGS-5	Available	TBD

(https://oc.ufz.de/index.php/s/nlph7bhfDkj6tC7)

$$\operatorname{div} \boldsymbol{\sigma} = 0. \tag{9.3.33}$$

Thermoelastic Plates

Peter Vogel and Jobst Maßmann

Plain strain problems on heated plates are the topic of this section. We focus on the closed form solutions. The associated simulation exercises have been checked by OGS; they may serve as verification tests. For the underlying theory of linear thermoelasticity see Carlson (1972), for more advanced examples see Nowacki (1986).

9.4 A Bilinear Temperature Distribution

Given length $L = 1$ m and thickness $H = 0.1$ m the domain represents the rectangular plate $[0, L] \times [0, L] \times [0, H]$. It is discretized by $120 \times 121 \times 1$ equally sized hexahedral elements. The solid material has been selected elastic with Young's modulus $E = 10,000$ MPa, Poisson's ratio $\nu = 0.2$, zero heat capacity, and thermal expansion $\alpha_T = 10^{-4}$ 1/K. Gravity is neglected via zero material density. The faces $x = L$ and $y = L$ are free of mechanical load by default. Fixities have been prescribed with zero z-displacement along the faces $z = 0$ and $z = H$, zero x-displacement along the face $y = 0$, and zero y-displacement along the face $x = 0$. The imposed thermal boundary conditions have top and bottom of the domain thermally insulated by default. Zero temperature prevails at the faces $x = 0$ and $y = 0$, respectively, the temperature distribution $T_0 y/L$ ($T_0 = 10\,°C$) has been assigned along the face $x = L$, the temperature distribution $T_0 x/L$ has been assigned along the face $y = L$. The simulation (Table 9.4) starts from zero initial temperature and comprises one timestep to establish the steady-state temperature distribution (Fig. 9.7) and to evaluate the associated displacements (Fig. 9.8), strains, and stresses.

Due to the setup the thermal problem does not depend on the z-coordinate and the Laplace equation governing the steady-state temperature distribution $T(x, y)$ becomes

$$\frac{\partial^2 T}{\partial x^2} + \frac{\partial^2 T}{\partial y^2} = 0. \tag{9.4.1}$$

The temperature distribution

Fig. 9.7 Deformations scaled up, temperature distribution

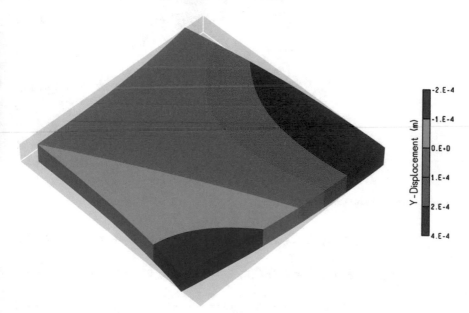

Fig. 9.8 Deformations scaled up, y-displacements

$$T(x, y) = T_0 \frac{x}{L}\frac{y}{L} \qquad (9.4.2)$$

satisfies the Laplace equation and the imposed thermal boundary conditions, hence, this is the solution of the thermal problem.

For the solution of the mechanical problem note that the applied temperature change is identical to the above $T(x, y)$. It will be shown that the displacements

$$u_x(x, y) = (1 + \nu)\alpha_T \frac{T_0}{L^2}\left(\frac{1}{2}x^2 y - \frac{1}{6}y^3\right), \qquad (9.4.3)$$

$$u_y(x, y) = (1 + \nu)\alpha_T \frac{T_0}{L^2}\left(\frac{1}{2}y^2 x - \frac{1}{6}x^3\right), \qquad (9.4.4)$$

$$u_z = 0 \qquad (9.4.5)$$

constitute the solution of the mechanical problem. The prescribed fixities are satisfied; there is zero x-displacement u_x along the face $y = 0$, zero y-displacement u_y along the face $x = 0$, and zero z-displacement u_z along the faces $z = 0$ and $z = H$. The strains read

$$\epsilon_{11} = \frac{\partial u_x}{\partial x} = (1 + \nu)\alpha_T T(x, y), \qquad (9.4.6)$$

$$\epsilon_{22} = \frac{\partial u_y}{\partial y} = (1 + \nu)\alpha_T T(x, y), \qquad (9.4.7)$$

$$\epsilon_{33} = \frac{\partial u_z}{\partial z} = 0, \qquad (9.4.8)$$

$$\epsilon_{12} = \frac{\partial u_x}{\partial y} + \frac{\partial u_y}{\partial x} = 0, \qquad (9.4.9)$$

$$\epsilon_{13} = \frac{\partial u_x}{\partial z} + \frac{\partial u_z}{\partial x} = 0, \qquad (9.4.10)$$

$$\epsilon_{23} = \frac{\partial u_y}{\partial z} + \frac{\partial u_z}{\partial y} = 0. \qquad (9.4.11)$$

The constitutive equations

$$\epsilon_{11} - \alpha_T T(x, y) = \nu\alpha_T T(x, y) = \frac{1}{E}[\sigma_{11} - \nu(\sigma_{22} + \sigma_{33})], \qquad (9.4.12)$$

$$\epsilon_{22} - \alpha_T T(x, y) = \nu\alpha_T T(x, y) = \frac{1}{E}[\sigma_{22} - \nu(\sigma_{11} + \sigma_{33})], \qquad (9.4.13)$$

$$\epsilon_{33} - \alpha_T T(x, y) = -\alpha_T T(x, y) = \frac{1}{E}[\sigma_{33} - \nu(\sigma_{11} + \sigma_{22})], \qquad (9.4.14)$$

$$\epsilon_{12} = 0 = \frac{2(1 + \nu)}{E}\sigma_{12}, \qquad (9.4.15)$$

$$\epsilon_{13} = 0 = \frac{2(1 + \nu)}{E}\sigma_{13}, \qquad (9.4.16)$$

Table 9.5 Benchmark deposit (https://docs.opengeosys.org/books/bmb-5)

BM code	Author	Code	Files	CTest
BMB5-9.5	Peter Vogel	OGS-5	Available	TBD

(https://oc.ufz.de/index.php/s/nlph7bhfDkj6tC7)

$$\epsilon_{23} = 0 = \frac{2(1+\nu)}{E}\sigma_{23} \tag{9.4.17}$$

yield the stress tensor

$$\sigma = \begin{pmatrix} \sigma_{11} & \sigma_{12} & \sigma_{13} \\ \sigma_{12} & \sigma_{22} & \sigma_{23} \\ \sigma_{13} & \sigma_{23} & \sigma_{33} \end{pmatrix} = -E\alpha_T T(x,y) \begin{pmatrix} 0 & 0 & 0 \\ 0 & 0 & 0 \\ 0 & 0 & 1 \end{pmatrix}, \tag{9.4.18}$$

which satisfies the equation of mechanical equilibrium

$$\text{div } \sigma = 0 \tag{9.4.19}$$

as well as the default mechanical boundary conditions along the faces $x = L$ and $y = L$.

9.5 A Temperature Distribution Represented by a Quadratic Form

Given length $L = 1\,\text{m}$ and thickness $H = 0.1\,\text{m}$ the domain represents the rectangular plate $[0, L] \times [0, L] \times [0, H]$. It is discretized by $40 \times 40 \times 1$ equally sized hexahedral elements. The solid material has been selected elastic with Young's modulus $E = 10,000\,\text{MPa}$, Poisson's ratio $\nu = 0.2$, zero heat capacity, thermal conductivity $k = 1\,\text{W/(m·K)}$, and thermal expansion $\alpha_T = 10^{-4}\,\text{1/K}$. Gravity is neglected via zero material density. The faces $x = L$ and $y = L$ are free of mechanical load by default. Fixities have been prescribed with zero z-displacement along the faces $z = 0$ and $z = H$, zero y-displacement along the face $y = 0$, and zero x-displacement along the face $x = 0$. The imposed thermal boundary conditions have top and bottom of the domain as well as the lateral faces on the coordinate planes thermally insulated by default. Zero temperature prevails at the origin. The heat flow $h = 20\,\text{W/m}^2$ has been prescribed at the face $y = L$, the heat flow $-h = -20\,\text{W/m}^2$ has been prescribed at the face $x = L$. The simulation (Table 9.5) starts from zero initial temperature and comprises one timestep to establish the steady-state temperature distribution (Fig. 9.9) and to evaluate the associated displacements (Fig. 9.10), strains, and stresses.

Fig. 9.9 Deformations scaled up, temperature distribution

Fig. 9.10 Deformations scaled up, y-displacements

Due to the setup the thermal problem does not depend on the z-coordinate and the Laplace equation governing the steady-state temperature distribution $T(x, y)$ becomes

$$\frac{\partial^2 T}{\partial x^2} + \frac{\partial^2 T}{\partial y^2} = 0. \tag{9.5.1}$$

The temperature distribution

$$T(x, y) = -\frac{h}{2kL}(x^2 - y^2) \tag{9.5.2}$$

satisfies the Laplace equation and the imposed thermal boundary conditions. There is zero temperature at the origin and Fourier's law gives for the heat flow across the face $x = 0$

$$k\frac{\partial T}{\partial x}(0, y) = 0 \quad \text{for} \ \ 0 \le y \le L, \tag{9.5.3}$$

across the face $y = 0$

$$k\frac{\partial T}{\partial y}(x, 0) = 0 \quad \text{for} \ \ 0 \le x \le L, \tag{9.5.4}$$

across the face $x = L$

$$k\frac{\partial T}{\partial x}(L, y) = -k\frac{h}{2kL}2L = -h \quad \text{for} \ \ 0 \le y \le L, \tag{9.5.5}$$

and across the face $y = L$

$$k\frac{\partial T}{\partial y}(x, L) = -k\frac{h}{2kL}(-2L) = h \quad \text{for} \ \ 0 \le x \le L. \tag{9.5.6}$$

Therefore, the temperature distribution $T(x, y)$ is the solution of the thermal problem.
For the solution of the mechanical problem note that the applied temperature change is identical to the above $T(x, y)$. It will be shown that the displacements

$$u_x(x, y) = -(1 + \nu)\alpha_T \frac{h}{2kL}\left(\frac{1}{3}x^3 - xy^2\right), \tag{9.5.7}$$

$$u_y(x, y) = -(1 + \nu)\alpha_T \frac{h}{2kL}\left(x^2 y - \frac{1}{3}y^3\right), \tag{9.5.8}$$

$$u_z = 0 \tag{9.5.9}$$

constitute the solution of the mechanical problem. The prescribed fixities are satisfied; there is zero x-displacement u_x along the face $x = 0$, zero y-displacement u_y along the face $y = 0$, and zero z-displacement u_z along the faces $z = 0$ and $z = H$. The strains read

$$\epsilon_{11} = \frac{\partial u_x}{\partial x} = (1 + \nu)\alpha_T T(x, y), \tag{9.5.10}$$

$$\epsilon_{22} = \frac{\partial u_y}{\partial y} = (1 + \nu)\alpha_T T(x, y), \tag{9.5.11}$$

$$\epsilon_{33} = \frac{\partial u_z}{\partial z} = 0, \tag{9.5.12}$$

$$\epsilon_{12} = \frac{\partial u_x}{\partial y} + \frac{\partial u_y}{\partial x} = 0, \tag{9.5.13}$$

$$\epsilon_{13} = \frac{\partial u_x}{\partial z} + \frac{\partial u_z}{\partial x} = 0, \tag{9.5.14}$$

$$\epsilon_{23} = \frac{\partial u_y}{\partial z} + \frac{\partial u_z}{\partial y} = 0. \tag{9.5.15}$$

The constitutive equations

$$\epsilon_{11} - \alpha_T T(x, y) = \nu\alpha_T T(x, y) = \frac{1}{E}[\sigma_{11} - \nu(\sigma_{22} + \sigma_{33})], \tag{9.5.16}$$

$$\epsilon_{22} - \alpha_T T(x, y) = \nu\alpha_T T(x, y) = \frac{1}{E}[\sigma_{22} - \nu(\sigma_{11} + \sigma_{33})], \tag{9.5.17}$$

$$\epsilon_{33} - \alpha_T T(x, y) = -\alpha_T T(x, y) = \frac{1}{E}[\sigma_{33} - \nu(\sigma_{11} + \sigma_{22})], \tag{9.5.18}$$

$$\epsilon_{12} = 0 = \frac{2(1 + \nu)}{E}\sigma_{12}, \tag{9.5.19}$$

$$\epsilon_{13} = 0 = \frac{2(1 + \nu)}{E}\sigma_{13}, \tag{9.5.20}$$

$$\epsilon_{23} = 0 = \frac{2(1 + \nu)}{E}\sigma_{23} \tag{9.5.21}$$

yield the stress tensor

$$\sigma = \begin{pmatrix} \sigma_{11} & \sigma_{12} & \sigma_{13} \\ \sigma_{12} & \sigma_{22} & \sigma_{23} \\ \sigma_{13} & \sigma_{23} & \sigma_{33} \end{pmatrix} = -E\alpha_T T(x, y) \begin{pmatrix} 0 & 0 & 0 \\ 0 & 0 & 0 \\ 0 & 0 & 1 \end{pmatrix}, \tag{9.5.22}$$

which satisfies the equation of mechanical equilibrium

$$\text{div } \sigma = 0 \tag{9.5.23}$$

as well as the default mechanical boundary conditions along the faces $x = L$ and $y = L$.

9.6 A Temperature Distribution Represented by a Fourier Series

Given length $L = 1\,\text{m}$ and thickness $H = 0.1\,\text{m}$ the domain represents the rectangular plate $[0, L] \times [0, L] \times [0, H]$. It is discretized by $80 \times 80 \times 1$ equally sized hexahedral elements. The solid material has been selected elastic with Young's modulus $E = 10,000\,\text{MPa}$, Poisson's ratio $\nu = 0.2$, zero heat capacity, thermal conductivity $k = 1\,\text{W/(m·K)}$, and thermal expansion $\alpha_T = 10^{-4}\,\text{1/K}$. Gravity is neglected via zero material density. The faces $x = 0$ and $x = L$ are free of mechanical load by default. Fixities have been prescribed with zero z-displacement along the faces $z = 0$ and $z = H$, zero y-displacement along the faces $y = 0$ and $y = L$, and zero x-displacement along the line $(0, L/2, z)$. The imposed thermal boundary conditions have top and bottom of the domain as well as the faces $y = 0$ and $y = L$ thermally insulated by default. Zero temperature prevails at the face $x = 0$, the heat flow $h(2y/L - 1)$ ($h = 40\,\text{W/m}^2$) has been prescribed at the face $x = L$. The simulation (Table 9.6) starts from zero initial temperature and comprises one timestep to establish the steady-state temperature distribution (Fig. 9.11) and to evaluate the associated displacements (Fig. 9.12), strains, and stresses.

Due to the setup the thermal problem does not depend on the z-coordinate and the Laplace equation governing the steady-state temperature distribution $T(x, y)$ becomes

$$\frac{\partial^2 T}{\partial x^2} + \frac{\partial^2 T}{\partial y^2} = 0. \tag{9.6.1}$$

The Laplace equation and the applied thermal boundary conditions pose a boundary value problem, which will be solved by separation of variables. Assuming a product solution

$$T(x, y) = F(x)G(y) \tag{9.6.2}$$

the Laplace equation gives

$$-\frac{1}{F}\frac{d^2 F}{dx^2} = \frac{1}{G}\frac{d^2 G}{dy^2}. \tag{9.6.3}$$

Since the left hand side depends only on x and the right hand side depends only on y, both sides are equal to some constant value $-\omega^2$. Thus the Laplace equation separates into two ordinary differential equations

Table 9.6 Benchmark deposit (https://docs.opengeosys.org/books/bmb-5)

BM code	Author	Code	Files	CTest
BMB5-9.6	Peter Vogel	OGS-5	Available	TBD

(https://oc.ufz.de/index.php/s/nlph7bhfDkj6tC7)

Fig. 9.11 Deformations scaled up, temperature distribution

Fig. 9.12 Deformations scaled up, x-displacements

$$\frac{d^2 F}{dx^2} = \omega^2 F, \tag{9.6.4}$$

$$\frac{d^2 G}{dy^2} = -\omega^2 G \tag{9.6.5}$$

with the general solutions

$$F(x) = C1 \sinh(\omega x) + C2 \cosh(\omega x), \tag{9.6.6}$$
$$G(y) = C3 \cos(\omega y) + C4 \sin(\omega y). \tag{9.6.7}$$

This yields

$$T(x, y) = A \, \sinh(\omega x) \cos(\omega y) + B \sinh(\omega x) \sin(\omega y)$$
$$+ C \cosh(\omega x) \cos(\omega y) + D \cosh(\omega x) \sin(\omega y). \tag{9.6.8}$$

The free constants A, B, C, D, and the eigenvalue ω will be determined from the boundary conditions. Zero temperature along the face $x = 0$

$$T(0, y) = 0 = C \cos(\omega y) + D \sin(\omega y) \tag{9.6.9}$$

is satisfied by

$$C = D = 0, \tag{9.6.10}$$

and the thermal boundary condition along the insulated face $y = 0$

$$\frac{\partial T}{\partial y}(x, 0) = 0 = \omega B \sinh(\omega x) \tag{9.6.11}$$

is satisfied by

$$B = 0. \tag{9.6.12}$$

The thermal boundary condition along the insulated face $y = L$

$$\frac{\partial T}{\partial y}(x, L) = 0 = \omega A \sinh(\omega x) \sin(\omega L) \tag{9.6.13}$$

yields the eigenvalues

$$\omega_n = \frac{n\pi}{L} \quad \text{for } n = 1, 2, \ldots \tag{9.6.14}$$

and the associated solutions

$$T_n(x, y) = A_n \sinh\left(n\pi\frac{x}{L}\right) \cos\left(n\pi\frac{y}{L}\right) \quad \text{for } n = 1, 2, \ldots . \tag{9.6.15}$$

The temperature distribution takes the form

$$T(x, y) = \sum_{n=1}^{\infty} A_n \sinh\left(n\pi\frac{x}{L}\right) \cos\left(n\pi\frac{y}{L}\right). \tag{9.6.16}$$

The specified heat flow across the face $x = L$ yields the remaining constants $A_1, A_2, \ldots$. By Fourier's law

$$\frac{\partial T}{\partial x}(L, y) = \frac{h}{k}\left(2\frac{y}{L} - 1\right) = \sum_{n=1}^{\infty} A_n \frac{n\pi}{L} \cosh(n\pi) \cos\left(n\pi\frac{y}{L}\right)$$

$$= \frac{a_0}{2} + \sum_{n=1}^{\infty} a_n \cos\left(n\pi\frac{y}{L}\right) \tag{9.6.17}$$

with the cosine series expansion of the imposed boundary condition on the last line. The Fourier coefficients $a_0, a_1, a_2, \ldots$ read

$$a_0 = \frac{2}{L}\int_0^L \frac{h}{k}\left(2\frac{y}{L} - 1\right) dy = 0, \tag{9.6.18}$$

and for $n = 1, 2, \ldots$

$$a_n = \frac{2}{L}\int_0^L \frac{h}{k}\left(2\frac{y}{L} - 1\right) \cos\left(n\pi\frac{y}{L}\right) dy = \frac{h}{k}\frac{4}{(n\pi)^2}[(-1)^n - 1]. \tag{9.6.19}$$

Comparing coefficients gives the constants $A_1, A_2, \ldots$ and the temperature distribution $T(x, y)$ becomes

$$T(x, y) = -\frac{8hL}{k\pi^3} \sum_{n=1}^{\infty} \frac{\sinh\left[(2n-1)\pi\frac{x}{L}\right]}{\cosh[(2n-1)\pi]} \frac{\cos\left[(2n-1)\pi\frac{y}{L}\right]}{(2n-1)^3}. \tag{9.6.20}$$

The series thus obtained is uniformly absolutely-convergent on the entire domain. It satisfies the Laplace equation and the thermal boundary conditions, hence, the temperature distribution $T(x, y)$ is the solution of the thermal problem.

For the solution of the mechanical problem note that the applied temperature change is identical to the above $T(x, y)$. It will be shown that the displacements

$$u_x(x, y) = -(1 + \nu)\alpha_T \frac{8hL^2}{k\pi^4} \sum_{n=1}^{\infty} \frac{\cosh\left[(2n-1)\pi\frac{x}{L}\right]}{\cosh[(2n-1)\pi]} \frac{\cos\left[(2n-1)\pi\frac{y}{L}\right]}{(2n-1)^4},$$

$$\tag{9.6.21}$$

$$u_y(x, y) = -(1 + \nu)\alpha_T \frac{8hL^2}{k\pi^4} \sum_{n=1}^{\infty} \frac{\sinh\left[(2n-1)\pi\frac{x}{L}\right]}{\cosh[(2n-1)\pi]} \frac{\sin\left[(2n-1)\pi\frac{y}{L}\right]}{(2n-1)^4}, \quad (9.6.22)$$

$$u_z = 0 \qquad\qquad (9.6.23)$$

constitute the solution of the mechanical problem. The prescribed fixities are satisfied; there is zero x-displacement u_x along the plane $y = L/2$, zero y-displacement u_y along the faces $y = 0$ and $y = L$, and zero z-displacement u_z along the faces $z = 0$ and $z = H$. The strains read

$$\epsilon_{11} = \frac{\partial u_x}{\partial x} = (1 + \nu)\alpha_T T(x, y), \qquad\qquad (9.6.24)$$

$$\epsilon_{22} = \frac{\partial u_y}{\partial y} = (1 + \nu)\alpha_T T(x, y), \qquad\qquad (9.6.25)$$

$$\epsilon_{33} = \frac{\partial u_z}{\partial z} = 0, \qquad\qquad (9.6.26)$$

$$\epsilon_{12} = \frac{\partial u_x}{\partial y} + \frac{\partial u_y}{\partial x} = 0, \qquad\qquad (9.6.27)$$

$$\epsilon_{13} = \frac{\partial u_x}{\partial z} + \frac{\partial u_z}{\partial x} = 0, \qquad\qquad (9.6.28)$$

$$\epsilon_{23} = \frac{\partial u_y}{\partial z} + \frac{\partial u_z}{\partial y} = 0. \qquad\qquad (9.6.29)$$

The constitutive equations

$$\epsilon_{11} - \alpha_T T(x, y) = \nu\alpha_T T(x, y) = \frac{1}{E}[\sigma_{11} - \nu(\sigma_{22} + \sigma_{33})], \qquad (9.6.30)$$

$$\epsilon_{22} - \alpha_T T(x, y) = \nu\alpha_T T(x, y) = \frac{1}{E}[\sigma_{22} - \nu(\sigma_{11} + \sigma_{33})], \qquad (9.6.31)$$

$$\epsilon_{33} - \alpha_T T(x, y) = -\alpha_T T(x, y) = \frac{1}{E}[\sigma_{33} - \nu(\sigma_{11} + \sigma_{22})], \qquad (9.6.32)$$

$$\epsilon_{12} = 0 = \frac{2(1 + \nu)}{E}\sigma_{12}, \qquad\qquad (9.6.33)$$

$$\epsilon_{13} = 0 = \frac{2(1 + \nu)}{E}\sigma_{13}, \qquad\qquad (9.6.34)$$

$$\epsilon_{23} = 0 = \frac{2(1 + \nu)}{E}\sigma_{23} \qquad\qquad (9.6.35)$$

yield the stress tensor

$$\boldsymbol{\sigma} = \begin{pmatrix} \sigma_{11} & \sigma_{12} & \sigma_{13} \\ \sigma_{12} & \sigma_{22} & \sigma_{23} \\ \sigma_{13} & \sigma_{23} & \sigma_{33} \end{pmatrix} = -E\alpha_T T(x, y) \begin{pmatrix} 0 & 0 & 0 \\ 0 & 0 & 0 \\ 0 & 0 & 1 \end{pmatrix}, \qquad (9.6.36)$$

which satisfies the equation of mechanical equilibrium

$$\text{div } \boldsymbol{\sigma} = 0 \qquad (9.6.37)$$

as well as the default mechanical boundary conditions along the faces $x = 0$ and $x = L$.

9.7 A Phase-Field Model for Brittle Fracturing of Thermo-Elastic Solids

Xing-Yuan Miao, Thomas Nagel

9.7.1 The Model

A phase-field model for fracture is derived and coupled to deformation as well as heat transport processes in order to simulate crack propagation in brittle materials under thermo-mechanical loads. The phase-field approach is a numerical treatment which is capable of describing the lower-dimensional crack boundaries in a continuous context. Smooth phase transition boundaries connect the fully damaged zones and the intact material and—in that sense—represent the fracture surfaces, see Fig. 9.13 and cf. Miehe et al. (2010b).

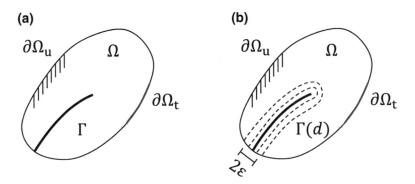

Fig. 9.13 Elastic body with a sharp crack (**a**) or a diffusive crack (**b**)

An energy functional of a fractured body can be formulated as

$$E\left(\boldsymbol{u}, \operatorname{grad} \boldsymbol{u}, d\right) = \int_{\Omega} \left[\left(d^2 + k\right) \psi_{\mathrm{e}}^+\left(\boldsymbol{\epsilon}_{\mathrm{el}}\right) + \psi_{\mathrm{e}}^-\left(\boldsymbol{\epsilon}_{\mathrm{el}}\right)\right] \mathrm{d}V$$
$$+ g_{\mathrm{c}} \int_{\Omega} \left[\frac{1}{4\varepsilon}\left(1 - d\right)^2 + \varepsilon \operatorname{grad} d \cdot \operatorname{grad} d\right] \mathrm{d}V$$

where k represents a residual stiffness used in numerical calculations to ensure a residual elastic energy density in the fully damaged state. The parameter g_{c} stands for the critical Griffith-type fracture energy release rate. The order parameter d is constrained between 0 (fully damaged) and 1 (intact material).

The total strain is composed of the elastic and thermal part

$$\boldsymbol{\epsilon} = \boldsymbol{\epsilon}_{\mathrm{el}} + \boldsymbol{\epsilon}_{\mathrm{th}} \tag{9.7.1}$$

with

$$\boldsymbol{\epsilon}_{\mathrm{th}} = \boldsymbol{\alpha} \Delta T \tag{9.7.2}$$

where $\boldsymbol{\alpha}$ stands for the linear thermal expansion tensor. Only the elastic part contributes to the elastic energy density

$$\psi_{\mathrm{e}}^+\left(\boldsymbol{\epsilon}_{\mathrm{el}}\right) := \frac{1}{2} K \left\langle \operatorname{tr}\left(\boldsymbol{\epsilon} - \boldsymbol{\epsilon}_{\mathrm{th}}\right)\right\rangle_+^2 + \mu \left(\boldsymbol{\epsilon} - \boldsymbol{\epsilon}_{\mathrm{th}}\right)^{\mathrm{D}} : \left(\boldsymbol{\epsilon} - \boldsymbol{\epsilon}_{\mathrm{th}}\right)^{\mathrm{D}} \tag{9.7.3}$$

$$\psi_{\mathrm{e}}^-\left(\boldsymbol{\epsilon}_{\mathrm{el}}\right) := \frac{1}{2} K \left\langle \operatorname{tr}\left(\boldsymbol{\epsilon} - \boldsymbol{\epsilon}_{\mathrm{th}}\right)\right\rangle_-^2 \tag{9.7.4}$$

where K is the bulk modulus, μ the shear modulus, and $\langle \bullet \rangle_\pm := \left(\bullet \pm | \bullet |\right)/2$.

The governing partial differential equations (derived in part from Eq. (9.7.1)) in the case of thermo-elastic deformation and brittle fracture along with the respective Neumann-type boundary conditions are

$$\operatorname{div}\left[\left(d^2 + k\right) \boldsymbol{\sigma}_0^+ + \boldsymbol{\sigma}_0^-\right] + \varrho \boldsymbol{b} = \boldsymbol{0} \tag{9.7.5}$$

$$2d\mathcal{H}\left(\boldsymbol{\epsilon}_{\mathrm{el}}\right) - \frac{1 - d}{2\varepsilon} g_{\mathrm{c}} - 2\varepsilon g_{\mathrm{c}} \operatorname{div}\left(\operatorname{grad} d\right) = 0 \tag{9.7.6}$$

$$\left(\varrho c_p\right)_{\mathrm{eff}} \frac{\partial T}{\partial t} - \operatorname{div}\left(\lambda_{\mathrm{eff}} \operatorname{grad} T\right) = 0 \tag{9.7.7}$$

$$\left[\left(d^2 + k\right) \boldsymbol{\sigma}_0^+ + \boldsymbol{\sigma}_0^-\right] \cdot \boldsymbol{n} - \bar{\boldsymbol{t}} = \boldsymbol{0} \quad \text{on} \quad \partial\Omega_t \tag{9.7.8}$$

$$\operatorname{grad} d \cdot \boldsymbol{n} = 0 \quad \text{on} \quad \partial\Omega \tag{9.7.9}$$

$$-\boldsymbol{q} \cdot \boldsymbol{n} - \bar{q}_n = 0 \quad \text{on} \quad \partial\Omega_{\mathrm{q}} \tag{9.7.10}$$

where $\left(\varrho c_p\right)_{\mathrm{eff}}$ represents the effective volumetric heat capacity and λ_{eff} the effective thermal conductivity which incorporate the degradation of both quantities due to the crack field (Kuhn 2013); $\boldsymbol{q}$ is the heat flux. $\mathcal{H}\left(\boldsymbol{\epsilon}_{\mathrm{el}}\right)$ represents a damage-driving

Fig. 9.14 Single-edge-
notched model

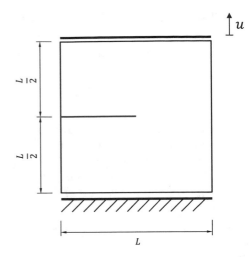

history field associated with the maximum local tensile strain energy density used
instead of its current value ψ_e^+ to ensure irreversibility of the crack propagation by
preventing crack healing (Miehe et al. 2010b).

9.7.2 Single-edge-notched Isothermal Tensile Test

A classical single-edge-notched tension test often used in literature for testing crack
propagation models (Ehlers and Luo 2017; Miehe et al. 2010b) was performed to
verify the implemented algorithm from a mechanical perspective. The mesh was
refined a priori with a minimum mesh size of approximately 0.001 mm in the region
through which the crack is expected to propagate and a total of 20,094 linear triangular
plane-strain elements. The incremental displacement loading of $\delta u = 1 \cdot 10^{-5}$ mm was
applied on the upper edge of the model to drive the propagation of the pre-existing
horizontal crack (see Fig. 9.14).

A direct solver (SparseLU) was used for solving the linear system. The Newton–
Raphson method was used as the nonlinear solver with an absolute tolerance of 10^{-3}.
The time-step size was set to 0.01 s.

The OGS-6 version used for the single-edge-notched tension benchmark can be
found in the official master branch available at

- https://github.com/ufz/ogs

along with the input files named `square_line_h` on available as CTests under

- https://github.com/ufz/ogs-data/tree/master/PhaseField.

The propagation path of the pre-existing crack driven by the tensile loading
through the material is illustrated in Fig. 9.15a. Figure 9.15b depicts the corresponding load-displacement curve for the upper edge. Both the crack propagation and

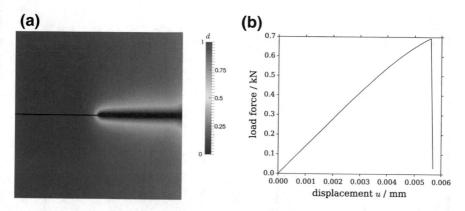

Fig. 9.15 Single-edge-notched tension test: **a** crack propagation pattern; **b** load - displacement curve

the load-displacement patterns are consistent with the literature, cf. Ehlers and Luo (2017), Miehe et al. (2010a). For a detailed description, derivation, as well as model validation of the described phase-field model, the interested reader can refer to Miao et al. (2017a, b).

9.7.3 Thermo-Mechanical Tests

To test the model under non-isothermal conditions, a series of models is set up for which analytical solutions for purely mechanical conditions can be found. The constraints of these problems are then modified such that the driving mechanical boundary conditions are replaced by a combination of mechanical constraints and thermal boundary conditions such that an equivalent stress state is created in the specimen but caused by thermally-induced deformations.

The first example is set up to exclude spatial gradients of the phase field and represents homogeneous damage. Specifically, a specimen compressed in a uniaxial stress field is modelled. This unconfined compression test was compared to a specimen which was heated up while its axial displacements were constrained. The compression test was simulated with a compressive displacement load applied to the upper surface of a three-dimensional cubic model, see Fig. 9.16a. The thermal expansion test was implemented by imposing a temperature increase to the domain while the top surface of the model was held in place. The temperature loading was chosen to achieve the same compressive load as that imposed in the unconfined compression test, Fig. 9.16b. Both models were simulated based on a single three-dimensional finite element. Due to the homogeneous evolution of the phase field (div grad $d = 0$), the analytical solutions for the two cases can be obtained

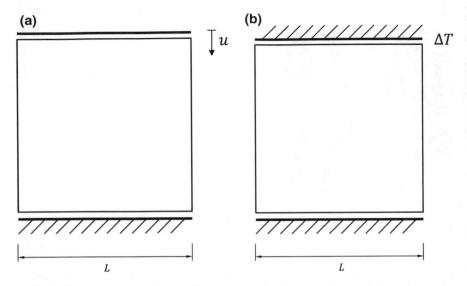

Fig. 9.16 Geometry and boundary conditions: **a** unconfined compression test; **b** thermal expansion test

$$d = \frac{g_c}{g_c + 4\epsilon\psi_e^+} \qquad (9.7.11)$$

where due to negative (elastic) volume strains only the deviatoric energy drives the phase field.

$$\psi_e^+ = \mu\,\epsilon^D : \epsilon^D = \frac{2\mu}{3}\left(\frac{u(1+\nu)}{L}\right)^2 \qquad (9.7.12)$$

for mechanical case, and

$$\psi_e^+ = \mu\,\epsilon_{el}^D : \epsilon_{el}^D = \frac{2\mu}{3}\,[\alpha\Delta T(1+\nu)]^2 \qquad (9.7.13)$$

for thermo-mechanical case. Where the Poisson's ratio is evolving with the degradation of the shear modulus

$$\nu(d) = \frac{3K - 2Gd^2}{2(3K + Gd^2)} \qquad (9.7.14)$$

The main parameters used in this benchmark are listed in Tables 9.7 and 9.8.

The phase-field evolution in the thermo-mechanical case follows the mechanical case, and both solutions correspond to the analytical solution, see Fig. 9.17. The increase of dilatant volume results from the compressive shearing, which is also an identification of the initiation and propagation of the cracks.

Table 9.7 Parameter values for numerical examples

Material properties	
Young's modulus E / GPa	210
Poison's ratio ν	0.3
Fracture toughness g_c / (N mm^{-1})	2.7
Thermal conductivity λ / (W m^{-1}K^{-1})	0.1
Thermal expansion α / K^{-1}	$1 \cdot 10^{-3}$
Specific heat capacity c_p / (J kg^{-1}K^{-1})	2000
Numerical settings	
Residual stiffness parameter k	$1 \cdot 10^{-3}$
Length scale 2ϵ / mm	0.002 (beam) / 0.01 (square) / 1 (cube)
Kinetic coefficient M / (mm^2 N^{-1} s^{-1})	$1 \cdot 10^6$

Table 9.8 Geometric parameters for numerical examples

Geometry	length/mm	width/mm	depth/mm
Beam	1	0.000999	0.000999
Square	1	1	–
Cube	1	1	1

Fig. 9.17 Phase-field - vertical elastic strain curve

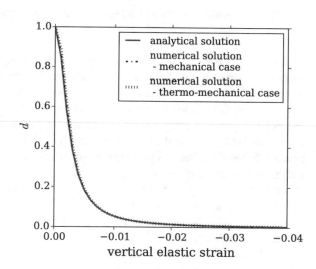

The second example entails the fracturing of a bar. The solution can be calculated analytically for a one-dimensional problem as outlined in Kuhn (2013). The uni-axial tension problem can be transformed to an axially constrained problem undergoing thermal shrinkage, see Figs. 9.18 and 9.19.

A discretisation of 1001 hexahedral elements with an element length of 0.999 m was applied. The tension test was simulated with equivalent displacement loads

Fig. 9.18 Bar under uniaxial tension. Mechanical model

Fig. 9.19 Bar under thermally induced uniaxial tension. Thermo-mechanical model

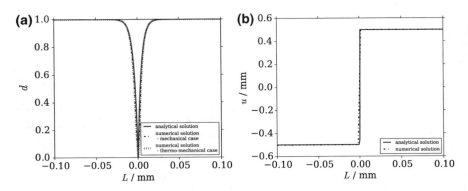

Fig. 9.20 Phase-field parameter and displacement field distribution along the bar

applied to both sides of the model, see Fig. 9.18. The top, bottom, front and back surfaces are constrained in their normal directions. The thermal shrinkage test was set to achieve the same tensile load as that imposed during the tension test by applying a temperature decline to the domain and holding the two ends of the bar in place, see Fig. 9.19. Figure 9.20 compares the two approximated phase-field and axial displacement distributions along the bar to the analytical solution. The crack forms initially in the middle of the bar and propagates towards the edges very fast, nearly oriented perpendicular to the direction of the applied loads.

In all thermo-mechanical simulations, a direct solver (SparseLU) was used and Newton–Raphson iterations performed to resolve non-linearities with an absolute tolerance of 10^{-3} in thermo-mechanical expansion case and of 10^{-2} in thermo-mechanical shrinkage case (Table 9.9).

For the thermal expansion test, the temperature lift was set to achieve 100 K within 10 s with a equidistant time-step size of 0.0001 s. The OGS-6 version used is available at:

Table 9.9 Benchmark deposit (https://docs.opengeosys.org/books/bmb-5)

BM code	Author	Code	Files	CTest
BMB5-9.7	Xing-Yuan Miao	OGS-6	See below	TBD

(https://oc.ufz.de/index.php/s/nlph7bhfDkj6tC7)

- https://github.com/xingyuanmiao/ogs/tree/tmphasefield,

with available commit hash https://github.com/xingyuanmiao/ogs/commit/810f5cf 04841f6e36f6c91166c20ca105789b024, while the input files named cube_1e0 can be found at:

- https://github.com/ufz/ogs-data/tree/986966acb8f0a82d966505a1dc502e144b86 e22d/TMPhaseField.

For the thermal shrinkage test (bar), the temperature decline was set to attain 500 K within 10 s with a equidistant time-step size of 0.001 s. The OGS-6 version used is available at:

- https://github.com/xingyuanmiao/ogs/tree/tmphasefield,

with the input files named beam3d available on:

- https://github.com/ufz/ogs-data/tree/986966acb8f0a82d966505a1dc502e144b86 e22d/TMPhaseField.

Chapter 10
THM Processes

Peter Vogel, Jobst Maßmann, Tianyuan Zheng, Xing-Yuan Miao,
Dmitri Naumov and Thomas Nagel

Porothermoelastic Beams

Peter Vogel and Jobst Maßmann

This section presents problems on permeable elastic beams subject to liquid pressure and temperature changes. We focus on the closed form solutions. The associated simulation exercises have been checked by OGS; they may serve as verification tests. For the underlying theory of porothermoelasticity and for more advanced examples see Cheng (2016).

10.1 Liquid Flow and Heat Transport in a Permeable Elastic Beam I

This example illustrates liquid flow and heat transport in a permeable elastic beam with constant inlet temperature. Given length $L = 10$ m and thickness $H = 1$ m the

P. Vogel (✉) · J. Maßmann
BGR, Federal Institute for Geosciences and Natural Resources, Hanover, Germany
e-mail: Peter.Vogel@bgr.de

T. Zheng · X.-Y. Miao · D. Naumov
UFZ, Helmholtz Centre for Environmental Research, Leipzig, Germany
e-mail: xing-yuan.miao@ufz.de

D. Naumov
e-mail: Dmitri.Naumov@ufz.de

T. Zheng · X.-Y. Miao
TU Dresden, Dresden, Germany

T. Nagel
UFZ, Helmholtz Centre for Environmental Research, Dresden, Germany

T. Nagel
Trinity College Dublin, Dublin 2, Ireland
e-mail: Thomas.Nagel@ufz.de

© Springer International Publishing AG 2018
O. Kolditz et al. (eds.), *Thermo-Hydro-Mechanical-Chemical Processes
in Fractured Porous Media: Modelling and Benchmarking*,
Terrestrial Environmental Sciences, https://doi.org/10.1007/978-3-319-68225-9_10

domain represents the rectangular beam $[0, L] \times [0, H] \times [0, H]$. It is discretized by $100 \times 1 \times 1$ equally sized hexahedral elements. The solid material has been selected elastic, various model parameters and the hydraulic properties are given in Table 10.1. The liquid is incompressible and has viscosity $\mu = 1$ mPa·s. For densities, heat capacities, and thermal conductivities of liquid and solid grain see Table 10.2, gravity has explicitly been neglected. Fixities have been prescribed on the entire surface of the domain with zero x-displacement along the faces $x = 0$ and $x = L$, zero y-displacement along the faces $y = 0$ and $y = H$, and zero z-displacement along the faces $z = 0$ and $z = H$. Pressure $p_0 = 10^5$ Pa at the liquid inlet ($x = 0$) and zero pressure at the liquid outlet ($x = L$) generate steady-state 1D flow along the x-axis. At the liquid inlet a constant temperature $T_0 = -10\,°C$ is specified for times $t > 0$, free outflow prevails at the liquid outlet by default. Starting from zero initial temperature the simulation (Table 10.3) evaluates the transient temperature distribution as well as the mechanical load with output after 10,000 and 20,000 s.

The formal solution proceeds in three steps, first to solve for pressure and specific discharge, next to evaluate the temperature distribution (Fig. 10.1), and finally to determine displacements (Fig. 10.2), strains, and stresses.

For incompressible liquids Darcy's law and continuity equation yield the Laplace equation as the governing equation describing the steady-state pressure distribution. It reads

$$\frac{d^2 p}{dx^2} = 0 \tag{10.1.1}$$

for 1D flow along the x-axis, hence, the pressure is given by

$$p(x) = p_0 \left(1 - \frac{x}{L}\right), \tag{10.1.2}$$

Table 10.1 Elastic and hydraulic parameters

Value	Quantity
$E = 5000$ MPa	Young's modulus
$\nu = 0.25$	Poisson's ratio
$\alpha_T = 10^{-6}$ 1/K	Thermal expansion
$k = 10^{-11}$ m^2	Isotropic permeability
$n = 0.1$	Porosity
1	Biot number

Table 10.2 Densities and thermal properties

Quantity	Liquid	Solid grain
Density	$\rho_l = 1000$ kg/m^3	$\rho_s = 2000$ kg/m^3
Heat capacity	$c_l = 1100$ J/(kg K)	$c_s = 250$ J/(kg K)
Thermal conductivity	$k_l = 10$ J/(s m K)	$k_s = 50$ J/(s m K)

Fig. 10.1 Temperature distribution after 20,000 s

Fig. 10.2 X-Displacements after 20,000 s

and the average pressure p_{av} becomes

$$p_{av} = \frac{1}{L} \int_0^L p(x)dx = \frac{p_0}{2}. \tag{10.1.3}$$

The specific discharge q is obtained by Darcy's law

$$q = \frac{k}{\mu} \frac{p_0}{L}. \tag{10.1.4}$$

Due to the setup the thermal problem depends on time and x-coordinate only and the heat transport equation governing the transient temperature distribution $T(x, t)$ reads

$$(n\rho_l c_l + (1-n)\rho_s c_s)\frac{\partial T}{\partial t} + (n\rho_l c_l)\frac{q}{n}\frac{\partial T}{\partial x} = (nk_l + (1-n)k_s)\frac{\partial^2 T}{\partial x^2}. \tag{10.1.5}$$

Introducing the notation

$$w = \frac{n\rho_l c_l}{n\rho_l c_l + (1-n)\rho_s c_s}\frac{q}{n}, \tag{10.1.6}$$

$$\chi = \frac{nk_l + (1-n)k_s}{n\rho_l c_l + (1-n)\rho_s c_s}, \tag{10.1.7}$$

the heat transport equation becomes

$$\frac{\partial T}{\partial t} + w\frac{\partial T}{\partial x} = \chi\frac{\partial^2 T}{\partial x^2}. \tag{10.1.8}$$

Due to free outflow at $x = L$ the formal problem is to determine the solution $T(x, t)$ of the above heat transport equation subject to the initial condition

$$T(x, 0) = 0 \quad \text{for } x > 0, \tag{10.1.9}$$

and the boundary conditions

$$\begin{aligned} T(0, t) &= T_0 \quad \text{for } t > 0, \\ \lim_{x \to \infty} T(x, t) &= 0 \quad \text{for } t > 0. \end{aligned} \tag{10.1.10}$$

Applying the Laplace transform with respect to t yields the ordinary differential equation

$$\chi\bar{T}'' - w\bar{T}' - s\bar{T} = 0, \tag{10.1.11}$$

where $\bar{T}$ is the transform of T, s is the transformation parameter, and the prime denotes the derivative with respect to x. This equation has to be solved with respect to the transformed boundary conditions. Hence,

$$\bar{T}(x, s) = \frac{T_0}{s} \exp\left(x\left[\frac{w}{2\chi} - \sqrt{(\frac{w}{2\chi})^2 + \frac{s}{\chi}} \right] \right). \tag{10.1.12}$$

The transform of the temperature may also serve to evaluate integrals of the temperature which will be required below.

$$\begin{aligned} \mathcal{L}\left\{ \int_0^x T(\hat{x}, t)d\hat{x} \right\} &= \int_0^x \bar{T}(\hat{x}, s)d\hat{x} \tag{10.1.13} \\ &= \frac{T_0}{s} \int_0^x \exp\left(\hat{x}\left[\frac{w}{2\chi} - \sqrt{(\frac{w}{2\chi})^2 + \frac{s}{\chi}} \right] \right) d\hat{x} \\ &= \frac{T_0}{s} \frac{\exp\left(x\left[\frac{w}{2\chi} - \sqrt{(\frac{w}{2\chi})^2 + \frac{s}{\chi}} \right] \right) - 1}{\frac{w}{2\chi} - \sqrt{(\frac{w}{2\chi})^2 + \frac{s}{\chi}}} \end{aligned}$$

and the average temperature

$$T_{av}(t) = \frac{1}{L} \int_0^L T(x, t)dx \tag{10.1.14}$$

has the transform

$$\mathscr{L}\{T_{av}\} = \frac{T_0}{sL} \frac{\exp\left(L\left[\frac{w}{2\chi} - \sqrt{(\frac{w}{2\chi})^2 + \frac{s}{\chi}}\right]\right) - 1}{\frac{w}{2\chi} - \sqrt{(\frac{w}{2\chi})^2 + \frac{s}{\chi}}}.$$

(10.1.15)

The Laplace transforms thus obtained are well suited for numerical inversion. The numerical inversion scheme by Crump (1976) may easily be applied to give the required values of temperature $T(x, t)$, average temperature $T_{av}(t)$, and the above integral of the temperature.

For the solution of the mechanical problem note that the applied temperature change is identical to the temperature distribution $T(x, t)$. Let $\boldsymbol{\sigma}$ denote the stress tensor, $\mathbf{I}$ the unit tensor, (u_x, u_y, u_z) the displacement vector, and

$$e = \frac{\partial u_x}{\partial x} + \frac{\partial u_y}{\partial y} + \frac{\partial u_z}{\partial z}$$

(10.1.16)

the volumetric strain. Employing Biot's simplified theory (i.e. Biot number equal one) the equations of equilibrium become

$$\frac{\partial^2 u_x}{\partial x^2} + \frac{\partial^2 u_x}{\partial y^2} + \frac{\partial^2 u_x}{\partial z^2} + \frac{1}{1-2\nu}\frac{\partial e}{\partial x} = \frac{2(1+\nu)}{1-2\nu}\alpha_T\frac{\partial T}{\partial x} + \frac{2(1+\nu)}{E}\frac{\partial p}{\partial x}$$

$$= \frac{2(1+\nu)}{1-2\nu}\alpha_T\frac{\partial T}{\partial x} - \frac{2(1+\nu)}{E}\frac{p_0}{L},$$

$$\frac{\partial^2 u_y}{\partial x^2} + \frac{\partial^2 u_y}{\partial y^2} + \frac{\partial^2 u_y}{\partial z^2} + \frac{1}{1-2\nu}\frac{\partial e}{\partial y} = \frac{2(1+\nu)}{1-2\nu}\alpha_T\frac{\partial T}{\partial y} + \frac{2(1+\nu)}{E}\frac{\partial p}{\partial y}$$

(10.1.17)

$$= 0,$$

$$\frac{\partial^2 u_z}{\partial x^2} + \frac{\partial^2 u_z}{\partial y^2} + \frac{\partial^2 u_z}{\partial z^2} + \frac{1}{1-2\nu}\frac{\partial e}{\partial z} = \frac{2(1+\nu)}{1-2\nu}\alpha_T\frac{\partial T}{\partial z} + \frac{2(1+\nu)}{E}\frac{\partial p}{\partial z}$$

$$= 0.$$

These equations and the imposed fixities are satisfied by

$$u_x = u_x(x, t),$$ (10.1.18)

$$u_y = 0,$$ (10.1.19)

$$u_z = 0,$$ (10.1.20)

if $u_x(x, t)$ satisfies the differential equation

$$\frac{\partial^2 u_x}{\partial x^2} + \frac{1}{1-2\nu}\frac{\partial^2 u_x}{\partial x^2} = \frac{2(1+\nu)}{1-2\nu}\alpha_T\frac{\partial T}{\partial x} - \frac{2(1+\nu)}{E}\frac{p_0}{L}$$

(10.1.21)

subject to the boundary conditions

$$u_x(0, t) = u_x(L, t) = 0 \quad \text{for } t > 0.$$

(10.1.22)

The x-displacement becomes

$$u_x(x, t) = \frac{1+\nu}{1-\nu}\left[\alpha_T \int_0^x T(\hat{x}, t)d\hat{x} - \alpha_T x T_{av}(t)\right.$$
$$\left. + \frac{1-2\nu}{2E}p_0\left(x - \frac{x^2}{L}\right)\right]. \qquad (10.1.23)$$

The strains read

$$\epsilon_{11} = \frac{\partial u_x}{\partial x} = \frac{1+\nu}{1-\nu}\left\{\alpha_T[T(x, t) - T_{av}(t)]\right.$$
$$\left. + \frac{1-2\nu}{E}[p(x) - p_{av}]\right\}, \qquad (10.1.24)$$

$$\epsilon_{22} = \frac{\partial u_y}{\partial y} = 0, \qquad (10.1.25)$$

$$\epsilon_{33} = \frac{\partial u_z}{\partial z} = 0, \qquad (10.1.26)$$

$$\epsilon_{12} = \frac{\partial u_x}{\partial y} + \frac{\partial u_y}{\partial x} = 0, \qquad (10.1.27)$$

$$\epsilon_{13} = \frac{\partial u_x}{\partial z} + \frac{\partial u_z}{\partial x} = 0, \qquad (10.1.28)$$

$$\epsilon_{23} = \frac{\partial u_y}{\partial z} + \frac{\partial u_z}{\partial y} = 0. \qquad (10.1.29)$$

The constitutive equations

$$\epsilon_{11} - \alpha_T T(x, t) = \frac{1+\nu}{1-\nu}\left\{\alpha_T[T(x, t) - T_{av}(t)]\right. \qquad (10.1.30)$$
$$\left. + \frac{1-2\nu}{E}[p(x) - p_{av}]\right\} - \alpha_T T(x, t)$$

$$= \frac{1}{E}[\sigma_{11} - \nu(\sigma_{22} + \sigma_{33})], \qquad (10.1.31)$$

$$\epsilon_{22} - \alpha_T T(x, t) = -\alpha_T T(x, t) = \frac{1}{E}[\sigma_{22} - \nu(\sigma_{11} + \sigma_{33})], \qquad (10.1.32)$$

$$\epsilon_{33} - \alpha_T T(x, t) = -\alpha_T T(x, t) = \frac{1}{E}[\sigma_{33} - \nu(\sigma_{11} + \sigma_{22})], \qquad (10.1.33)$$

$$\epsilon_{12} = 0 = \frac{2(1+\nu)}{E}\sigma_{12}, \qquad (10.1.34)$$

$$\epsilon_{13} = 0 = \frac{2(1+\nu)}{E}\sigma_{13}, \qquad (10.1.35)$$

$$\epsilon_{23} = 0 = \frac{2(1+\nu)}{E}\sigma_{23} \qquad (10.1.36)$$

yield the stress tensor

$$\sigma = \begin{pmatrix} \sigma_{11} & \sigma_{12} & \sigma_{13} \\ \sigma_{12} & \sigma_{22} & \sigma_{23} \\ \sigma_{13} & \sigma_{23} & \sigma_{33} \end{pmatrix} = \left[\frac{p(x) - p_{av}}{1 - \nu} - \frac{E\alpha_T T_{av}(t)}{(1 - 2\nu)(1 - \nu)} \right] \begin{pmatrix} 1 - \nu & 0 & 0 \\ 0 & \nu & 0 \\ 0 & 0 & \nu \end{pmatrix}$$

$$- \frac{E\alpha_T}{1 - \nu} T(x, t) \begin{pmatrix} 0 & 0 & 0 \\ 0 & 1 & 0 \\ 0 & 0 & 1 \end{pmatrix}, \qquad (10.1.37)$$

which satisfies the equation of mechanical equilibrium

$$\operatorname{div}(\sigma - p(x)\mathbf{I}) = 0. \qquad (10.1.38)$$

10.2   Liquid Flow and Heat Transport in a Permeable Elastic Beam II

The setup of this exercise has been adopted from the previous example, however, the inlet temperature is now discontinuous. Given length $L = 10$ m and thickness $H = 1$ m the domain represents the rectangular beam $[0, L] \times [0, H] \times [0, H]$. It is discretized by $100 \times 1 \times 1$ equally sized hexahedral elements. The solid material has been selected elastic, various model parameters and the hydraulic properties are given in Table 10.1. The liquid is incompressible and has viscosity $\mu = 1$ mPa·s. For densities, heat capacities, and thermal conductivities of liquid and solid grain see Table 10.2, gravity has explicitly been neglected. Fixities have been prescribed on the entire surface of the domain with zero x-displacement along the faces $x = 0$ and $x = L$, zero y-displacement along the faces $y = 0$ and $y = H$, and zero z-displacement along the faces $z = 0$ and $z = H$. Pressure $p_0 = 10^5$ Pa at the liquid inlet ($x = 0$) and zero pressure at the liquid outlet ($x = L$) generate steady-state 1D flow along the x-axis. At the liquid inlet temperature $T_0 = -12\,°C$ is specified for times less than $t_0 = 17,000$ s and zero afterwards. Free outflow prevails at the liquid outlet by default. Starting from zero initial temperature the simulation (Table 10.4) evaluates the transient temperature distribution as well as the mechanical load with output after 10,000 and 20,000 s.

The formal solution proceeds in three steps, first to solve for pressure and specific discharge, next to evaluate the temperature distribution (Fig. 10.3), and finally to determine displacements (Fig. 10.4), strains, and stresses.

Table 10.3 Benchmark deposit (https://docs.opengeosys.org/books/bmb-5)

BM code	Author	Code	Files	CTest
BMB5-10.1	Peter Vogel	OGS-5	Available	ToDo

(https://oc.ufz.de/index.php/s/nlph7bhfDkj6tC7)

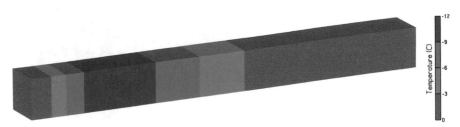

Fig. 10.3 Temperature distribution after 20,000 s

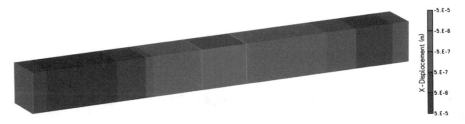

Fig. 10.4 X-Displacements after 20,000 s

For incompressible liquids Darcy's law and continuity equation yield the Laplace equation as the governing equation describing the steady-state pressure distribution. It reads

$$\frac{d^2 p}{dx^2} = 0 \qquad (10.2.1)$$

for 1D flow along the x-axis, hence, the pressure is given by

$$p(x) = p_0 \left(1 - \frac{x}{L} \right), \qquad (10.2.2)$$

and the average pressure p_{av} becomes

$$p_{av} = \frac{1}{L} \int_0^L p(x) dx = \frac{p_0}{2}. \qquad (10.2.3)$$

The specific discharge q is obtained by Darcy's law

$$q = \frac{k}{\mu} \frac{p_0}{L}. \qquad (10.2.4)$$

Due to the setup the thermal problem depends on time t and x-coordinate only and the heat transport equation governing the transient temperature distribution $T(x, t)$ reads

$$(n\rho_l c_l + (1-n)\rho_s c_s)\frac{\partial T}{\partial t} + (n\rho_l c_l)\frac{q}{n}\frac{\partial T}{\partial x} = (nk_l + (1-n)k_s)\frac{\partial^2 T}{\partial x^2}. \quad (10.2.5)$$

Introducing the notation

$$w = \frac{n\rho_l c_l}{n\rho_l c_l + (1-n)\rho_s c_s}\frac{q}{n}, \quad (10.2.6)$$

$$\chi = \frac{nk_l + (1-n)k_s}{n\rho_l c_l + (1-n)\rho_s c_s}, \quad (10.2.7)$$

the heat transport equation becomes

$$\frac{\partial T}{\partial t} + w\frac{\partial T}{\partial x} = \chi\frac{\partial^2 T}{\partial x^2}. \quad (10.2.8)$$

Due to free outflow at $x = L$ the formal problem is to determine the solution $T(x, t)$ of the above heat transport equation subject to the initial condition

$$T(x, 0) = 0 \quad \text{for } x > 0, \quad (10.2.9)$$

and the boundary conditions

$$T(0, t) = \begin{cases} T_0 & \text{for } t_0 > t > 0, \\ 0 & \text{for } t > t_0, \end{cases} \quad (10.2.10)$$

$$\lim_{x\to\infty} T(x, t) = 0 \quad \text{for } t > 0. \quad (10.2.11)$$

Applying the Laplace transform with respect to t yields the ordinary differential equation

$$\chi\bar{T}'' - w\bar{T}' - s\bar{T} = 0, \quad (10.2.12)$$

where $\bar{T}$ is the transform of T, s is the transformation parameter, and the prime denotes the derivative with respect to x. This equation has to be solved with respect to the transformed boundary conditions. Hence,

$$\bar{T}(x, s) = T_0\frac{1 - \exp(-t_0 s)}{s}\exp\left(x\left[\frac{w}{2\chi} - \sqrt{(\frac{w}{2\chi})^2 + \frac{s}{\chi}}\right]\right). \quad (10.2.13)$$

The transform of the temperature may also serve to evaluate integrals of the temperature which will be required below.

$$\mathscr{L}\left\{\int_0^x T(\hat{x}, t)d\hat{x}\right\} = \int_0^x \bar{T}(\hat{x}, s)d\hat{x} \tag{10.2.14}$$

$$= T_0 \frac{1 - \exp(-t_0 s)}{s} \int_0^x \exp\left(\hat{x}\left[\frac{w}{2\chi} - \sqrt{(\frac{w}{2\chi})^2 + \frac{s}{\chi}}\right]\right) d\hat{x}$$

$$= T_0 \frac{1 - \exp(-t_0 s)}{s} \frac{\exp\left(x\left[\frac{w}{2\chi} - \sqrt{(\frac{w}{2\chi})^2 + \frac{s}{\chi}}\right]\right) - 1}{\frac{w}{2\chi} - \sqrt{(\frac{w}{2\chi})^2 + \frac{s}{\chi}}}$$

and the average temperature

$$T_{av}(t) = \frac{1}{L}\int_0^L T(x, t)dx \tag{10.2.15}$$

has the transform

$$\mathscr{L}\{T_{av}\} = \frac{T_0}{L}\frac{1 - \exp(-t_0 s)}{s}\frac{\exp\left(L\left[\frac{w}{2\chi} - \sqrt{(\frac{w}{2\chi})^2 + \frac{s}{\chi}}\right]\right) - 1}{\frac{w}{2\chi} - \sqrt{(\frac{w}{2\chi})^2 + \frac{s}{\chi}}}. \tag{10.2.16}$$

The Laplace transforms thus obtained are well suited for numerical inversion. The numerical inversion scheme by Crump (1976) may easily be applied to give the required values of temperature $T(x, t)$, average temperature $T_{av}(t)$, and the above integral of the temperature.

For the solution of the mechanical problem we adopt the ideas outlined in the context of the previous example. Note that the applied temperature change is identical to the temperature distribution $T(x, t)$. Let σ denote the stress tensor, $\mathbf{I}$ the unit tensor, (u_x, u_y, u_z) the displacement vector, and

$$e = \frac{\partial u_x}{\partial x} + \frac{\partial u_y}{\partial y} + \frac{\partial u_z}{\partial z} \tag{10.2.17}$$

the volumetric strain. Employing Biot's simplified theory (i.e. Biot number equal one) the equations of equilibrium become

$$\frac{\partial^2 u_x}{\partial x^2} + \frac{\partial^2 u_x}{\partial y^2} + \frac{\partial^2 u_x}{\partial z^2} + \frac{1}{1 - 2\nu}\frac{\partial e}{\partial x} = \frac{2(1 + \nu)}{1 - 2\nu}\alpha_T \frac{\partial T}{\partial x} + \frac{2(1 + \nu)}{E}\frac{\partial p}{\partial x}$$

$$= \frac{2(1 + \nu)}{1 - 2\nu}\alpha_T \frac{\partial T}{\partial x} - \frac{2(1 + \nu)}{E}\frac{p_0}{L},$$

$$\frac{\partial^2 u_y}{\partial x^2} + \frac{\partial^2 u_y}{\partial y^2} + \frac{\partial^2 u_y}{\partial z^2} + \frac{1}{1 - 2\nu}\frac{\partial e}{\partial y} = \frac{2(1 + \nu)}{1 - 2\nu}\alpha_T \frac{\partial T}{\partial y} + \frac{2(1 + \nu)}{E}\frac{\partial p}{\partial y}$$

$$= 0,$$

$$\frac{\partial^2 u_z}{\partial x^2} + \frac{\partial^2 u_z}{\partial y^2} + \frac{\partial^2 u_z}{\partial z^2} + \frac{1}{1 - 2\nu}\frac{\partial e}{\partial z} = \frac{2(1 + \nu)}{1 - 2\nu}\alpha_T \frac{\partial T}{\partial z} + \frac{2(1 + \nu)}{E}\frac{\partial p}{\partial z}$$

$$= 0. \tag{10.2.18}$$

These equations and the imposed fixities are satisfied by

$$u_x = u_x(x, t), \tag{10.2.19}$$
$$u_y = 0, \tag{10.2.20}$$
$$u_z = 0, \tag{10.2.21}$$

if $u_x(x, t)$ satisfies the differential equation

$$\frac{\partial^2 u_x}{\partial x^2} + \frac{1}{1 - 2\nu} \frac{\partial^2 u_x}{\partial x^2} = \frac{2(1 + \nu)}{1 - 2\nu} \alpha_T \frac{\partial T}{\partial x} - \frac{2(1 + \nu)}{E} \frac{p_0}{L} \tag{10.2.22}$$

subject to the boundary conditions

$$u_x(0, t) = u_x(L, t) = 0 \quad \text{for } t > 0. \tag{10.2.23}$$

The x-displacement becomes

$$u_x(x, t) = \frac{1 + \nu}{1 - \nu} \left[\alpha_T \int_0^x T(\hat{x}, t) d\hat{x} - \alpha_T x T_{av}(t) \right.$$
$$\left. + \frac{1 - 2\nu}{2E} p_0 \left(x - \frac{x^2}{L} \right) \right]. \tag{10.2.24}$$

The strains read

$$\epsilon_{11} = \frac{\partial u_x}{\partial x} = \frac{1 + \nu}{1 - \nu} \left\{ \alpha_T [T(x, t) - T_{av}(t)] \right.$$
$$\left. + \frac{1 - 2\nu}{E} [p(x) - p_{av}] \right\}, \tag{10.2.25}$$

$$\epsilon_{22} = \frac{\partial u_y}{\partial y} = 0, \tag{10.2.26}$$

$$\epsilon_{33} = \frac{\partial u_z}{\partial z} = 0, \tag{10.2.27}$$

$$\epsilon_{12} = \frac{\partial u_x}{\partial y} + \frac{\partial u_y}{\partial x} = 0, \tag{10.2.28}$$

$$\epsilon_{13} = \frac{\partial u_x}{\partial z} + \frac{\partial u_z}{\partial x} = 0, \tag{10.2.29}$$

$$\epsilon_{23} = \frac{\partial u_y}{\partial z} + \frac{\partial u_z}{\partial y} = 0. \tag{10.2.30}$$

The constitutive equations

Table 10.4 Benchmark deposit (https://docs.opengeosys.org/books/bmb-5)

BM code	Author	Code	Files	CTest
BMB5-10.2	Peter Vogel	OGS-5	Available	TBD

(https://oc.ufz.de/index.php/s/nlph7bhfDkj6tC7)

$$\epsilon_{11} - \alpha_T T(x, t) = \frac{1 + \nu}{1 - \nu} \Big\{ \alpha_T [T(x, t) - T_{av}(t)] \tag{10.2.31}$$

$$+ \frac{1 - 2\nu}{E} [p(x) - p_{av}] \Big\} - \alpha_T T(x, t)$$

$$= \frac{1}{E} [\sigma_{11} - \nu(\sigma_{22} + \sigma_{33})], \tag{10.2.32}$$

$$\epsilon_{22} - \alpha_T T(x, t) = -\alpha_T T(x, t) = \frac{1}{E} [\sigma_{22} - \nu(\sigma_{11} + \sigma_{33})], \tag{10.2.33}$$

$$\epsilon_{33} - \alpha_T T(x, t) = -\alpha_T T(x, t) = \frac{1}{E} [\sigma_{33} - \nu(\sigma_{11} + \sigma_{22})], \tag{10.2.34}$$

$$\epsilon_{12} = 0 = \frac{2(1 + \nu)}{E} \sigma_{12}, \tag{10.2.35}$$

$$\epsilon_{13} = 0 = \frac{2(1 + \nu)}{E} \sigma_{13}, \tag{10.2.36}$$

$$\epsilon_{23} = 0 = \frac{2(1 + \nu)}{E} \sigma_{23} \tag{10.2.37}$$

yield the stress tensor

$$\sigma = \begin{pmatrix} \sigma_{11} & \sigma_{12} & \sigma_{13} \\ \sigma_{12} & \sigma_{22} & \sigma_{23} \\ \sigma_{13} & \sigma_{23} & \sigma_{33} \end{pmatrix} = \left[\frac{p(x) - p_{av}}{1 - \nu} - \frac{E \alpha_T T_{av}(t)}{(1 - 2\nu)(1 - \nu)} \right] \begin{pmatrix} 1 - \nu & 0 & 0 \\ 0 & \nu & 0 \\ 0 & 0 & \nu \end{pmatrix}$$

$$- \frac{E \alpha_T}{1 - \nu} T(x, t) \begin{pmatrix} 0 & 0 & 0 \\ 0 & 1 & 0 \\ 0 & 0 & 1 \end{pmatrix}, \tag{10.2.38}$$

which satisfies the equation of mechanical equilibrium

$$\text{div} (\sigma - p(x)\mathbf{I}) = 0. \tag{10.2.39}$$

10.3 Liquid Flow and Heat Transport in a Permeable Elastic Beam III

The setup of this exercise has been adopted from the previous examples, however, we now consider constant heat flow at the liquid inlet. Given length $L = 10$ m and thickness $H = 1$ m the domain represents the rectangular beam $[0, L] \times [0, H] \times [0, H]$. It is discretized by $80 \times 1 \times 1$ equally sized hexahedral elements. The solid material has been selected elastic, various model parameters and the hydraulic properties are given in Table 10.1. The liquid is incompressible and has viscosity $\mu = 1$ mPa·s. For densities, heat capacities, and thermal conductivities of liquid and solid grain see Table 10.2, gravity has explicitly been neglected. Fixities have been prescribed on the entire surface of the domain with zero x-displacement along the faces $x = 0$ and $x = L$, zero y-displacement along the faces $y = 0$ and $y = H$, and zero z-displacement along the faces $z = 0$ and $z = H$. Pressure $p_0 = 10^5$ Pa at the liquid inlet $(x = 0)$ and zero pressure at the liquid outlet $(x = L)$ generate steady-state 1D flow along the x-axis. At the liquid inlet a constant heat flow $h = 100$ W/m^2 is specified for times $t > 0$; it acts as a heat source to the domain. Free outflow prevails at the liquid outlet by default. Starting from zero initial temperature the simulation (Table 10.5) evaluates the transient temperature distribution as well as the mechanical load with output after 5000 and 10,000 s.

The formal solution proceeds in three steps, first to solve for pressure and specific discharge, next to evaluate the temperature distribution (Fig. 10.5), and finally to determine displacements (Fig. 10.6), strains, and stresses.

For incompressible liquids Darcy's law and continuity equation yield the Laplace equation as the governing equation describing the steady-state pressure distribution. It reads

$$\frac{d^2 p}{dx^2} = 0 \qquad (10.3.1)$$

for 1D flow along the x-axis, hence, the pressure is given by

Fig. 10.5 Temperature distribution after 10,000 s

Fig. 10.6 X-Displacements after 10,000 s

$$p(x) = p_0 \left(1 - \frac{x}{L}\right), \tag{10.3.2}$$

and the average pressure p_{av} becomes

$$p_{av} = \frac{1}{L} \int_0^L p(x)dx = \frac{p_0}{2}. \tag{10.3.3}$$

The specific discharge q is obtained by Darcy's law

$$q = \frac{k}{\mu} \frac{p_0}{L}. \tag{10.3.4}$$

Due to the setup the thermal problem depends on time and x-coordinate only and the heat transport equation governing the transient temperature distribution $T(x, t)$ reads

$$(n\rho_l c_l + (1 - n)\rho_s c_s)\frac{\partial T}{\partial t} + (n\rho_l c_l)\frac{q}{n}\frac{\partial T}{\partial x} = (nk_l + (1 - n)k_s)\frac{\partial^2 T}{\partial x^2}. \tag{10.3.5}$$

Introducing the notation

$$w = \frac{n\rho_l c_l}{n\rho_l c_l + (1 - n)\rho_s c_s}\frac{q}{n}, \tag{10.3.6}$$

$$\chi = \frac{nk_l + (1 - n)k_s}{n\rho_l c_l + (1 - n)\rho_s c_s}, \tag{10.3.7}$$

the heat transport equation becomes

$$\frac{\partial T}{\partial t} + w\frac{\partial T}{\partial x} = \chi\frac{\partial^2 T}{\partial x^2}. \tag{10.3.8}$$

The initial condition reads

$$T(x, 0) = 0 \quad \text{for } x > 0. \tag{10.3.9}$$

By Fourier's law and due to free outflow at $x = L$ the boundary conditions become

$$\frac{\partial T}{\partial x}(0, t) = -\frac{h}{nk_l + (1 - n)k_s} \quad \text{for } t > 0, \tag{10.3.10}$$

$$\lim_{x \to \infty} T(x, t) = 0 \quad \text{for } t > 0. \tag{10.3.11}$$

The formal problem is to determine the solution $T(x, t)$ of the heat transport equation subject to the above initial and boundary conditions. Applying the Laplace transform with respect to t yields the ordinary differential equation

$$\chi \bar{T}'' - w \bar{T}' - s \bar{T} = 0, \tag{10.3.12}$$

where $\bar{T}$ is the transform of T, s is the transformation parameter, and the prime denotes the derivative with respect to x. This equation has to be solved with respect to the transformed boundary conditions. Hence,

$$\bar{T}(x, s) = -\frac{h}{nk_l + (1-n)k_s} \frac{\exp\left(x\left[\frac{w}{2\chi} - \sqrt{(\frac{w}{2\chi})^2 + \frac{s}{\chi}}\right]\right)}{s\left[\frac{w}{2\chi} - \sqrt{(\frac{w}{2\chi})^2 + \frac{s}{\chi}}\right]}. \tag{10.3.13}$$

The transform of the temperature may also serve to evaluate integrals of the temperature which will be required below.

$$\mathscr{L}\left\{\int_0^x T(\hat{x}, t)d\hat{x}\right\} = \int_0^x \bar{T}(\hat{x}, s)d\hat{x}$$

$$= -\frac{h}{nk_l + (1-n)k_s} \frac{\exp\left(x\left[\frac{w}{2\chi} - \sqrt{(\frac{w}{2\chi})^2 + \frac{s}{\chi}}\right]\right) - 1}{s\left[\frac{w}{2\chi} - \sqrt{(\frac{w}{2\chi})^2 + \frac{s}{\chi}}\right]^2}$$

$$\tag{10.3.14}$$

and the average temperature

$$T_{av}(t) = \frac{1}{L}\int_0^L T(x, t)dx \tag{10.3.15}$$

has the transform

$$\mathscr{L}\{T_{av}\} = -\frac{h}{nk_l + (1-n)k_s} \frac{\exp\left(L\left[\frac{w}{2\chi} - \sqrt{(\frac{w}{2\chi})^2 + \frac{s}{\chi}}\right]\right) - 1}{sL\left[\frac{w}{2\chi} - \sqrt{(\frac{w}{2\chi})^2 + \frac{s}{\chi}}\right]^2}. \tag{10.3.16}$$

The Laplace transforms thus obtained are well suited for numerical inversion. The numerical inversion scheme by Crump (1976) may easily be applied to give the required values of temperature $T(x, t)$, average temperature $T_{av}(t)$, and the above integral of the temperature.

For the solution of the mechanical problem we adopt the ideas outlined in the context of the previous examples. Note that the applied temperature change is identical to the temperature distribution $T(x, t)$. Let σ denote the stress tensor, $\mathbf{I}$ the unit tensor, (u_x, u_y, u_z) the displacement vector, and

$$e = \frac{\partial u_x}{\partial x} + \frac{\partial u_y}{\partial y} + \frac{\partial u_z}{\partial z} \tag{10.3.17}$$

the volumetric strain. Employing Biot's simplified theory (i.e. Biot number equal one) the equations of equilibrium become

$$\frac{\partial^2 u_x}{\partial x^2} + \frac{\partial^2 u_x}{\partial y^2} + \frac{\partial^2 u_x}{\partial z^2} + \frac{1}{1-2\nu}\frac{\partial e}{\partial x} = \frac{2(1+\nu)}{1-2\nu}\alpha_T\frac{\partial T}{\partial x} + \frac{2(1+\nu)}{E}\frac{\partial p}{\partial x}$$

$$= \frac{2(1+\nu)}{1-2\nu}\alpha_T\frac{\partial T}{\partial x} - \frac{2(1+\nu)}{E}\frac{p_0}{L},$$

$$\frac{\partial^2 u_y}{\partial x^2} + \frac{\partial^2 u_y}{\partial y^2} + \frac{\partial^2 u_y}{\partial z^2} + \frac{1}{1-2\nu}\frac{\partial e}{\partial y} = \frac{2(1+\nu)}{1-2\nu}\alpha_T\frac{\partial T}{\partial y} + \frac{2(1+\nu)}{E}\frac{\partial p}{\partial y}$$

$$= 0,$$

$$\frac{\partial^2 u_z}{\partial x^2} + \frac{\partial^2 u_z}{\partial y^2} + \frac{\partial^2 u_z}{\partial z^2} + \frac{1}{1-2\nu}\frac{\partial e}{\partial z} = \frac{2(1+\nu)}{1-2\nu}\alpha_T\frac{\partial T}{\partial z} + \frac{2(1+\nu)}{E}\frac{\partial p}{\partial z}$$

$$= 0. \tag{10.3.18}$$

These equations and the imposed fixities are satisfied by

$$u_x = u_x(x, t), \tag{10.3.19}$$
$$u_y = 0, \tag{10.3.20}$$
$$u_z = 0, \tag{10.3.21}$$

if $u_x(x, t)$ satisfies the differential equation

$$\frac{\partial^2 u_x}{\partial x^2} + \frac{1}{1-2\nu}\frac{\partial^2 u_x}{\partial x^2} = \frac{2(1+\nu)}{1-2\nu}\alpha_T\frac{\partial T}{\partial x} - \frac{2(1+\nu)}{E}\frac{p_0}{L} \tag{10.3.22}$$

subject to the boundary conditions

$$u_x(0, t) = u_x(L, t) = 0 \quad \text{for } t > 0. \tag{10.3.23}$$

The x-displacement becomes

$$u_x(x, t) = \frac{1+\nu}{1-\nu}\left[\alpha_T \int_0^x T(\hat{x}, t)d\hat{x} - \alpha_T x T_{av}(t) \right.$$
$$\left. + \frac{1-2\nu}{2E}p_0\left(x - \frac{x^2}{L}\right)\right]. \tag{10.3.24}$$

The strains read

$$\epsilon_{11} = \frac{\partial u_x}{\partial x} = \frac{1+\nu}{1-\nu}\left\{\alpha_T[T(x,t) - T_{av}(t)]\right.$$
$$\left. + \frac{1-2\nu}{E}[p(x) - p_{av}]\right\}, \qquad (10.3.25)$$

$$\epsilon_{22} = \frac{\partial u_y}{\partial y} = 0, \qquad (10.3.26)$$

$$\epsilon_{33} = \frac{\partial u_z}{\partial z} = 0, \qquad (10.3.27)$$

$$\epsilon_{12} = \frac{\partial u_x}{\partial y} + \frac{\partial u_y}{\partial x} = 0, \qquad (10.3.28)$$

$$\epsilon_{13} = \frac{\partial u_x}{\partial z} + \frac{\partial u_z}{\partial x} = 0, \qquad (10.3.29)$$

$$\epsilon_{23} = \frac{\partial u_y}{\partial z} + \frac{\partial u_z}{\partial y} = 0. \qquad (10.3.30)$$

The constitutive equations

$$\epsilon_{11} - \alpha_T T(x,t) = \frac{1+\nu}{1-\nu}\left\{\alpha_T[T(x,t) - T_{av}(t)]\right. \qquad (10.3.31)$$
$$\left. + \frac{1-2\nu}{E}[p(x) - p_{av}]\right\} - \alpha_T T(x,t)$$
$$= \frac{1}{E}[\sigma_{11} - \nu(\sigma_{22} + \sigma_{33})], \qquad (10.3.32)$$

$$\epsilon_{22} - \alpha_T T(x,t) = -\alpha_T T(x,t) = \frac{1}{E}[\sigma_{22} - \nu(\sigma_{11} + \sigma_{33})], \qquad (10.3.33)$$

$$\epsilon_{33} - \alpha_T T(x,t) = -\alpha_T T(x,t) = \frac{1}{E}[\sigma_{33} - \nu(\sigma_{11} + \sigma_{22})], \qquad (10.3.34)$$

$$\epsilon_{12} = 0 = \frac{2(1+\nu)}{E}\sigma_{12}, \qquad (10.3.35)$$

$$\epsilon_{13} = 0 = \frac{2(1+\nu)}{E}\sigma_{13}, \qquad (10.3.36)$$

$$\epsilon_{23} = 0 = \frac{2(1+\nu)}{E}\sigma_{23} \qquad (10.3.37)$$

yield the stress tensor

$$\boldsymbol{\sigma} = \begin{pmatrix} \sigma_{11} & \sigma_{12} & \sigma_{13} \\ \sigma_{12} & \sigma_{22} & \sigma_{23} \\ \sigma_{13} & \sigma_{23} & \sigma_{33} \end{pmatrix} = \left[\frac{p(x) - p_{av}}{1-\nu} - \frac{E\alpha_T T_{av}(t)}{(1-2\nu)(1-\nu)}\right]\begin{pmatrix} 1-\nu & 0 & 0 \\ 0 & \nu & 0 \\ 0 & 0 & \nu \end{pmatrix}$$
$$- \frac{E\alpha_T}{1-\nu}T(x,t)\begin{pmatrix} 0 & 0 & 0 \\ 0 & 1 & 0 \\ 0 & 0 & 1 \end{pmatrix}, \qquad (10.3.38)$$

Table 10.5 Benchmark deposit (https://docs.opengeosys.org/books/bmb-5)

BM code	Author	Code	Files	CTest
BMB5-10.3	Peter Vogel	OGS-5	Available	TBD

(https://oc.ufz.de/index.php/s/nlph7bhfDkj6tC7)

which satisfies the equation of mechanical equilibrium

$$\text{div}\,(\sigma - p(x)\mathbf{I}) = 0. \tag{10.3.39}$$

10.4 Mass Conservation, Thermal Pressurization and Stress Distribution in Coupled Thermo-Hydro-Mechanical Processes

Xing-Yuan Miao, Tianyuan Zheng, Thomas Nagel

10.4.1 Governing Equations

The numerical analysis of multi-field problems in porous media is an important task for different geo-engineering subjects (e.g., geothermal energy, oil and gas reservoirs, energy storage and nuclear waste management). In particular, the coupling between heat transport and biphasic consolidation in saturated porous media is of high practical relevance.

To simulate the thermo-hydro-mechanical processes, the basic set of governing equations is given as:

- Mass (volume) balance

$$\text{div}\left[(\mathbf{u}_S)'_S + \phi_F \mathbf{w}_{FS}\right] = \underbrace{\beta_T^{\text{eff}} T'_S}_{\text{first term}} + \underbrace{\phi_F \beta_{TF} \text{grad}\, T \cdot \mathbf{w}_{FS}}_{\text{second term}} \tag{10.4.1}$$

with $\phi_F \mathbf{w}_{FS} = -\kappa_F/\mu_{FR}\left[\text{grad}\, p - \varrho_{FR} g\right]$ and $\beta_T^{\text{eff}} = \phi_F \beta_{TF} + 3(1 - \phi_F)\alpha_{TS}$
- Momentum balance

$$\text{div}\left[\sigma_S^E - \alpha_B p \mathbf{I}\right] + \varrho^{\text{eff}} g = 0 \tag{10.4.2}$$

with $\sigma_S^E = \mathcal{C} : (\epsilon - \epsilon_{\text{th}})$ and $\epsilon_{\text{th}} = \alpha_{TS}\Delta T$
- Energy balance

$$(\varrho c_p)^{\text{eff}} \frac{\partial T}{\partial t} + \phi_F \varrho_{FR} c_{pF} \text{grad}\, T \cdot \mathbf{w}_{FS} - \text{div}\left[\lambda^{\text{eff}} \text{grad}\, T\right] \tag{10.4.3}$$

where α_{TS} is the linear coefficient of thermal expansion of the solid phase, β_{TF} is the volumetric coefficient of thermal expansion of the fluid phase. ϕ_F is the porosity. κ_F is the intrinsic permeability, μ_{FR} is the viscosity, α_B is the Biot coefficient, σ_S^E are the effective Cauchy stresses, T is the absolute temperature, p is the pore pressure, λ the heat conductivity and $(\bullet)'_\alpha$ denotes a material time derivative following the motion of the αth constituent.

In OGS-5, the mass balance and momentum balance equations are solved monolithically, while the energy balance is solved sequentially. In OGS-6, a model is available in which the three governing equations are assembled and solved monolithically. Different constitutive relations (e.g. elasto-plasticity, visco-elasticity) can be used to describe the material behaviour of the porous solid. For initial verification, the case of a linearly elastic solid matrix is considered here.

10.4.2   OGS-5

Mass Conservation During Non-isothermal Flow

A two-dimensional channel-like model (Fig. 10.7) was set up to check the mass conservation for a fluid flow flowing through a porous solid. The solid phase is assumed to be homogeneous and incompressible. The chosen parameters can be found in Table 10.6 and boundary conditions are illustrated in Fig. 10.7. Consider flow entering the domain through the left boundary: after reaching steady-state conditions the same mass flux should exit the right-hand boundary (i.e. no more mass accumulation in the model):

$$\Delta \dot{m} = \dot{m}_{in} - \dot{m}_{out} = 0 \qquad (10.4.4)$$

and $\dot{m} = \int_{\partial \Omega} \varrho_F(T) \mathbf{w}_{FS} \cdot \mathbf{n} dA$.

Two scenarios (isothermal and non-isothermal) are considered to check the accuracy of the FEM implementation. In the non-isothermal case, the second term of the

Fig. 10.7 Single-material domain

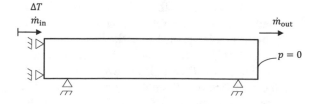

Table 10.6 Parameters used in the benchmark

Parameter	Value	Unit
Initial temperature	283.15	°C
Porosity	0.1	–
Water specific heat capacity	4280	J kg^{-1} K^{-1}
Water thermal conductivity	0.56	W m^{-1} K^{-1}
Water real density	1000	kg m^{-3}
Solid specific heat capacity	1714	J kg^{-1} K^{-1}
Solid thermal conductivity	2.5	W m^{-1} K^{-1}
Solid real density	1750	kg m^{-3}
Intrinsic permeability	10^{-6}	cm^2
Viscosity	$1.278 \cdot 10^{-3}$	Pa s
Time step size	10000	s
Young's modulus	21	GPa
Poisson's ratio	0.3	–
Fluid volumetric thermal expansion coefficient	$2.07 \cdot 10^{-4}$	K^{-1}
Solid linear thermal expansion coefficient	$0.7 \cdot 10^{-5}$	K^{-1}

right hand side of Eq. (10.4.1) represents the influence of the spatial temperature gradient on flow and is commonly omitted from the set of governing equations. Its relevance for an accurate mass balance is tested in this benchmark by including or excluding it.

The considered has a length of 1 mm, and a height of 0.2 mm. The inlet flow left is $2 \cdot 10^{-4}$ kg/s and A temperature gradient of 200 K was imposed. Meshes with triangular and quadrilateral elements are used for discretization, respectively. Shape functions are linear for the primary variables T and p, and quadratic for **u**. The direct solver PARDISO for the linear system was combined with a Newton-Raphson iterative scheme with a relative error tolerance of 10^{-12} for all primary variables. We check the accuracy by calculating the relative deviation between the inlet mass flow and the outlet mass flow, i.e., $\epsilon_{err} = |\Delta \dot{m}/\dot{m}_{in}|$. A space discretisation of 100 quadratic elements and time discretisation 10 s with 10 time steps.

For comparison, an isothermal process was simulated under otherwise unchanged boundary conditions. Its relative mass flux deviation was found to be $\epsilon_{err} = 1.5 \cdot 10^{-6}$. In the non-isothermal case, the fluid density is no longer constant but changes in a linear fashion proportional to β_{TF}. Results for the different element types are listed in Table 10.7. The temperature distribution is illustrated in Fig. 10.8. We can see that the relative error does not vary much with different mesh density. Considering the gradient term on the RHS of Eq. (10.4.1) does indeed improve the mass balance.

Table 10.7 Relative error in % depending on whether the second term is considered on the RHS of Eq. (10.4.1) or not

Mesh type	Number of nodes	With second term	Without second term
△ (coarse)	196	0.18	4.36
□ (coarse)	561	0.46	4.71
△ (fine)	716	0.11	4.36
□ (fine)	1440	0.28	4.55

Fig. 10.8 Temperature distribution of the domain

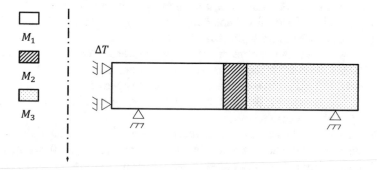

Fig. 10.9 Bi-material domain. Texture indicates material domains 1, 2 (interface) and 3

However, this effect only becomes relevant if the product of temperature gradient and the flow velocity is very high. In most practical applications, this term can indeed be neglected.

Bi-Material Interface

In many application, different material domains with contrasting properties are adjacent to each other. This can pose numerical difficulties and model accuracy needs to be tested before application. For this purpose, a composite material composed of a cylindrical core and an annular domain of a different material were modelled as an axisymmetric domain with three materials was set up as illustrated in Fig. 10.9. The third material is the interface between the two material domains by which we control the mechanical and hydraulic connectivity of the two adjacent domains.

The properties were set such that two cases could be studied:

Case 1 (sealed test):   **Case 2 (unsealed test):**

$$E_1 = 2E_3 \qquad\qquad\qquad\qquad E_1 = 2E_3$$

$$\nu_1 = \nu_2 = \nu_3 \qquad\qquad\qquad \nu_1 = \nu_2 = \nu_3$$

$$\alpha_{\mathrm{TS}1} = \frac{1}{2}\alpha_{\mathrm{TS}3} \qquad\qquad\qquad \alpha_{\mathrm{TS}1} = \frac{1}{2}\alpha_{\mathrm{TS}3}$$

$$\kappa_1 = \kappa_3 \qquad\qquad\qquad\qquad \kappa_1 = \kappa_3$$

$$E_2 = \frac{1}{100}E_3 \qquad\qquad\qquad E_2 = \frac{1}{100}E_3$$

$$\alpha_{\mathrm{TS}2} = \frac{1}{2}\alpha_{\mathrm{TS}3} \qquad\qquad\qquad \alpha_{\mathrm{TS}2} = \frac{1}{2}\alpha_{\mathrm{TS}3}$$

$$\kappa_2 \approx 0 \qquad\qquad\qquad\qquad \kappa_2 = \kappa_3$$

The porosity and the Biot coefficient of the intermediate layer were both set to 0 to block the effect of temperature and pressure on solid deformation in this region and make the interface passive.

Due to the different thermal expansion coefficients between material 1 and material 3, the thermally induced volume changes of these two materials are different, which produces different fluid pressures in the materials.

For the two cases, the bottom of the model is constrained in vertical direction and the left side is constrained in horizontal direction. Temperature of $363.15\,\mathrm{K}$ is applied to the left side of the domain. The initial temperature of the whole area is set to $283.15\,\mathrm{K}$. For case 1, the domain is sealed and for case 2, there is a Dirichlet boundary $p = 0$ on the right side.

Case 1 would be expected to capture this feature as if the two materials were separated from each other. The impermeable intermediate layer ($\kappa_2 \approx 0$) would prevent the flow moving mass from regions of higher pressure to those of lower pressure. The thermally induced volume change in material 1 is smaller than that of material 2 which leads to $p_1 > p_3$ given an identical pore fluid. Assuming the intermediate layer allowed free expansion of the neighbouring two materials, the analytical solution for the fluid pressure in each domain is

$$p = -K_{\mathrm{S}}\left(e_{\mathrm{M}} - e^{\mathrm{th}}\right) \tag{10.4.5}$$

with $e_{\mathrm{M}} = \beta_{\mathrm{TM}}\Delta T$ and $e^{\mathrm{th}} = \beta_{\mathrm{TS}}\Delta T$, where K_{S} is the bulk modulus of the solid skeleton obtained from

$$K_{\mathrm{S}} = \frac{E_{\mathrm{S}}}{3(1 - 2\nu_{\mathrm{S}})} \tag{10.4.6}$$

and β_{TM} is the effective volumetric coefficient of thermal expansion:

$$\beta_{\mathrm{TM}} = \phi_{\mathrm{S}}\beta_{\mathrm{TS}} + \phi_{\mathrm{F}}\beta_{\mathrm{TF}} \tag{10.4.7}$$

The same solver settings were used as in the previous example. The space discretization is 927 quadratic elements and the time discretization is 10 s with 10 time steps.

By comparing the analytical solution of the fluid pressure in the two domains, $p_1^{\text{analytical}} = 0.02604\,\text{MPa}$ and $p_3^{\text{analytical}} = 0.01155\,\text{MPa}$ to the numerical approximation, $p_1^{\text{numerical}} = 0.0278\,\text{MPa}$ and $p_3^{\text{numerical}} = 0.0123\,\text{MPa}$, a good correspondence is found. Note that due to $\kappa_2 \approx 0 \neq 0$ fluid mass is indeed moved from one domain to the other given enough time. The simulation thus needs to take this time scale into account if hydraulic isolation is to be modelled (Fig. 10.10).

Case 2 established a hydraulic connection between the two reservoirs. To estimate the average modulus of the bi-material domain, the rule of mixtures was applied. Here, $E_S = \phi_S^{S1} E_{S1} + (1 - \phi_S^{S1}) E_{S2}$ served as an upper-bound (in the direction of parallel to the "springs") and $E_S = \left(\frac{\phi_S^{S1}}{E_{S1}} + \frac{1 - \phi_S^{S1}}{E_{S2}} \right)^{-1}$ as a lower-bound estimate (considering springs in series). The Poisson's ratios in all domains were equal $\nu_S = \nu_1 = \nu_2$, and $\phi_S^{S1} = \frac{V_{S1}}{V_{S1} + V_{S2}}$ etc.

The same method can be used to obtain the overall thermal expansion coefficient of the solid phase.

Thus, the analytical solution of the fluid pressure caused by thermal expansion evaluates to $p_{\text{upper}}^{\text{analytical}} = 0.01879\,\text{MPa}$ and $p_{\text{lower}}^{\text{analytical}} = 0.01638\,\text{MPa}$. By numerical simulation, $p^{\text{numerical}} = 0.01634\,\text{MPa}$ are found which is close to the lower bound. This is reminiscent of the fact that due to the externally unconstrained expansion and the internal fluid pressure equilibration, this set-up corresponds to the spring-in-series analogy (Fig. 10.11).

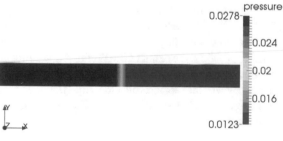

Fig. 10.10 Pressure distribution of the sealed test (case 1)

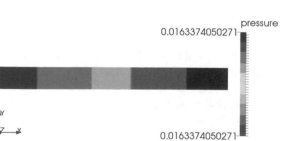

Fig. 10.11 Pressure distribution of the unsealed test (case 2)

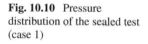

Fig. 10.12 2D domain and
FE mesh. Square 1.0 ×
1.0 mm; 10 × 10 elements

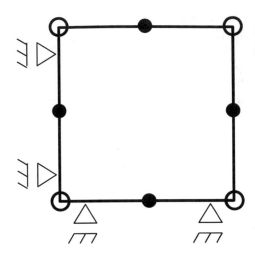

10.4.3 OGS-6

On the aspects of OGS-6, three numerical benchmarks were performed to verify
the numerical results of thermo-hydro-mechanical process in OGS-6 with analytical
solutions. A homogeneous square model, a bi-material beam model and a point heat
source consolidation model were set up for the verification.

Homogeneous Square Model

In this case, an axisymmetric homogeneous domain (see Fig. 10.12) was set up with
the length of 1 mm. The whole domain was heated up from the left boundary from
273.15 to 353.15 K and the liquid was sealed in the area from the surrounding Neumann no flow boundary.

After 10,000 s, the result of the numerical solution reached steady state with homogeneous pressure of 0.1042 MPa. Then the analytical solution of the fluid pressure
created by volume expansion (see Eq. 10.4.8) is derived through the definition of
general volumetric thermal expansion coefficient which can be found in the OGS-
5 section.

$$p = -K_S(e_M - e^{th})$$
$$= -K_S \phi_F(\beta_{TF} - \beta_{TS}) \Delta T \qquad (10.4.8)$$

where $e^{th} = \beta_{TS} \Delta T$. Using the analytical solution, the steady state pressure in the
whole domain is 0.1042 MPa with Young's Modulus of 21 MPa, Possion's ratio of 0.3,
porosity of 0.4, volumetric thermal expansion coefficient of 2.07×10^{-4} for fluid and
linear thermal expansion coefficient of 0.7×10^{-5} for solid. The numerical solution
used newton-rapson method for nonlinear solver with tolerance of 10^{-7}, 10^{-5}, 10^{-5},
10^{-5} for primary variables and BiCGSTAB for the linear solver. The OGS-6 version performed for this benchmark can be found on github with links https://github.
com/grubbymoon/ogs/THM_PR and the input files can be found on https://github.
com/ufz/ogs-data/THM_PR/ThermoHydroMechanics/Linear. The numerical solu-

Fig. 10.13 Heterogeneous model setup

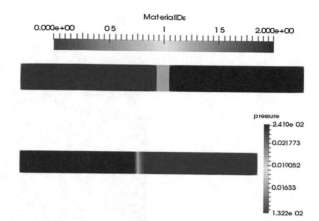

Fig. 10.14 Pressure distribution in the composite beam

tion reaches steady state within 10,000 s and equals 0.1042 MPa which fits very well with the analytical solution.

Bi-Material Beam Model

In this case, a composite beam model of axisymmetry with three different materials was built (See Fig. 10.13). The relationship of the different properties of the composite beam was set up as follows (same with the composite beam case 1 in the OGS-5 section): The approximate numerical result can be found in Fig. 10.14. $p_1^{\text{numerical}} = 0.0241$ MPa and $p_3^{\text{numerical}} = 0.01322$ MPa, With the analytical solution of different regions from Sect. 10.4.2, $p_1^{\text{analytical}} = 0.02604$ MPa and $p_3^{\text{analytical}} = 0.01155$ MPa, the relative error is 7.45 percent for the maximum value and 14.45 percent for the minimum value. While in OGS-5, the relative error is 6.7 percent for the maximum value and 6.5 percent for the minimum value.

Point Heat Source Consolidation Model

When a heat source such as a canister of radioactive waste is buried in a saturated porous medium, the variation of temperature will casue the pore water to expand a greater amount than the voids of the porous material. The temperature lift will thus usually be accompanied by an increase in pore pressure. If the domain is sufficiently permeable, these pore pressures will dissipate. The derivation of analytical solution can be found in Kolditz et al. (2016c).

A 2D axisymmertic model is set up for the verification. The model domain and meshes can be found in Fig. 10.15. A line source is seleted to represent the injection source (0.00204357 m in this case) which is located between center point of the quarter and the closest node. After the axisymmetric rotation around the vertical direction, the line source has converted into a circular source and the Neumann boundary heat flux can thus be calculated by $300/2/(\pi r^2)$W. The radius of the domain is 10 m and the initial temperature and pore pressure are 273 K and 0 Pa respectively. The model parameters can be found in Table 10.8. Three different observation locations are selected for the analytical and numerical solutions (0.25, 0.5 and 1 m from the injection source).

Fig. 10.15 Mesh
distribution and domain

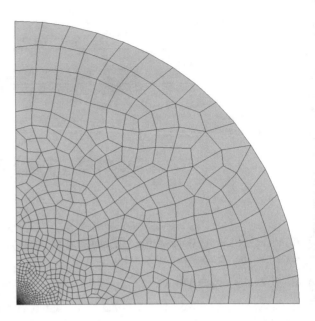

Table 10.8 Parameters for the point heat source consolidation

Parameter	Value	Unit
Initial temperature	283.15	°C
Porosity	0.16	–
Water specific heat capacity	4280	$J\,kg^{-1}\,K^{-1}$
Water thermal conductivity	0.56	$W\,m^{-1}\,K^{-1}$
Water real density	1000	$kg\,m^{-3}$
Solid specific heat capacity	1000	$J\,kg^{-1}\,K^{-1}$
Solid thermal conductivity	1.64	$W\,m^{-1}\,K^{-1}$
Solid real density	2450	$kg\,m^{-3}$
Intrinsic permeability	$2 \cdot 10^{-20}$	m^2
Viscosity	$1 \cdot 10^{-3}$	Pa s
Time step size	10,000	s
Young's modulus	5	GPa
Poisson's ratio	0.3	–
Biot coefficient	1	–
Fluid volumetric thermal expansion coefficient	$4 \cdot 10^{-4}$	K^{-1}
Solid linear thermal expansion coefficient	$4.5 \cdot 10^{-5}$	K^{-1}

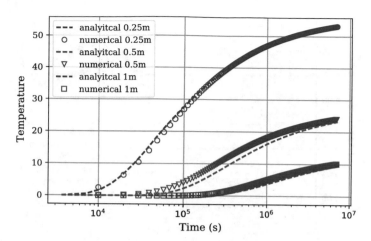

Fig. 10.16 The numerical solution compared with analytical solution (temperature)

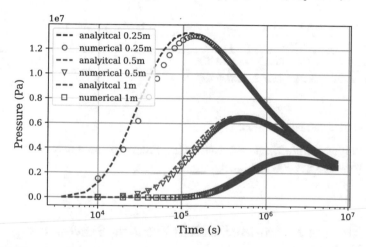

Fig. 10.17 The numerical solution compared with analytical solution (pressure)

From Figs. 10.16 and 10.17, we can find a generally good match of the numerical solution and analytical solution. In Fig. 10.16, the difference gets larger with the further observation location because the density of the mesh is very fine around the injection point and quickly becomes coarser and coarser along the radius. Since the chosen point is on the displacment boundary and does not have displacement in the y direction, the displacement on the x direction is plotted (Fig. 10.18).

The OGS-6 version performed for this benchmark can be found on github with links (Table 10.9):

• https://github.com/grubbymoon/ogs/tree/THM_PR.

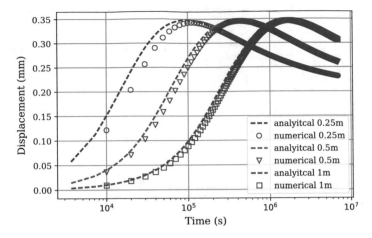

Fig. 10.18 The numerical solution compared with analytical solution (displacement)

Table 10.9 Benchmark deposit (https://docs.opengeosys.org/books/bmb-5)

BM code	Author	Code	Files	CTest
BMB5-10.4	Xing-Yuan Miao	OGS-6	See below	TBD

(https://oc.ufz.de/index.php/s/nlph7bhfDkj6tC7)

The input files can be found on:

- https://github.com/ufz/ogs-data/tree/THM_Freezing/ThermoHydroMechanics/
  Linear/Point_injection_r0.00204357.

10.5 Thermo-Hydro-Mechanical Freezing Benchmark (CIF Test)

Tianyuan Zheng, Xing-Yuan Miao, Dmitri Naumov, Thomas Nagel

The thermal-hydro-mechanical behaviour of freezing in Thermo-hydro-mechanical coupling fluid-saturated porous media is of great interest in soil construction, geotechnics, energy storage and geothermal applications. A fully coupled thermo-hydro-mechanical model of freezing is implemented into OGS-6 (Zheng et al. 2016) and (Zheng et al. 2017). The system of equations is treated numerically by a monolithically coupled incremental-iterative Newton-Raphson scheme. A numerical example of CIF (Capillary suction, Internal damage and Freezing-thawing) is performed.

10.5.1   Governing Equations

The governing equations are given by
(i) the mixture volume balance

$$0 = \text{div } (\mathbf{v_S} + \phi_L \mathbf{w}_{LS}) + \hat{\varrho}_I \left(\varrho_{LR}^{-1} - \varrho_{IR}^{-1} \right) - \beta_T T_S'$$

$$\text{with } \phi_L \mathbf{w}_{LS} = -\frac{\mathbf{K}}{\mu_{LR}} \, (\text{grad } p_{LR} - \varrho_{LR} \mathbf{b}_L) \tag{10.5.1}$$

where ϱ_R is the real density of phase α, $\hat{\varrho}_I$ is phase transition term between ice and water, $\beta_T = \sum_\alpha \phi_\alpha \beta_{T\alpha}$ is the mixed volumetric thermal expansion coefficient, $\mathbf{K}$ is the Intrinsic permeability tensor, $\mathbf{w}_{LS}$ is the seepage velocity, $\mathbf{v}_\alpha$ is the velocity of phase α and ϕ_α is the volume fraction of phase α.
(ii) the mixture momentum balance.

$$\text{div } [-p_{LR}\mathbf{I} + \lambda_S \text{tr}(\epsilon_S)\mathbf{I} + 2\mu_S \epsilon_S - 3\alpha_{TS} k_S (T - T_{S0})\mathbf{I} + \lambda_I \text{tr}(\epsilon_I)\mathbf{I} + 2\mu_I \epsilon_I$$
$$- 3\alpha_{TI} k_I (T - T_{I0})\mathbf{I} - 3\alpha_{FI} k_I (\phi_I - \phi_{I0} - \phi_I \epsilon_I : \mathbf{I})\mathbf{I}] + \varrho \mathbf{b} = \mathbf{0} \tag{10.5.2}$$

where λ_α and μ_α are the Lamé coefficients of phase α and α_{TS} is the linear thermal expansion coefficient for solid phase, T_0 is the reference temperature for phase α, k_S is the bulk modulus of the solid phase, $\mathbf{b}$ is the body force and α_{FI} is the freezing expansion coefficient.
(iii) the mixture energy balance

$$\left((\varrho c_p)^{\text{eff}} - \frac{\partial \varrho_I^{\text{eq}}}{\partial T} \Delta h_I \right) \frac{\partial T}{\partial t} - \text{div } (\boldsymbol{\lambda}^{\text{eff}} \text{ grad } T) + \varrho_L c_{pL} \mathbf{w}_{LS} \cdot \text{grad } T = 0$$

$$\tag{10.5.3}$$

where $\boldsymbol{\lambda}^{\text{eff}}$ is the heat conductivity tensor and Δh_I is the specific enthalpy of fusion.
   The ice volume fraction is determined based on an equilibrium approach and follows the relation (Zheng et al. 2016)

$$\phi_I \equiv \phi_I^{\text{eq}} = \phi \frac{1}{1 + e^{-k(T - T_m)}} \tag{10.5.4}$$

10.5.2   Benchmark Description

A cuboid model with a cross section of 15 cm^2 and a height of 7.5 cm^2 is used for the numerical example. Material properties and relevant numerical parameters are listed in Table 10.10 (Setzer et al. 2001) and (Bluhm et al. 2014).

Table 10.10 Parameters used in the numerical example

Parameter	Value	Unit
Initial temperature	20	°C
Initial solid volume fraction	0.5	–
Water specific heat capacity	4179	J kg^{-1} K^{-1}
Water thermal conductivity	0.58	W m^{-1} K^{-1}
Water real density	1000	kg m^{-3}
Ice specific heat capacity	2052	J kg^{-1} K^{-1}
Ice thermal conductivity	2.2	W m^{-1} K^{-1}
Ice real density	920	kg m^{-3}
Solid specific heat capacity	5900	J kg^{-1} K^{-1}
Solid thermal conductivity	1.1	W m^{-1} K^{-1}
Solid real density	2000	kg m^{-3}
Initial intrinsic permeability	10^{-8}	cm^2
Viscosity	$1.278 \cdot 10^{-3}$	Pa s
Time step size	300	s
Lamé constant μ_I	4.17	GPa
Lamé constant λ_I	2.78	GPa
Lamé constant μ_S	12.5	GPa
Lamé constant λ_S	8.33	GPa

Fig. 10.19 Temperature load for the bottom boundary

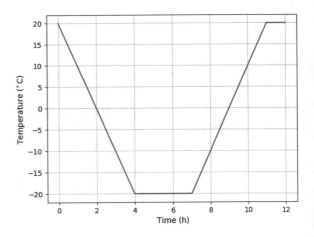

The ice formation is simulated in a 2D setting. A Dirichlet boundary condition for temperature was applied at the bottom of the domain. The temperature profile is illustrated in Fig. 10.19. The bottom of the domain was cooled in the first 7 h and heated up for the final 4 h. The remaining boundaries were considered adiabatic.

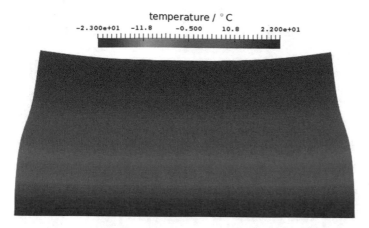

Fig. 10.20 Temperature after 5 h

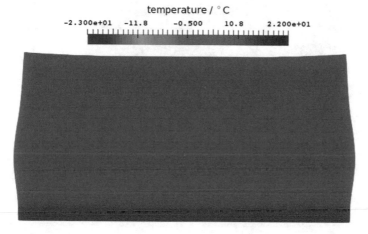

Fig. 10.21 Temperature after 11.8 h

The bottom surface was constrained in the vertical direction and was sealed while all others were treated as free displacement and draining boundaries. The time step size was set to 2400 s. The SparseLU solver (Guennebaud et al. 2010) was chosen to solve the linear system of equations, while for the nonlinear Newton-Raphson solver absolute tolerances were set to 10^{-6} for displacements, 10^{-3} for temperature, 10^{-4} for pressure. The OGS-6 version used for this benchmark can be found under https://github.com/grubbymoon/ogs/THM_Freezing_Stiffness and the input files at https://github.com/ufz/ogs-data/Benchmark_chapter/ThermoHydroMechanics/Linear/CIF_test.

Figures 10.20, 10.21, 10.22, 10.23, 10.24 and 10.25 show the temperature distribution (with unit of °C), ice volume fraction and volume ratio at different times of the freeze-thaw cycle. The cooling process lasted 4 h to decrease the temperature below

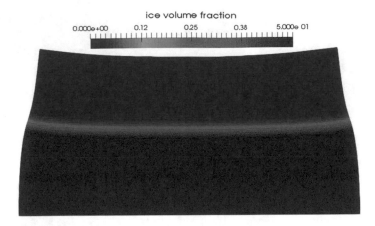

Fig. 10.22 Ice volume fraction after 5 h

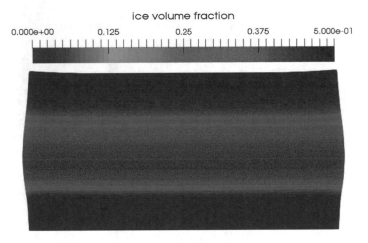

Fig. 10.23 Ice volume fraction after 11.8 h

the freezing point and trigger the phase change from liquid to ice. For pure water, the volume deformation due to phase change is 9%. With the setting of the initial porosity to 0.5, the volume deformation of the calculated numerical model is 4.5%.

Later on, the temperature increased above the onset temperature for phase change, causing the volume to contract due to thawing. Volume deformation and ice volume fractions are clearly co-localized (Figs. 10.20, 10.21, 10.22, 10.23, 10.24 and 10.25). The results show a satisfactory match with those presented in Bluhm et al. (2014) and Ricken et al. (2010). More details on these models can be found in Zheng et al. (2016) and Zheng et al. (2017) (Table 10.11).

The OGS-6 version used for this benchmark can be found under:

- https://github.com/grubbymoon/ogs/tree/THM_Freezing_Stiffness.

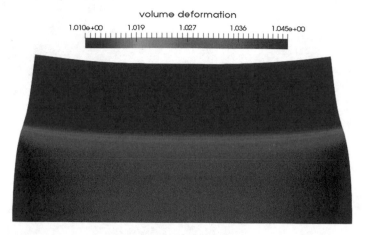

Fig. 10.24 Volume deformation after 5 h. Displacements magnified ten-fold

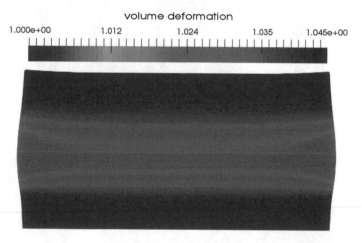

Fig. 10.25 Volume deformation after 11.8 h. Displacements magnified ten-fold

Table 10.11 Benchmark deposit (https://docs.opengeosys.org/books/bmb-5)

BM code	Author	Code	Files	CTest
BMB5-10.5	Tianyuan Zheng	OGS-6	See below	TBD

(https://oc.ufz.de/index.php/s/nlph7bhfDkj6tC7)

The input files are available at:

- https://github.com/ufz/ogs-data/tree/THM_Freezing/ThermoHydroMechanics/ Linear/CIF_test.

Chapter 11
RTM Processes

Renchao Lu, Norihiro Watanabe, Eunseon Jang and Haibing Shao

11.1 Reactive Mass Transport in a Compacted Granite Fracture with Pressure Solution Acting upon Grain Contacts

Renchao Lu, Norihiro Watanabe, Eunseon Jang, Haibing Shao

Fluid-mineral interaction, taking place over the fracture surface, renders a permanent change of fracture surface geometry. Hydraulic characteristic of fracture is consequently altered in response to the geometric change. For the sake of simplicity, the hydraulic process is assumed to be decoupled from the surface weathering. The dissolution-induced change of fracture surface geometry is neglected accordingly. Following this presumption, a benchmark is carried out. In the benchmark, we focus on the water-granite interaction in a flow-through undeformable fracture under confining stress, with highlights on the enhanced mineral dissolution at grain contacts where pressure solution is operative.

Furthermore, this benchmark is also part of the international DECOVALEX-2015 Project Task C1.[1] A full specification of the present benchmark can be referred to Lu et al. (2017b).

---

[1] http://decovalex.org/D-2015/task-c1.html.

R. Lu (✉) · E. Jang · H. Shao
Helmholtz Centre for Environmental Research-UFZ, Leipzig, Germany
e-mail: renchao.lu@ufz.de

R. Lu
TU Dresden, Dresden, Germany

H. Shao
TU Freiberg, Freiberg, Germany

N. Watanabe
National Institute of Advanced Industrial Science and Technology (AIST),
Renewable Energy Research Center, Koriyama, Fukushima, Japan

© Springer International Publishing AG 2018
O. Kolditz et al. (eds.), *Thermo-Hydro-Mechanical-Chemical Processes in Fractured Porous Media: Modelling and Benchmarking*,
Terrestrial Environmental Sciences, https://doi.org/10.1007/978-3-319-68225-9_11

11.1.1 Theory

Mineral dissolution reactions spontaneously take place over the compacted rough-walled fracture surface amid fluid-mineral interaction. Unlike the chemical weathering at open pores, the weathering at compacted grain contacts is dominated by pressure solution. In particular, the stress-induced chemical potential difference between center and periphery of the contact drives enhanced mineral dissolution over the rough grain surface. The dissolved components radially diffuse from the center into the surrounding open pores afterwards through the embedded water film. Precipitates have the potential to be formed in the course, and eventually cemented around the periphery.

Although the weathering mechanism at open pores is distinguished from that at grain contacts, the mass removal, arising from the dissolution at either, can be quantitatively described with a general rate equation. The rate equation is expressed as a summation over rates in various aqueous environments (modified from Palandri and Kharaka 2004, with reference to the work of Taron and Elsworth 2010.)

$$\dot{m}_{\bar{X}} = \sum_n k_n^+ a \left(1 - \frac{Q}{a K_{eq}} \right), \quad n = \text{acid, neutral, and base,} \tag{11.1.1}$$

where $\bar{X}$ denotes the mineral reactant, $\dot{m}_{\bar{X}}$ [mol/m^2/s] is the mass removal rate per reactive surface area, k^+ [mol/m^2/s] is the dissolution rate constant, a [-] is the activity of mineral, Q [-] is the ion activity product, and K_{eq} [-] is the equilibrium constant.

The activity of mineral a is equal to unity in the case of chemical weathering at open pores. In the other situation, the stress-induced chemical potential difference between center and periphery of the contact $\Delta\mu$ [J/mol] results in the elevation of the activity (Heidug 1995)

$$a = \exp\left(\frac{\Delta\mu}{RT}\right) = \exp\left[\frac{(\sigma_{con} - p_w) V_m}{R_c RT}\right], \tag{11.1.2}$$

where R [J/K/mol] is the ideal gas constant, T [K] is the temperature, σ_{con} [Pa] is the confining stress, p_w [Pa] is the pore pressure, V_m [m^3/mol] is the molar volume of mineral, and R_c [-] is the contact area ratio.

The dissolution rate constant k^+ can be given from the Arrhenius equation as (Palandri and Kharaka 2004)

$$k^+ = A e^{E/RT} a_{H^+}^{n_{H^+}}, \tag{11.1.3}$$

where A [mol/m^2/s] is the pre-exponential factor, E [J/mol] is the activation energy, a_{H^+} [-] is the activity of hydrogen ion, and n_{H^+} [-] is the H$^+$ catalysis constant.

Reactive transport, in terms of a particular dissolved component X_j, along the flow path in a fracture follows the advection-dispersion-reaction equation

$$b_{\mathrm{m}}^{\mathrm{p}}\frac{\partial C_j}{\partial t} + \nabla \cdot \left(-b_{\mathrm{m}}^{\mathrm{p}}\mathbf{D}\nabla C_j + b_{\mathrm{m}}^{\mathrm{p}} v C_j\right) = 2 f_r' \sum_{i=1}^{i} s_{ij}\left(\phi_i^{\mathrm{p}} \dot{m}_i^{\mathrm{p}} + \frac{R_{\mathrm{c}}}{1 - R_{\mathrm{c}}} f_{r,i} \phi_i^{\mathrm{c}} \dot{m}_i^{\mathrm{c}}\right),$$

(11.1.4)

where C_j [mol/m^3] is the concentration of the dissolved component, $b_{\mathrm{m}}^{\mathrm{p}}$ [m] is the mean of the mechanical aperture at open pores (simply called mechanical aperture at open pores in the following text), $\mathbf{D}$ [m^2/s] is the dispersion tensor, v [m/s] is the flow velocity, s_{ij} is the component of the stoichiometric coefficient matrix of reaction products $\mathbf{s}$, f_r' [-] is the fracture surface roughness factor, i.e., the area ratio of apparent fracture surface over fracture surface plane, f_r [-] is the intragranular roughness factor, defined as the ratio of true (total) grain surface area over apparent (geometric) grain surface area, ϕ_i^{p} and ϕ_i^{c} [-] are the area fractions of the mineral reactant $\bar{X}_i$ with respect to the pore surface and contact surface. $\dot{m}_i^{\mathrm{p}}$ and $\dot{m}_i^{\mathrm{c}}$ [mol/m^2/s] are the normalized mass removal rates on pore walls and at grain contacts, respectively, which can be determined by Eq. (11.1.1). The explicit multiplication of the reaction terms by 2 shows the dissolved component sources from chemical weathering on the opposed fracture surfaces.

The dispersion tensor $\mathbf{D}$ is given as (Scheidegger 1961)

$$D_{ij} = D_{\mathrm{m}}\delta_{ij} + \alpha_T|\mathbf{v}|\delta_{ij} + (\alpha_L - \alpha_T)\frac{v_i v_j}{|\mathbf{v}|},$$

(11.1.5)

with the molecular diffusion coefficient D_{m} [m^2/s], the longitudinal and transversal dispersion coefficients α_L, α_T [m].

The intragranular roughness factor f_r can be estimated via Brunauer–Emmet–Teller (BET) analysis of N$_2$ adsorption experiment (Tester et al. 1994)

$$f_r = \frac{S_{\mathrm{BET}}}{S_{\mathrm{GEO}}} = \frac{S_{\mathrm{BET}}\, d\, \rho_{\mathrm{m}}}{6},$$

(11.1.6)

where S_{BET} [m^2/kg] is the specific surface area of grain measured from nitrogen adsorption isotherms, S_{GEO} [m^2/kg] is the geometric surface area of ideally smooth spherical grain, d [m] is the grain diameter, and ρ_{m} [kg/m^3] is the mineral density.

Fluid flow through a rough-walled fracture can be treated as steady-state laminar flow without gravitational effect, which is constructed on the cubic law (Witherspoon et al. 1980)

$$\nabla \cdot \left(-\frac{b_{\mathrm{h}}^3}{12\eta}\nabla p_w\right) = 0,$$

(11.1.7)

where η [Pa·s] is the fluid dynamic viscosity, and b_{h} [m] is the hydraulic aperture which can be estimated from the mean mechanical aperture and contact area ratio (Walsh 1981)

$$b_{\mathrm{h}}^3 = \frac{1 - R_{\mathrm{c}}}{1 + R_{\mathrm{c}}} b_{\mathrm{m}}^3.$$

(11.1.8)

Flow velocity v [m/s] can be calculated once the pore pressure is obtained

$$v = -\frac{b_h^2}{12\eta}\,\nabla p_w.$$

(11.1.9)

11.1.2 Example

With the developed 1-D reactive transport model, we aim to reproduce part of the flow-through experiment reported in Yasuhara et al. (2011). The flow-through experiment is conducted on a centrally bisected granite core sample (see Fig. 11.1). The fractured granite core sample, consisting of 50% quartz, 25% k-feldspar, 10% albite, 10% anorthite, and 5% biotite (see Table 11.1), is subjected to confining stress of 5 MPa, differential hydraulic pressure of 0.04 MPa, and temperature of 25 °C at the stage of concern. A stream of deionized water (pH = 7) is injected into the artificially induced fracture with a length of 61.2 mm and a width of 29.4 mm. The outflow rate and effluent element concentrations are regularly measured in the experiment.

The chemical reactions, involved in the water-granite interaction, comprise the mineral surface reactions (see Table 11.2) and a series of aqueous reactions. It is noteworthy that amorphous silica and gibbsite act as precipitates in the calibrated geochemical system. The dissolution kinetics of the surface reactions are listed in Table 11.3. The equilibrium constants of all the reactions source from the LLNL thermodynamic database (Parkhurst et al. 1999).

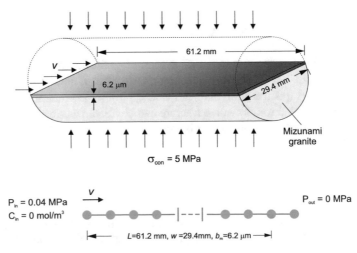

Fig. 11.1 Schematic diagram of the fractured Mizunami granite core sample subjected to the prescribed experimental conditions. A 1-D reactive transport model, taking account of the chemical weathering both at open pores and at grain contacts, is set up accordingly

Table 11.1 Mineralogical composition of Mizunami granite and mineral properties

Mineral	ϕ [%]	[a]ϕ^p [%]	[a]ϕ^c [%]	ρ_m [g/cm^3]	V_m [m^3/mol]	[b]f_r [-]
Quartz	50	50	50	2.65	2.27×10^{-5}	40.10
K-feldspar	25	25	25	2.55	1.09×10^{-4}	38.58
Albite	10	10	10	2.62	1.00×10^{-4}	39.64
Anorthite	10	10	10	2.75	1.01×10^{-4}	41.61
Biotite	5	5	5	2.10	1.40×10^{-4}	31.77

[a]Area fractions of the minerals at the non-contact and contact zones are assumed to follow the overall distribution on the fracture surface.
[b]Mineral-specific intragranular roughness factors are calculated by Eq. (11.1.6), with the measured grain diameter d of 178 μm and the BET surface area S_{BET} of 0.51 m^2/g (Yasuhara et al. 2011).

Table 11.2 Mineral surface reactions and precipitation reactions involved in the calibrated geochemical system, and logarithmic mineral dissolution rates log k at pH $= 7$ and T $= 25\,°C$

Mineral	Chemical reaction	log k
Quartz	$SiO_2 + 2\,H_2O \rightarrow H_4SiO_4$	-13.99
K-feldspar	$KAlSi_3O_8 + 4\,H^+ + 4\,H_2O \rightarrow Al^{3+} + K^+ + 3\,H_4SiO_4$	-12.38
Albite	$NaAlSi_3O_8 + 4\,H^+ + 4\,H_2O \rightarrow Al^{3+} + Na^+ + 3\,H_4SiO_4$	-11.96
Anorthite	$CaAl_2(SiO_4)_2 + 8\,H^+ \rightarrow 2\,Al^{3+} + Ca^{2+} + 2\,H_4SiO_4$	-11.05
Biotite[a]	[5 Phlogopite : 1 Annite] $+ 10\,H^+ \rightarrow$ $Al^{3+} + K^+ + 2.5\,Mg^{2+} + 0.5\,Fe^{2+} + 3\,H_4SiO_4$	-12.51
Amorphous silica	$SiO_2 + 2\,H_2O \leftrightarrow H_4SiO_4(am)$	
Gibbsite	$Al^{3+} + 3\,H_2O \leftrightarrow Al(OH)_3 + 3\,H^+$	

[a]The chemical formulas of phlogopite and annite are $KAlMg_3Si_3O_{10}(OH)_2$ and $KFe_3AlSi_3O_{10}(OH)_2$.

Table 11.3 Dissolution kinetics of the rock-forming minerals

Mineral	Acid			Neutral		Base		
	log k_1	n_1	E_1	log k_2	E_2	log k_3	n_3	E_3
Quartz	–	–	–	-13.99	87.7	–	–	–
K-feldspar	-10.06	0.500	51.7	-12.41	38.0	-21.20	-0.823	94.1
Albite	-9.87	0.457	65.0	-12.04	69.8	-16.98	-0.572	71.0
Anorthite[a]	-3.32	1.5	18.4	-11.6	18.4	-13.5	-0.33	18.4
Biotite	-9.84	0.525	22.0	-12.55	22.0	–	–	–

[a]Source: anorthite - from Li et al. (2006); the others - from Palandri and Kharaka (2004).

Furthermore, pressure solution is presumed to act upon grain contacts apart from the anorthite ones under experimental conditions. Model parameters are summarized in Table 11.4.

Figure 11.2 compares the predictions with the measured effluent element concentrations. Apart from the overestimated Si concentration, the other simulated element concentrations are in good agreement with the measurements.

Table 11.4 Model parameters used in the mass transport simulation

Parameter	Value	Unit
Confining stress σ_{con}	5	MPa
Contact area ratio R_{c0}	3.16%	–
Mean mechanical aperture b_{m0}	6.2	μm
Mechanical aperture at open pores b_{m0}^{p}	6.402	μm
Fracture surface roughness factor f_r'	1.0	–
[a]Intragranular roughness factor f_r	$31.77 - 41.61$	–
Water dynamic viscosity η	9×10^{-4}	Pa·s
Molecular diffusion coefficient D_m	2.24×10^{-10}	m^2/s
Longitudinal dispersion coefficient α_L	0.001	m
Element length Δx	1.02	mm
Time step size Δt	0.004	s

[a]Specific values are listed in Table 11.1.

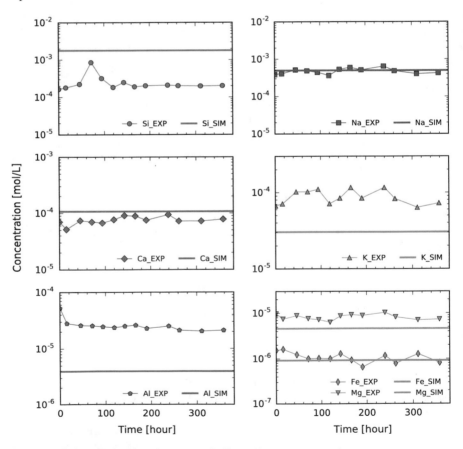

Fig. 11.2 Comparison against the measured effluent element concentrations

Chapter 12
THC-Processes

**Thomas Nagel, Peter Ostermeier, Gabriele Seitz, Holger Class
and Rainer Helmig**

12.1 A Benchmark Case for the Simulation of Thermochemical Heat Storage

12.1.1 Introduction

The operation of thermochemical heat storage devices in open mode by permeating a reactive porous body or particle bed by a compressible heat-transfer fluid which in turn transports a reactive fluid component constitutes a strongly coupled problem of multiple physical processes. High reaction rates associated with a rapid release of significant amounts of heat and a complicated dependence of processes and material properties on the physical and chemical state of the system add to the level of complexity. The development of suitable computational models capturing the salient features of such a problem is a challenging task (Nagel et al. 2016a). Verification of the implementation is usually only possible in parts by the use of analytical solutions under simplified conditions (Nagel et al. 2013). To build confidence in the numerical implementation under more complex conditions it is thus beneficial to perform a comparison of simulation results obtained by the use of different software packages/codes.

T. Nagel (✉)
UFZ, Helmholtz Centre for Environmental Research, Leipzig, Germany
e-mail: Thomas.Nagel@ufz.de

P. Ostermeier
Institute for Energy Systems, Technical University of Munich, Munich, Germany

G. Seitz · H. Class · R. Helmig
Deptartment of Hydromechanics and Modelling of Hydrosystems, University of Stuttgart,
Stuttgart, Germany

T. Nagel
Trinity College Dublin, Dublin, Ireland

© Springer International Publishing AG 2018
O. Kolditz et al. (eds.), *Thermo-Hydro-Mechanical-Chemical Processes
in Fractured Porous Media: Modelling and Benchmarking*,
Terrestrial Environmental Sciences, https://doi.org/10.1007/978-3-319-68225-9_12

It is the purpose of this chapter to (i) define a suitable benchmark and (ii) perform such a comparison by using the simulation platforms OpenGeoSys (Kolditz et al. 2012a), DuMu^x (Flemisch et al. 2011) and ANSYS Fluent (Ansys Inc., Canonsburg, PA, USA. 2013).

This chapter begins by briefly summarizing the specifics of the respective numerical implementations, continues with the definition of a common benchmark and closes with a comparison of simulation results.

12.1.2 Implementational Details

12.1.2.1   OpenGeoSys

In OpenGeoSys, the following set of equations was used for the simulation of the benchmark.

Gas mass balance (note: $\varrho_\alpha = \phi_\alpha \varrho_{\alpha R}$ for any phase α):

$$\frac{\partial \varrho_G}{\partial t} + \mathrm{div}\,(\varrho_G \mathbf{v}_G) = -\hat{\varrho}_S \tag{12.1.1}$$

Mass balance for the reactive component in the gas phase:

$$\varrho_G \frac{\partial x_{m\zeta}}{\partial t} - \mathrm{div}\,(\varrho_{GR} \mathbf{D}_\zeta\,\mathrm{grad}\,x_{m\zeta}) + \varrho_G\,\mathrm{grad}\,x_{m\zeta} \cdot \mathbf{v}_G - \hat{\varrho}_S x_{m\zeta} = -\hat{\varrho}_S \tag{12.1.2}$$

Energy balance for the mixture:

$$\left(\varrho_G c_{pG} + \varrho_S c_{pS}\right) \frac{\partial T}{\partial t} - \mathrm{div}\,(\boldsymbol{\lambda}_{\mathrm{eff}}\,\mathrm{grad}\,T) + \varrho_G c_{pG}\,\mathrm{grad}\,T \cdot \mathbf{v}_G - \frac{\partial(\phi_G p)}{\partial t} =$$
$$= \hat{\varrho}_S \Delta h \tag{12.1.3}$$

This is a simplified version of the model presented in references Nagel et al. (2014, 2013), Shao et al. (2013) where local thermal non-equilibrium has been taken into account. The deviations from local thermal equilibrium in the system considered here are sufficiently small to justify the use of only one energy balance in order to limit computing times. This assumption will be tested implicitly by the comparison to the ANSYS Fluent results which are based on local thermal non-equilibrium, see Sect. 12.1.2.3.

In the above, Darcy's law was used to describe the flow of the gas phase:

$$\phi_G \mathbf{v}_G = -\frac{\mathbf{k}}{\mu_{GR}}(\,\mathrm{grad}\,p - \varrho_{GR}\mathbf{b}) \tag{12.1.4}$$

Furthermore, ϱ_{GR} of the gas mixture followed from Dalton's law for ideal gases; λ_{eff} is the volumetric average of gas and solid contributions; to account for a variable gas composition, mixing laws were used for gas viscosity μ_{GR} (Model: Wilke in Poling et al. (2001)), heat capacity c_{pG} and heat conductivity λ_{GR} (Model: Wassilijewa, Maso and Saxena in Poling et al. (2001)). The specific heat capacity of the individual gases followed Eq. (12.1.13) with parameters from Tables 12.1 and 12.2, while heat conductivity and viscosity of nitrogen and water vapour were calculated following reference Stephan et al. (1987) and the IAPWS formulation from Wagner et al. (2000), respectively.

To arrive at a finite element formulation, the Bubnov–Galerkin weighted residual method (Bathe 2001; Zienkiewicz et al. 2006) is used to derive the weak forms of the governing equation of Eqs. (12.1.1)–(12.1.3). The weak forms are then discretized using an isoparametric finite element approach. The spatially discretized equations are discretized in time by a fully implicit backward Euler scheme and monolithically assembled into one equation system. Picard iterations were used to resolve the nonlinearities in the system. Details regarding the implementation can be found in Nagel et al. (2013).

Note, that the density production term

$$\hat{\rho}_S = \phi_S \hat{\rho}_{SR} \qquad (12.1.5)$$

determining the rate of the chemical reaction is a strictly local quantity. The corresponding ordinary differential equation (ODE) for $\hat{\rho}_{SR}$ was solved in each global iteration in each integration point of the finite element mesh by a method suitable for stiff ODEs (details of the approach used here can be found in Shao et al. (2013), Nagel et al. (2014)).

12.1.2.2 DuMu^x

The numerical model in DuMu^x consists of the following balance equations:

Each component of the gas phase, i.e. water and nitrogen, is balanced by the general mass balance equation, where Darcy's Law (12.1.4) is included to account for the porous medium flow. Here, ϱ_{GR} denotes the overall gas phase density determined according to Dalton's law.

$$\frac{\partial(\varrho_{GR} X^\kappa \phi_G)}{\partial t} - \operatorname{div}\left\{\varrho_{GR} X^\kappa \frac{\mathbf{k}}{\mu_{GR}} (\operatorname{grad} p - \varrho_{GR}\mathbf{g}) + \mathbf{D}_{pm}\varrho_{GR}\operatorname{grad} X^\kappa\right\} -$$
$$- q^\kappa = 0 \quad \text{with} \quad \kappa \in \{w, N_2\}$$

$$(12.1.6)$$

The components' mass or mole fractions X^κ, respectively, sum up to 1. The balance equations of the solid components are formulated in terms of their volume fractions Φ_λ considering the two solid species CaO and Ca(OH)$_2$:

$$\frac{\partial(\varrho_{R\lambda}\phi_\lambda)}{\partial t} = q_\lambda \qquad (12.1.7)$$

Thus, it is possible to account for the porosity change due to the volume change of the solid phase during the chemical reaction. However, in order to allow the comparison of the different models, the porosity ϕ_G in Eqs. (12.1.6), (12.1.8) is set as a constant. Under the assumption of local thermal equilibrium only one energy balance equation is needed:

$$\frac{\partial \phi_G \varrho_{GR} u_G}{\partial t} + \sum_\lambda \frac{\partial \phi_\lambda \varrho_{R\lambda} c_{p\lambda} T}{\partial t} - \operatorname{div}\left\{\varrho_{GR} h_G \frac{\mathbf{k}}{\mu_\alpha}(\operatorname{grad} p - \varrho_{GR}\mathbf{g})\right\} -$$
$$- \operatorname{div}(\lambda_{\text{eff}} \operatorname{grad} T) - q^h = 0. \qquad (12.1.8)$$

The reaction rate $\hat{\varrho}_{SR}$ is incorporated in the source terms as follows:[1]

$$q^w = -\hat{\varrho}_{SR} M_{CaO}^{-1} \qquad (12.1.9)$$
$$q_{CaO} = -\hat{\varrho}_{SR} M_{CaO}^{-1} \qquad (12.1.10)$$
$$q_{Ca(OH)_2} = \hat{\varrho}_{SR} M_{CaO}^{-1} \qquad (12.1.11)$$
$$q^h = \hat{\varrho}_{SR} M_{CaO}^{-1} \Delta h \qquad (12.1.12)$$

This set of equations is solved for the pressure p, the mole fraction of vapor X_{H_2O}, the two solid volume fractions ϕ_λ and the temperature as primary variables. The remaining variables are determined subsequently by the additional relations.

For the gas phase properties of the water component, the IAPWS formulation (IAPWS 1997) was used. The heat conductivity of nitrogen λ_{N_2} was calculated using Eq. (12.1.13) with parameters from Table 12.2. The compositional heat capacity and conductivity of the gas phase is weighted by the components' mole fractions to the overall term. For the compositional gas viscosity μ_{GR}, the mixing law of Wilke was used Poling et al. (2001). As in 12.1.2.1, λ_{eff} is the volumetric average of gas and solid contributions.

The equations are discretized in space using a fully coupled vertex-centered finite-volume-scheme, the so-called box-scheme (Helmig 1997) and by the implicit Euler method in time. The resulting system of equations is linearized by the Newton-Raphson method and solved using the linear solver Umfpack (Davis 2004).

---

[1] The source term formulations have been modified in an updated model formulation which will be presented in a follow-up publication of this chapter within this series.

Table 12.1 Polynomial coefficients for water vapour property calculations, see Eq. (12.1.13). Units of the coefficients are such that resulting units of the quantities are J/(kg K) for c_{pG}, W/(m K) for λ_{GR} and Pa s for μ_{GR}. Values from ANSYS Fluent database

	A_0	A_1	A_2	A_3	A_4
c_{pG}	$1.5630770 \cdot 10^3$	1.6037550	$-2.9327840 \cdot 10^{-3}$	$3.2161010 \cdot 10^{-6}$	$-1.1568270 \cdot 10^{-9}$
λ_{GR}	$-7.9679960 \cdot 10^{-3}$	$6.8813320 \cdot 10^{-5}$	$4.4904600 \cdot 10^{-8}$	$-9.0999370 \cdot 10^{-12}$	$6.1733140 \cdot 10^{-16}$
μ_{GR}	$-4.4189440 \cdot 10^{-6}$	$4.6876380 \cdot 10^{-8}$	$-5.3894310 \cdot 10^{-12}$	$3.2028560 \cdot 10^{-16}$	$4.9191790 \cdot 10^{-22}$

Table 12.2 Polynomial coefficients for nitrogen property calculations, see Eq. (12.1.13). Units of the coefficients are such that resulting units of the quantities are J/(kg K) for c_{pG}, W/(m K) for λ_{GR} and Pa s for μ_{GR}

	A_0	A_1	A_2	A_3	A_4
c_{pG}	$9.7904300 \cdot 10^2$	$4.1796390 \cdot 10^{-1}$	$-1.1762790 \cdot 10^{-3}$	$1.6743940 \cdot 10^{-6}$	$-7.2562970 \cdot 10^{-10}$
λ_{GR}	$4.7371090 \cdot 10^{-3}$	$7.2719380 \cdot 10^{-5}$	$-1.1220180 \cdot 10^{-8}$	$1.4549010 \cdot 10^{-12}$	$-7.8717260 \cdot 10^{-17}$
μ_{GR}	$7.4733060 \cdot 10^{-6}$	$4.0836890 \cdot 10^{-8}$	$-8.2446280 \cdot 10^{-12}$	$1.3056290 \cdot 10^{-15}$	$-8.1779360 \cdot 10^{-20}$

12.1.2.3 ANSYS Fluent

The same constitutive laws as mentioned above in Sect. 12.1.2.1 model are used for the calculation of the properties of the gas phase in the ANSYS Fluent models. Further details and results regarding the models not addressed below can be found in Ostermeier et al. (2016). Two different model implementations were used which are described in the following sections.

The individual gas-phase properties were calculated according to the following polynomial equations of state

$$P = \sum_{i=0}^{4} A_i \left(\frac{T}{T_0} \right)^i \quad \text{with} \quad P \in \{\lambda_{GR}, c_{pG}, \mu_{GR}\} \text{ and } T_0 = 1 \text{ K} \qquad (12.1.13)$$

and the coefficients listed in Tables 12.1 and 12.2 taken from the ANSYS Fluent database.

Two-Fluid Model (TFM)

In this Eulerian multiphase approach both gas and solid phase are treated as interpenetrating continua. The following set of equations is used in ANSYS Fluent. Equations for both models are presented in the form adapted to the benchmark. For the general form of the equations refer to the ANSYS Fluent Theory Guide (Ansys Inc., Canonsburg, PA, USA. 2013).

Saturation condition:

$$\phi_G + \phi_S = 1 \qquad (12.1.14)$$

Conservation of mass for the gas and solid phase:

$$\frac{\partial}{\partial t}(\phi_G \varrho_{GR}) + \text{div}\,(\phi_G \varrho_{GR} \mathbf{v}_G) = -\hat{\varrho}_S \qquad (12.1.15)$$

$$\frac{\partial}{\partial t}(\phi_S \varrho_{SR}) = \hat{\varrho}_S \tag{12.1.16}$$

The density production term $\hat{\rho}_S$ is determined in a User Defined Function (UDF) and incorporated into the simulation using the Heterogeneous Stiff Chemistry Solver.

Conservation of momentum for the gas phase:

$$\frac{\partial}{\partial t}(\phi_G \varrho_{GR} \mathbf{v}_G) + \operatorname{div}(\phi_G \varrho_{GR} \mathbf{v}_G \otimes \mathbf{v}_G) = \tag{12.1.17}$$
$$= -\phi_G \operatorname{grad} p + \operatorname{div} \bar{\bar{\tau}}_G + K \mathbf{v}_G - \hat{\varrho}_S \mathbf{v}_G$$

The gas-solid exchange coefficient K, governing the pressure drop in the gas phase, is calculated according to the Ergun equation $K = 150 \frac{(1-\phi_G)^2 \mu_{GR}}{\phi_G d_p^2} + 1.75 \frac{(1-\phi_G)\varrho_{GR}|\mathbf{v}_G|}{d_p}$ (Ergun 1952). The calculation of the fluid extra-stress tensor $\bar{\bar{\tau}}_G$ due to intra-fluid viscous dissipation can be found in Ansys Inc., Canonsburg, PA, USA. (2013).

Conservation of energy for the gas and solid phase:

$$\frac{\partial}{\partial t}(\phi_G \varrho_{GR} h_G) + \operatorname{div}(\phi_G \varrho_{GR} \mathbf{v}_G h_G) = \phi_G \frac{\partial p_G}{\partial t} + \bar{\bar{\tau}}_G : \operatorname{grad} \mathbf{v}_G + \tag{12.1.18}$$
$$+ \operatorname{div}(\phi_G \lambda_{GR} \operatorname{grad} T_G) + \dot{Q}_{GS} - \hat{\varrho}_S x_{m,G,H_2O} h_G$$

$$\frac{\partial}{\partial t}(\phi_S \varrho_{SR} h_S) = \phi_S \frac{\partial p_S}{\partial t} + \operatorname{div}(\phi_S \lambda_{SR} \nabla T_S) + \hat{\varrho}_S \Delta h - \tag{12.1.19}$$
$$- \dot{Q}_{GS} + \hat{\varrho}_S x_{m,G,H_2O} h_G$$

For calculating the interphase heat transfer $\dot{Q}_{GS}$, the model of Gunn (1978) is used, leading to almost identical gas and solid temperatures throughout the reactor for the small particles considered in the benchmark case ($d_{50} \approx 8 \, \mu m$).

Species transport equations for the components in the gas and solid phase:

$$\frac{\partial}{\partial t}(\phi_G \varrho_{GR} x_{m,G,H_2O}) + \operatorname{div}(\phi_G \varrho_{GR} x_{m,G,H_2O} \mathbf{v}_G) =$$
$$= -\operatorname{div}(-\phi_G \varrho_{GR} \mathbf{D}_G \operatorname{grad} x_{m,G,H_2O}) - \hat{\varrho}_S \tag{12.1.20}$$
$$\frac{\partial}{\partial t}(\phi_G \varrho_{GR} x_{m,G,N_2}) + \operatorname{div}(\phi_G \varrho_{GR} x_{m,G,N_2} \mathbf{v}_G) =$$
$$= -\operatorname{div}(-\phi_G \varrho_{GR} \mathbf{D}_G \operatorname{grad} x_{m,G,N_2}) \tag{12.1.21}$$
$$\frac{\partial}{\partial t}(\phi_S \varrho_{SR} x_{m,S,CaO}) = -\hat{\varrho}_S \tag{12.1.22}$$
$$\frac{\partial}{\partial t}(\phi_S \varrho_{SR} x_{m,S,Ca(OH)_2}) = \hat{\varrho}_S \tag{12.1.23}$$

Porous-Zone Model (PZM)

This model is a single-phase approach, considering only the pressure drop and the volume occupation caused by the particles. The conservation equations of mass, momentum and species for the gas phase are analogous to the TFM (Eqs. (12.1.15), (12.1.17), (12.1.20) and (12.1.21)). The model solves only one energy equation for the mixture, which is a valid assumption in this benchmark and is also validated by the results of the TFM ($T_G \approx T_S = T$).

Conservation of energy for the mixture:

$$\frac{\partial}{\partial t}[\phi_G \varrho_{GR} e_G + (1 - \phi_G)\varrho_{SR} e_S] + \text{div}\left[\phi_G \mathbf{v}_G(\varrho_{GR} e_G + p)\right] =$$

$$= \bar{\bar{\tau}}_G : \text{grad} \, \mathbf{v}_G + \text{div}\left[(\phi_G \lambda_{GR} + (1 - \phi_G)\lambda_{SR}) \, \text{grad} \, T\right] + \qquad (12.1.24)$$

$$+ \hat{\varrho}_S\left[\Delta h - \frac{\phi_G}{1 - \phi_G}\left(\frac{p}{\varrho_{GR}} - \frac{\mathbf{v}_G^2}{2}\right)\right]$$

Since this model uses the total specific energy $e = h - \frac{p}{\varrho} + \frac{\mathbf{v}^2}{2}$ instead of the specific enthalpy h, the source term for the reaction enthalpy on the right hand side of Eq. (12.1.24) has been adapted accordingly.

Virtual mass balance for the solid phase:

$$(1 - \phi_G)\varrho_{SR,t+\Delta t} = \varrho_{SR,t} + \hat{\varrho}_S \Delta t \qquad (12.1.25)$$

The mass balance for the solid phase is solved in a User Defined Function (UDF). It receives the local reactor conditions (pressure, temperature,...) from the simulations and furthermore calculates the reaction rate and the resulting source terms on the right hand sides of the conservation equations (Eqs. (12.1.15), (12.1.17), (12.1.20) and (12.1.24)).

Spatial and Temporal Discretization

ANSYS Fluent applies the finite volume method (FVM) to divide the domain into a computational grid. The governing equations are integrated on the individual control volumes (weak forms) and linearized and solved using a pressure-based solver. The pressure-velocity coupling is achieved by the phase coupled SIMPLE (semi-implicit method for pressure-linked equations) scheme in the TFM and PISO (pressure-implicit with splitting of operators) scheme in the PZM. (Ansys Inc., Canonsburg, PA, USA. 2013; Laurien and Oertel Jr 2013)

Spatial discretization is performed with second-order upwind schemes (except density and volume fraction handled with first-order upwind). Gradients are evaluated with the least-squares cell-based approach.

Temporal discretization is achieved by first-order fully implicit time integration.

Table 12.3 Material parameters of the laboratory scale reactor. Details see (Shao et al. 2013; Nagel et al. 2014)

Property	Symbol	Value
Porosity	ϕ_G	0.8
Effective density Ca(OH)$_2$	$\varrho_{Ca(OH)_2 R}$	$2200\,\mathrm{kg\,m^{-3}}$
Effective density CaO	$\varrho_{CaO R}$	$1665\,\mathrm{kg\,m^{-3}}$
Intrinsic permeability	k_S	$8.53\cdot10^{-12}\,\mathrm{m^2}$
Specific heat capacity Ca(OH)$_2$	$c_{p\,Ca(OH)_2}$	$1530\,\mathrm{J\,kg^{-1}\,K^{-1}}$
Specific heat capacity CaO	$c_{p\,CaO}$	$934\,\mathrm{J\,kg^{-1}\,K^{-1}}$
Solid heat conductivity	λ_{SR}	$0.4\,\mathrm{W\,m^{-1}K^{-1}}$
Diffusion coefficient	D_V	$9.65\cdot10^{-5}\,\mathrm{m^2s^{-1}}$
Reaction enthalpy	Δh_{iso}	$108.3\,\mathrm{kJ\,mol^{-1}}$

Table 12.4 Boundary conditions for the reference reactor during the discharge reaction. Axial direction z, radial direction r; outer radius of reactor R, length of reactor L

	Eq. (12.1.1)	Eq. (12.1.2)	Eq. (12.1.3)
$z = 0$:	$\varrho_G w_n = 0.25\,\mathrm{g/s}$	$T = 573.15\,\mathrm{K}$	$x_{mV} = 0.35$
$z = L$:	$p = 2\cdot10^5\,\mathrm{Pa}$	$q_n = 0$	$d_n = 0$
$r = R$:	$\varrho_G w_n = 0\,\mathrm{kg/s}$	$T = 573.15\,\mathrm{K}$	$d_n = 0$

12.1.3 Benchmark Description

A one-dimensional set-up is chosen to represent a cylindrical reactor filled with calcium oxide in which the following reaction occurs:

$$CaO + H_2O\,(g) \rightleftharpoons Ca(OH)_2 \; ; \; -\Delta H$$

This test case is similar to the analyses performed in Shao et al. (2013), Nagel et al. (2014). A discharge operation is simulated by pushing a mixture of water vapour (reactive and heat transfer fluid) and nitrogen (inert heat transfer fluid) through the reactor. Parameters and boundary conditions were as described Tables 12.3 and 12.4. The reaction kinetics $\hat{\rho}_{SR}$ were modelled with the full kinetic model described in references Nagel et al. (2014), Schaube et al. (2012). The van't Hoff equilibrium relation was given by

$$T_{eq} = \frac{A}{\ln\left(\frac{p_\varsigma}{p_0}\right) - B} \tag{12.1.26}$$

where $B = 16.508$, $A = -12845\,\mathrm{K}$ and $p_0 = 1\,\mathrm{bar}$. The reactor had a diameter of $5.5\,\mathrm{cm}$ and a length of $8\,\mathrm{cm}$. Its central axis was discretised into

- 418 linear finite elements in OpenGeoSys

- 1000 rectangular elements in DuMu^x
- 1000 linear rectangular cells in ANSYS Fluent

Time step sizes of

- 0.1 s (first 10 s), 0.2 s (next 20 s) and 0.5 s (remaining time); OpenGeoSys
- 0.5 s; DuMu^x
- 0.01 s; ANSYS Fluent

were chosen. Both temporal and spatial discretisation were confirmed by convergence analyses.

12.1.4   Results

12.1.4.1   Balance Checks

The respective models were checked for internal consistency by evaluating the mass and energy balance relations. This can be done via the gas phase (mass and energy transport at outlet vs. inlet; cf. e.g. Shao et al. (2013)) or the solid phase (integral over local reaction rate and associated heat release).

The mass of water remaining in the system (theoretical value 20.32 g) balanced via component mass flux (difference between inlet and outlet) yielded the following results:

Software Package	Mass Difference	Relative Error
OGS	20.30 g	0.12%
DuMu^x	20.61 g	1.42%
ANSYS TFM	20.75 g	2.09%
ANSYS PZM	20.30 g	0.11%

The total enthalpy released (theoretical value 122.24 kJ) balanced via the advective enthalpy flux (difference between inlet and outlet) yielded the following results:

Software Package	Enthalpy Difference	Relative Error
OGS	122.62 kJ	0.31%
DuMu^x	123.70 kJ	1.19%
ANSYS TFM	121.72 kJ	0.42%
ANSYS PZM	124.00 kJ	1.44%

With respect to these results, all models appear to be internally consistent.

12.1.4.2   Comparison of Selected Results

All simulators qualitatively capture the reactor behaviour according to different relevant variables (Figs. 12.1, 12.2, 12.3, 12.4 and 12.5). The best agreement is achieved

Fig. 12.1 Reaction rate
$q = \hat{\rho}_{SR}$ at the inlet of the
reactor

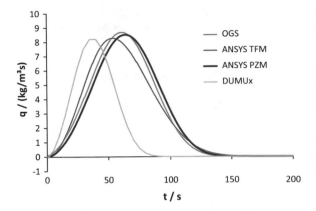

Fig. 12.2 Reaction rate
profile $q = \hat{\rho}_{SR}$ along the
reactor length after 1350 s of
simulated time

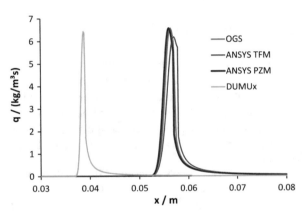

Fig. 12.3 Temperature
profile along the reactor
length after 1350 s of
simulated time

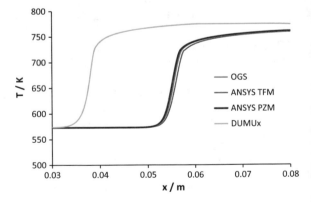

Fig. 12.4 Temperature at the outlet of the reactor corresponding to the heat output of the reactor

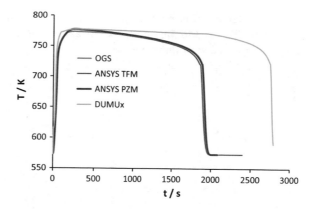

Fig. 12.5 Vapour mass fraction at the outlet of the reactor

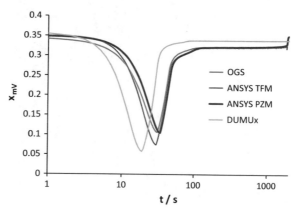

between the Porous-Zone Model in ANSYS Fluent and the implementation in Open-GeoSys. The Two-Fluid Model in ANSYS Fluent shows slight deviations, especially regarding the reaction rate (Figs. 12.1 and 12.2). There are quite significant deviations in all results obtained from the DuMu^x implementation. These differences are likely rooted in the treatment of the reaction rate and its integration into the balance equations. The reaction rate already deviates significantly at the reactor inlet (Fig. 12.1). The net effect is apparent especially when comparing the reaction rate profiles along the reactor length (Fig. 12.2) and the outlet temperature as it develops over time (Fig. 12.4). The reaction front propagates through the reactor much more slowly. As the associated heat release keeps the temperature at equilibrium conditions, this slow front propagation leads to a prolonged temperature rise at the reactor outlet. By visual comparison it becomes clear that the simulation in DuMu^x predicts a much higher heat release assuming similar gas properties (Fig. 12.4). As this is not seen in the balance checks presented before, the evaluation of the balances and the

model implementation both incorporate the same deviation from the other implementation. This is currently under investigation and the results will be published in
a follow-up version of this chapter.[2]

---

[2]A current implementation using different source-term formulations and a modification of the reaction enthalpy by the volume work suited for the formulation of the energy balance as implemented
in DuMux provides results that match the results of the other codes more closely. The specifics
require further testing, the results of which will be published as a separate chapter within this book
series.

Appendix A
OpenGeoSys-6

by Dmitri Naumov, Lars Bilke, Thomas Fischer, Yonghui Huang, Christoph Lehmann, Xing-Yuan Miao, Thomas Nagel, Francesco Parisio, Karsten Rink, Haibing Shao, Wenqing Wang, Norihiro Watanabe, Tianyuan Zheng, Olaf Kolditz

A.1 OGS-6: Development and Challenges

A.1.1 Introduction

Current development of the general multi-physics simulation platform OpenGeoSys concentrates mainly on the new version 6 of the software. OGS-6 is a major rework of the previous version OGS-5 motivated by several improvement possibilities described in detail in the previous benchmark book edition Kolditz et al. (2016b) Appendix A.

The software development itself occurs on an open shared platform, making both amenable to scrutiny by the scientific community and public authorities. Besides describing development workflows that assure code quality, we highlight how code performance and accuracy are routinely tested in an automated test environment as well in this benchmarking initiative.

In the following sections are an overview of the current implementation, a description of the testing and continuous integration environment, and finally an overview of the high-performance-computing setups and their parallel performance.

A.1.2 Overview of Processes

Current overview of the capabilities is partially covered by benchmarks in this book. and includes following process implementations:

© Springer International Publishing AG 2018
O. Kolditz et al. (eds.), *Thermo-Hydro-Mechanical-Chemical Processes in Fractured Porous Media: Modelling and Benchmarking*, Terrestrial Environmental Sciences, https://doi.org/10.1007/978-3-319-68225-9

- variations of heat and component transport equations: ground water flow (elliptic steady-state equation), heat conduction, hydro-thermal process, pressure based liquid flow process, inert component-transport, Richards flow, two-phase flow (pressure-pressure and pressure-saturation formulations), temperature dependet two-phase flow (pressure-pressure formulation), (all of the parabolic kind),
- coupled mechanics processes: small deformation process and hydro-mechanical process, both with Lower-Interface-Element extensions, thermo-mechanical and thermo-hydro-mechanical processes, and a phase-field formulation,
- thermal energy storage process

A.1.3 Implementation of Workflows

For an implementation of a new process and working with the existing implementations a developer has several flexibilities in the choice of the non-linear solver, the linear solver availabilities, and the pre- and post-processing options.

There are two implementation possibilities for the processes: monolithic and staggered schemes. Independent of this choice, to resolve the non-linearities of the equations a fixed-point iteration and Newton-Raphson method are available. Most processes are implemented using the Newton-Raphson scheme and therefore are implemented monolithically.

The staggered scheme (together with fixed-point iterations) allow for quick working implementations of coupled multi-field processes, while choosing the Newton-Raphson scheme makes it possible to solve highly non-linear equations.

For the solution of the linear equation systems currently there are three major libraries supported: Eigen (Guennebaud et al. 2010), LIS (2017), and PETSc (Balay et al. 2016). These give a big variety of linear equations solvers, both direct and iterative, for the users to experiment. The Eigen library, furthermore, provides wrappers to external solvers from the PaStiX, SuiteSparse, SuperLU, and Intel MKL libraries.

For the input and output of the mesh geometries the Visualization Toolkit library (VTK) with its rich set of manipulation and visualization tools; primarily Paraview—a graphical front-end to the VTK libraries, and the Python bindings to the VTK giving powerful scripting options.

A.1.4 Transition from OGS-5

The benchmarking of the code is continued in the new version of OGS. Comparisons with the analytical solutions is one of the requirements for the newly implemented processes, where it is possible, and comparison to the OGS-5 (or other codes like FEFLOW FEFLOW 2017 or FEBio FEBio 2017) if same implementation is available. The OGS-6 development aims to introduce new processes and solution techniques, which were not possible in the former version.

A.1.5 Heterogeneous Computing

Another challenge is the adaptation to heterogeneous computing, where different kinds of processors are utilized to achieve higher efficiencies. Examples of heterogeneous systems are, for example, computer systems using CPU and GPU cores simultaneously.

A.2 Software Engineering and Continuous Integration

by Lars Bilke

OGS-6 is developed as an open-source project by using the GitHub platform for code hosting, code review, and bug and issue tracking (OGS 2017c). A Jenkins continuous integration (CI) server (OGS 2017b) builds and tests the software on every proposed change (pull request) on supported platforms (Windows, Linux, Mac Desktop systems and Linux HPC cluster systems). See Kolditz et al. (2015) Appendix-B for a general introduction.

The QA process is automated and defined in *Jenkins Pipeline Domain Specific Language* (DSL, Jenkins 2017a) scripts which are part of the source code to allow the concurrent testing of multiple development branches. We aim to identify reusable pipeline definitions, such as CI steps typical for a C++/CMake-based project (configuration, compilation, testing, artifact deploying), which can be shared with other software projects, and integrate those CI steps as a standalone *Jenkins Pipeline Shared Library* (Jenkins 2017b).

Submitting a proposed change (in the form of a GitHub pull request) triggers a CI-job which consist of several sub-jobs with several stages each. A sub-job is created for every supported platform. We test both on bare-metal hardware (e.g. a Linux cluster system) as well on containerized environments using Docker (2017). Docker container providing Linux environments with different compiler setups are defined via the Dockerfile DSL in code as well (OGS 2017a). Sub-job stages group complex steps into logical parts such as configuring, building and testing. After a CI-job finishes the developer gets immediate feedback which sub-job or even stage failed with its corresponding log output. Web-based automated code documentation (with Doxygen) and interactive benchmark descriptions with embedded 3D visualizations (with VTK 2017) are generated. Binaries for all supported platforms are generated and provided for further manual testing if required.

With this setup large parts of the whole software engineering infrastructure and processes are formalized and defined via DSLs in version controllable code-repositories allowing for easy contributing and peer-review on the code, infrastructure and process levels. For citing GitHub-based source code, digital object identifiers (DOIs) can be generated with the Zenodo platform.

A.3 High-Performance-Computing

by Thomas Fischer, Wenqing Wang, Christoph Lehmann, Dmitri Naumov

A.3.1 Why is High-Performance-Computing Necessary?

- For the solution of non-stationary models many time steps are necessary.
- For coupled approaches in cross-compartment topics many iterations between for the particular compartments specialized software may be necessary.
- Large scale domains or complex geological structures lead to meshes containing many cells. Also, in order to improve the accuracy of simulation results or to resolve small scale effects, a finer discretization of the domain is necessary. Doubling the model resolution for d-dimensional domains in each space direction results in 2^d times more mesh elements.
- Multi-physics with complex equations of state and the modeling of chemical reactions requires high numerical effort for both the assembly and for solving the system of linear equations.

Consequently, a high demand of computational resources (CPU time, large memory requirements and fast input/output operations) for the numerical analysis of large scale or complex problems arises.

A.3.2 Parallelization Approaches

Nowadays the computational power of a single core CPU hits its physical limits. Most of the growth of computational power is attributed to parallelization. As a consequence the usual way to tackle the huge computational effort for solving high resolution models is parallel computing. In principle there are two programming approaches for parallel computing, the shared memory approach and the distributed memory approach.

OpenMP (Open Multi-Processing) is a technique that makes use of the shared memory approach and accelerates the program mostly by parallelization of loops. This technique is relatively easy to use. OpenMP benefits from the multi-core architectures. Typically such architectures have a limited number of cores.

One representative of the distributed memory parallelization technique is MPI (Message-Passing Interface). Programs using MPI can utilize many compute nodes in one computation, where each compute node typically has multiple CPU cores. Efficient implementation of algorithms for distributed memory architectures (MPI) is more challenging than the implementations for shared memory (OpenMP) systems.

So far, OGS-6 is parallelized implementing the distributed memory approach, i.e., it uses MPI.

A.3.3 Results

A.3.3.1 Description of the Benchmark Example

As a test, a groundwater flow equation, an elliptic PDE, in a homogeneous medium is solved

$$\nabla \cdot (\kappa \nabla h) = 0, \quad \text{in } \Omega = [0, 1]^3,$$

with Dirichlet-type boundary conditions

$$h = 1, \quad \text{for } x = 0, \quad \text{and} \quad h = -1, \quad \text{for } x = 1.$$

The conductivity κ is set to 1. The analytical solution is $h(x, y, z) = 1 - 2x$.

For the numerical solution of the elliptic PDE the domain was discretized in different resolutions from 10^6 to 10^8 hexahedra, see Tables A.1 and A.2.

A.3.3.2 Run Times and Speedup for IO, Assembly and Linear Solver

All run time tests were conducted on the EVE cluster at the Helmholtz Centre for Environmental Research GmbH – UFZ. EVE is a Linux-based cluster with 65 compute nodes where each compute node has 20 compute cores. It has 11.2 TB RAM in total. The compute nodes are connected via a 40 Gbit Infiniband network. The computations are performed on the compute nodes that have 120 GB of RAM.

Since such a cluster is often used in parallel by multiple users, the measurement environment is not always in the same state. In order to avoid that other simulations on the cluster influence the computations, the test problem was partitioned so that always complete compute nodes are occupied. However, the total network traffic and the total IO to external storage on the cluster can not be controlled. For this reason, each run time test is repeated 10 times to exclude or at least minimize external environment influences.

Each compute core participating on the computations, outputs the time measurements for the particular tasks, i.e., mesh reading, assembling, linear solving, writing results. The maximum time for each particular task is chosen to average over the 10 simulations.

Strong scaling: Run times and speedup for IO, assembly, linear solver

The scaling is normally defined as $s(N) = \frac{t(1)}{t(N)}$, where $t(1)$ is the run time using one compute core and $t(N)$ is the run time using N cores. Because the problem is too large to fit in the memory associated with one compute core in Fig. A.1 the slightly changed definition $s_{20}(N) = \frac{t(20)}{t(N)}$ is depicted. The figure is based on the data from the Table A.1. As expected, when more compute cores are deployed the run times decrease for the assembly and for the linear solver.

Run times for reading the mesh drops down until the employment of 60 cores and stay around 2 s for higher number of cores. This could be an effect of the GPFS file

Table A.1 Fixed discretization $236 \times 236 \times 236 (= 13\ 144\ 256)$ hexahedra, all run times in seconds

#cores	Reading mesh	Output time steps 0 and 1	Assembly	Linear solver
20	6.32	3.99 + 4.62	3.81	56.94
40	4.74	4.01 + 5.60	1.93	29.19
60	1.78	7.12 + 7.31	1.33	18.01
80	2.15	11.77 + 12.46	1.03	14.42
100	1.72	14.39 + 18.71	0.83	11.59
120	2.31	23.98 + 31.50	0.69	9.73
140	2.80	13.85 + 19.47	0.61	8.41

Fig. A.1 Scaling for different tasks of the simulation

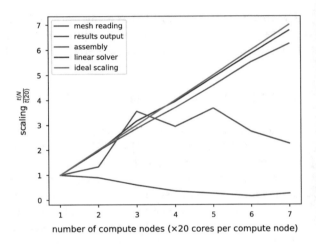

system of the cluster that has to distribute smaller and smaller parts of the binary input files to more and more compute nodes. The overhead connected to this could be one reason why the reading of the mesh does not scale as expected.

While the input consists of a small number of binary files, the output of the simulation results is done in a different way at the moment. Each process opens its own file and writes its sub-domain specific data. For a large number of processes this results in opening a large number of files which is typically a time consuming operation on GPFS file systems. Figure A.1 indicates some potential for improvement, especially for the output of simulation results.

Weak scaling: Run times for IO, assembly, linear solver

To investigate the weak scaling of OGS-6 the discretization and the number of compute cores for the simulation is chosen such that the workload on a core is almost the same, as illustrated in the third column of Table A.2. The run time measurements are done with two different versions of OGS-6, one from February 2017 (last three rows in the table) and one from September 2017 (first three rows in the table).

Table A.2 All run times in seconds

Discretization #hexahedra	#cores	DOFs per core	Reading mesh	Output time steps 0 and 1	Assembly	Linear solver
184 × 184 × 184	20	311 475	3.55	2.51+2.53	1.86	21.59
232 × 232 × 232	40	312 179	4.14	4.89+5.91	1.90	27.34
292 × 292 × 292	80	311 214	4.80	10.71 + 12.43	1.89	33.58
292 × 292 × 292	80	311 214	3.84	3.98+4.42	2.39	37.39
368 × 368 × 368	160	311 475	5.01	4.74+5.48	2.39	46.21
465 × 465 × 465	320	314 202	6.73	5.52+6.41	2.38	69.19

In a preprocessing step the mesh is partitioned and stored in a small number of binary files. The time for reading of the mesh for the simulation slightly decreases. This could be caused by distribution of the data via the GPFS file system. There is a significant variation in the run times of the output which needs further investigation.

The run times for the assembly, being almost the same for different number of process, demonstrate that this part scales weakly very well. Furthermore, the jump in the assembly times between the two version (02/2017 vs. 09/2017) is caused by switching between dynamically allocated local matrices (02/2017) and statically allocated local matrices (09/2017).

The run times for solving the systems of linear equations shows that it becomes more difficult to solve the system the bigger it is, i.e., the number of iterations increases with the number of degrees of freedom.

A.4 Data Integration and Visualisation

by Karsten Rink, Carolin Helbig, Lars Bilke

In order to cope with increasing amounts of heterogeneous data from various sources, workflows combining data integration and data analytics methods become more and more important. Therefore, data integration and visual data analysis are important elements within the OpenGeoSys workflow concept. Figure A.2 shows a generic workflow for scientific visualisation of environmental data including categories and examples for functionality.

Some general challenges for data integration and visual data analysis for the field of environmental science are summarized in the following list and described in detail in Helbig et al. (2015):

- Multifaceted data: High dimensional parameter spaces
- Multimodal data: Correlation of observation and simulation
- Incorporation of multiple scales: Catchment scale vs sewage network

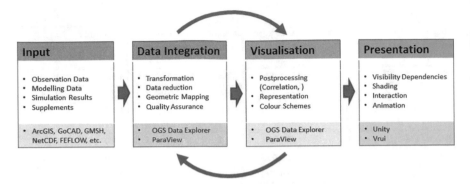

Fig. A.2 Elements of data integration and visual data analysis (Helbig et al. 2017)

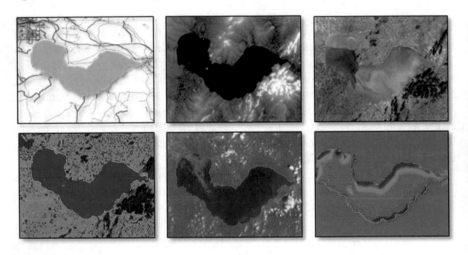

Fig. A.3 Example for data integration tasks: Environmental Information System Chaohu

- Multirun data: Ensembles or multiple scenarios
- Supplemental information: Adding Images/Videos/Diagrams/Papers/etc.

  Typical tasks for data integration include, e.g.:

- Visualisation of input- and modelling data as well as simulation results within unified context,
- Data exploration to understand complex data, find correlations and dependencies,
- Detection of potential problems, inconsistencies or missing data.

  Figure A.3 shows an example for heterogeneous data sets to be integrated for typical applications in environmental sciences. This includes urban infrastructures, digital elevation models (DEM), surface water quality, land use pattern and modeling results. Data integration is an essential prerequisite for building comprehensive Environmental Information Systems (EIS) (Rink et al. 2016).

Fig. A.4 Environmental Information System - Chaohu combining urban infrastructures (left top), integrating model results (left bottom), land use information (right)

Figure A.4 depicts elements of the Environmental Information System – Chaohu combining observation data and modeling results within a unified geographical context. Municipal waste water and fertilizers from agricultural production are the main sources for the pollution of Lake Chaohu. This information is the basis for building predictive models concerning the water quality of Lake Chaohu and assessing the effect of remediation concepts.

Data integration and analysis requires appropriate tools for specific disciplines. The OpenGeoSys (OGS) DataExplorer is particularly suited for environmental applications (Fig. A.5). Typical requirements for data analysis are:

- Data Conversion
- Data Representation
- Mesh Generation
- Error Detection
- Transformations
- Detecting and resolving inconsistencies
- Mapping
- Scalar Information
- Visual Metaphors
- Export to Graphics Frameworks

The OGS DataExplorer provides numerous interfaces for data import and export functions mainly for visualization. Several interface to complementary modeling softwares (preferable open source products) are available (Fig. A.6).

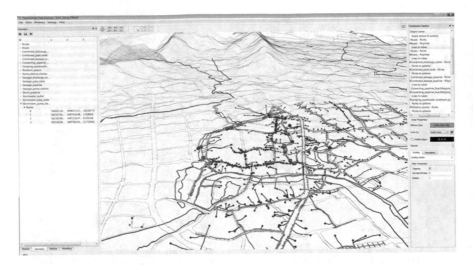

Fig. A.5 OpenGeoSys (OGS) DataExplorer

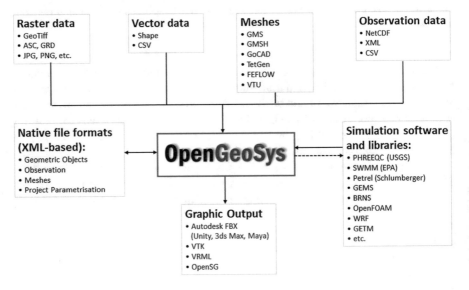

Fig. A.6 OGS Interfaces

Complex data collections for environmental applications benefit from embedding into Virtual Geographic Environments (VGE) (Fig. A.7) The Unity3D environment is used for interactive visualization applications providing a large variety of functionality, e.g.:

- Transfer functions
- Graphics shaders for visual appearance of data objects

Fig. A.7 Unity3D environment for building Virtual Geographic Environments (VGE)

- Semantic grouping and visibility management
- Viewpoints showing aspects of interest
- Animations of time-dependent data
- View path animations
- Performance optimisation
- Tracking user interaction
- Adding supplemental information (images, videos)

Data integration and visual data analytics provide numerous added values for environmental research, such as:

Data integration within the geographical context:

- Integrated data frame for observation and simulation data
- Visualisation complex and heterogeneous data bases
- Error and inconsistency detection

Data exploration and knowledge transfer:

- Exploration and understanding of complex data sets
- Visualisation of multifaceted/multimodal/multirun data
- Supports discussions between scientists and communication with stakeholders or for public participation

Recent research and progress in Environmental Visualization are presented and discussed in a related workshop series "Visualization in Environmental Sciences (EnvirVis)" which is a regular part of the annual European visualization conference EUROVIS.

Appendix B
GINA_OGS: A Tutorial of a Geotechnical Application

by Herbert Kunz (BGR)

The program GINA_OGS is developed as a pre- and postprocessing tool for the FE program OpenGeoSys (Fig. B.1). Extensive functions for the definition of geometry, initial/boundary conditions, physical/chemical parameter, and numerical parameters have been implemented in the GINA_OGS (Kolditz et al. 2012c, 2015, 2016b). As part of mesh generation, features for creating structured quad- and hexahedron elements are implemented. For unstructured triangles and tetrahedron elements interfaces to the freeware programs GMSH and TetGen are available. According to the definition of task, different mesh types can be freely selected.

Fig. B.1 Graphical user interface (GUI) of GINA_OGS

© Springer International Publishing AG 2018
O. Kolditz et al. (eds.), *Thermo-Hydro-Mechanical-Chemical Processes in Fractured Porous Media: Modelling and Benchmarking,* Terrestrial Environmental Sciences, https://doi.org/10.1007/978-3-319-68225-9

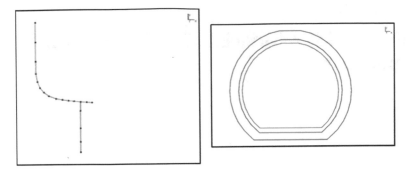

Fig. B.2 Path as line elements (left) and profiles (right)

As a tutorial of a geotechnical application, we introduce a mesh generation for an underground opening for a so-called mine-by experiment. For this purpose, four steps are needed and described in detail as follows.

1. Gallery route: GINA_OGS has implemented a function, which creates galleries along a given path. A path from line elements is required for the path (Fig. B.2(left)). Points along the gallery route can be set as geometry points and connected with polyline objects. Afterwards the geometry objects will be converted to (finite element) line elements. Different colors can be predefined for different feature, e.g. niche, existing gallery, or gallery be excavated.
2. Gallery profile: For a detailed investigation of the near-field of the gallery, a 3D model is needed. The real geometry of a gallery or niche must be therefore defined. In this application, the cross-section of the gallery is a horse shoe shape. We defined the gallery profile also by using line elements using the same method as for the gallery route. Then GINA_OGS extruded the profile along the given path. If side roads or cross roads in the model, the intersections must be calculated. GINA_OGS calculated the intersections automatically. Different material properties can be considered around the gallery. In this example, three Profiles were created: the real gallery surface (black), the boundary of the concrete supporting system (blue) and the boundary of the excavation disturbed zone (red) (Fig. B.2(right)). Figure B.3 shows the gallery with a side route.
3. Geological features: Around the gallery, a block was created. This block are divided into four zones to characterize different geological features. All block elements are defined as geometrical surface objects. The gallery structure, described before, was also converted to surface objects. This geometry can then be exported to a GMSH file. A 2D triangle mesh can be generated on the entire surface of the block and the gallery with the help of the program GMSH.

Fig. B.3 Gallery with a side route

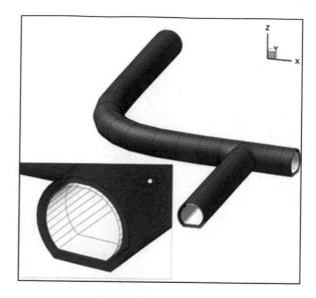

Fig. B.4 Model with 207000 tetrahedron elements

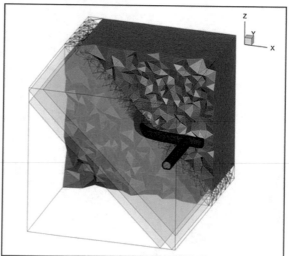

4. <u>Model block:</u> The volumes of the model will then be filled with tetrahedral elements. The program TetGen is very useful for this procedure. No volumes must be defined. Tetgen finds the volumes automatically and fills this with tetrahedral elements (Fig. B.4).

   Contact: e-mail: Herbert.Kunz@bgr.de

Appendix C
OGS: Input Files

The input files are available under:

- https://docs.opengeosys.org/books/bmb-5

as an archive - file name indicates the corresponding book section. Individual examples are available under:

- https://oc.ufz.de/index.php/s/nlph7bhfDkj6tC7 (pwd: BMB5)

    OpenGeoSys codes (excecutables and sources) are available under:

- General: http://www.opengeosys.org/resources/downloads
- OGS-5: https://github.com/ufz/ogs5
- OGS-6: https://github.com/ufz/ogs

© Springer International Publishing AG 2018
O. Kolditz et al. (eds.), *Thermo-Hydro-Mechanical-Chemical Processes in Fractured Porous Media: Modelling and Benchmarking*,
Terrestrial Environmental Sciences, https://doi.org/10.1007/978-3-319-68225-9

BM code	Author	Code	Files	CTest
H Processes				
BMB5-2.1	Peter Vogel	OGS-5	Available	TBD
BMB5-2.2	Peter Vogel	OGS-5	Available	TBD
BMB5-2.3	Peter Vogel	OGS-5	Available	TBD
BMB5-2.4	Peter Vogel	OGS-5	Available	TBD
BMB5-2.5	Peter Vogel	OGS-5	Available	TBD
BMB5-2.6	Tao Chen	OGS-5.7.0 (Mac)	Available	TBD
M Processes				
BMB5-3.1	Peter Vogel	OGS-5	Available	TBD
BMB5-3.2	Peter Vogel	OGS-5	Available	TBD
BMB5-3.3	Peter Vogel	OGS-5	Available	TBD
BMB5-3.4	Peter Vogel	OGS-5	Available	TBD
BMB5-3.5	Peter Vogel	OGS-5	Available	TBD
BMB5-3.6	Peter Vogel	OGS-5	Available	TBD
BMB5-3.7	Peter Vogel	OGS-5	Available	TBD
BMB5-3.8	Peter Vogel	OGS-5	Available	TBD
BMB5-3.9	Peter Vogel	OGS-5	Available	TBD
BMB5-3.10	Peter Vogel	OGS-5	Available	TBD
BMB5-3.11	Peter Vogel	OGS-5	Available	TBD
BMB5-3.12	Peter Vogel	OGS-5	Available	TBD
BMB5-3.13	Peter Vogel	OGS-5	Available	TBD
BMB5-3.14	Peter Vogel	OGS-5	Available	TBD
BMB5-3.15	Peter Vogel	OGS-5	Available	TBD
BMB5-3.16	Peter Vogel	OGS-5	Available	TBD
BMB5-3.17	Xing-Yuan Miao	OGS-6	Available	Below
https://github.com/ufz/ogs-data/tree/master/Mechanics/Burgers				
BMB5-3.18	Xing-Yuan Miao	OGS-6	Available	Below
https://github.com/ufz/ogs-data/tree/master/Mechanics/Ehlers				
BMB5-3.19	Francesco Parisio	OGS-6	Available	Below
https://github.com/ufz/ogs-data/tree/master/Mechanics/Linear				
T Processes				
BMB5-4.1	Vinay Kumar	OGS-5.7.1	Available	TBD
https://github.com/VinayGK/ogs5/tree/thermal-cond-averaging				
HH Processes				
BMB5-5.1	Aaron Peche	OGS-5	Available	NA
BMB5-5.2	Jobst Maßmann	OGS-5	Available	NA
BMB5-5.3	Miao Jing	OGS-5	Available	NA
H^2 Processes				
BMB5-6.1	Yonghui Huang	OGS-6	Available	PR
https://github.com/ufz/ogs/blob/master/ProcessLib/RichardsFlow				
https://github.com/ufz/ogs-data/tree/master/Parabolic/Richards				
BMB5-6.2	Yonghui Huang	OGS-6	Available	PR
https://github.com/ufz/ogs/tree/master/ProcessLib/TwoPhaseFlowWithPP				
https://github.com/ufz/ogs-data/tree/master/Parabolic/TwoPhaseFlowPP/Liakopoulos				
BMB5-6.3	Yonghui Huang	OGS-6	Available	PR
https://github.com/ufz/ogs/tree/master/ProcessLib/TwoPhaseFlowWithPP				
https://github.com/ufz/ogs-data/tree/master/Parabolic/TwoPhaseFlowPP/McWorts				

HT Processes				
BMB5-7.1	Fabien Magri	OGS-5	Available	TBD
	Fabien Magri	FEFLOW		NA
	Mauro Cacace	MOOSE		NA
BMB5-7.2	Tianyuan Zheng	OGS-6	Available	Below
https://github.com/ufz/ogs-data/tree/master/Parabolic/HT/ConstViscosity				
BMB5-7.3	Tianyuan Zheng	OGS-6	Available	Below
https://github.com/ufz/ogs-data/tree/TH-monolithic/Parabolic/TH				
HM Processes				
BMB5-8.1	Gesa Ziefle	OGS-6	Available	Below
https://github.com/endJunction/ogs/tree/LIE3D_0				
TM Processes				
BMB5-9.1	Peter Vogel	OGS-5	Available	TBD
BMB5-9.2	Peter Vogel	OGS-5	Available	TBD
BMB5-9.3	Peter Vogel	OGS-5	Available	TBD
BMB5-9.4	Peter Vogel	OGS-5	Available	TBD
BMB5-9.5	Peter Vogel	OGS-5	Available	TBD
BMB5-9.6	Peter Vogel	OGS-5	Available	TBD
BMB5-9.7	Xing-Yuan Miao	OGS-6	Available	PR
https://github.com/xingyuanmiao/ogs/tree/phasefield_wip				
THM Processes				
BMB5-10.1	Peter Vogel	OGS-5	Available	TBD
BMB5-10.2	Peter Vogel	OGS-5	Available	TBD
BMB5-10.3	Peter Vogel	OGS-5	Available	TBD
BMB5-10.4	Xing-Yuan Miao	OGS-6	Available	PR
https://github.com/grubbymoon/ogs/tree/THM_PR				
BMB5-10.5	Tianyuan Zheng	OGS-6	Available	PR
https://github.com/grubbymoon/ogs/tree/THM_Freezing_Stiffness				
RTM Processes				
BMB5-11.1	Renchao Lu	OGS-5	Available	TBD
THC Processes				
BMB5-12.1	Thomas Nagel	OGS-5	Available	TBD
	Peter Ostermeier	ANSYS		NA
	Holger Class	DUMUX		NA

References

A. Peche, T. Graf, L. Fuchs, I. Neuweiler, A coupled approach for the three-dimensional simulation of pipe leakage in variably saturated soil. J. Hydrol. (2017). https://doi.org/10.1016/j.jhydrol.2017.10.050

A.A. Abed, M. Laitinen, J. Lämsä, T. Harjupatana, W.T. Sołowski, M. Kataja, Hydro-mechanical modelling of mx-80 bentonite: One dimensional study, vol. 9 (2016), https://www.scopus.com/inward/record.uri?eid=2-s2.0-85013659903&doi=10.1051%2fe3sconf%2f20160918005&partnerID=40&md5=c7b00010560f8e2b145a1eb5b9b2bfbd

M. Abramowitz, I.A. Stegun, *Handbook of Mathematical Functions*, 10th edn. (Dover, New York, 1972)

S. Alevizos, T. Poulet, S.D.C. Walsh, T. Mohr Durdez, M. Veveakis, K. Regenauer-Lieb, The dynamics of multiscale, multiphysics faults: Part ii - episodic stick-slip can turn the jelly sandwich into a crème brûlée. Tectonophysics (2017), https://www.scopus.com/inward/record.uri?eid=2-s2.0-85020931717&doi=10.1016%2fj.tecto.2017.06.010&partnerID=40&md5=5074aa8ec491978216f9812dacad1646

Altair, *HyperMesh*. Altair Engineering, Inc., Troy, Michigan, United States, 10 (2016), www.altairhyperworks.com/product/HyperMesh. Accessed 18 Oct 2016

Ansys Inc., Canonsburg, PA, USA. *ANSYS Fluent Theory Guide 15.0* (2013)

A.C. Aplin, A.J. Fleet, J.H.S. Macquaker, *Muds and mudstones: Physical and fluid-flow properties*, vol. 158 (Geological Society special publication, Geological Society and USA distributor (AAPG Bookstore, London and Tulsa, OK, 1999). ISBN 1-86239-044-4

C. Aristides, LIAKOPOULOS. Retention and distribution of moisture in soils after infiltration has ceased. Hydrol. Sci. J. **10**(2), 58–69 (1965)

S. Balay, S. Abhyankar, M.F. Adams, J. Brown, P. Brune, K. Buschelman, L. Dalcin, V. Eijkhout, W.D. Gropp, D. Kaushik, M.G. Knepley, L. Curfman McInnes, K. Rupp, B.F. Smith, S. Zampini, H. Zhang, H. Zhang. PETSc (2016), http://www.mcs.anl.gov/petsc, http://www.mcs.anl.gov/petsc

K.J. Bathe, *Finite-Elemente-Methoden (German Edition)* (Springer, 2001), http://www.amazon.com/Finite-Elemente-Methoden-German-Edition-K-J-Bathe/dp/3540668063%3FSubscriptionId%3D0JYN1NVW651KCA56C102%26tag%3Dtechkie-20%26linkCode%3Dxm2%26camp%3D2025%26creative%3D165953%26creativeASIN%3D3540668063. ISBN 3540668063

J. Bear, *Dynamics of Fluids in Porous Media* (Dover, Mineola, 1972). ISBN 978-0-486-65675-5

© Springer International Publishing AG 2018
O. Kolditz et al. (eds.), *Thermo-Hydro-Mechanical-Chemical Processes in Fractured Porous Media: Modelling and Benchmarking*,
Terrestrial Environmental Sciences, https://doi.org/10.1007/978-3-319-68225-9

S. Bente, V. Krase, U. Kowalsky, D. Dinkler, Model for degradation-induced settlements as part of a coupled landfill model. Int. J. Numer. Anal. Methods Geomech. (2017), https://www.scopus.com/inward/record.uri?eid=2-s2.0-85014686862&doi=10.1002%2fnag.2687&partnerID=40&md5=2939c0782f17e7ca314a276a97ed7f1e

F. Bernier, F. Lemy, P. De Cannière, V. Detilleux. Implications of safety requirements for the treatment of thmc processes in geological disposal systems for radioactive waste. J. Rock Mech. Geotech. Eng. **9**(3), 428–434 (2017), https://www.scopus.com/inward/record.uri?eid=2-s2.0-85018956448&doi=10.1016%2fj.jrmge.2017.04.001&partnerID=40&md5=8c0d8127332a457125a8e537e42fc878

BGR, TC ACSAD: a decision support system for integrated water resources management (IWRM). web page of the Federal Institute for Geosciences and Natural Resources (2017), https://www.bgr.bund.de/IWRM-DSS. Accessed 11 July 2017

G. Blöcher, T. Reinsch, J. Henninges, H. Milsch, S. Regenspurg, J. Kummerow, H. Francke, S. Kranz, A. Saadat, G. Zimmermann, E. Huenges, Hydraulic history and current state of the deep geothermal reservoir groß schönebeck. Geothermics **63**, 27–43 (2016), https://www.scopus.com/inward/record.uri?eid=2-s2.0-84979263748&doi=10.1016%2fj.geothermics.2015.07.008&partnerID=40&md5=70c6824de64c8b11288a02e917b07f4d

J. Bluhm, W.M. Bloßfeld, T. Ricken, Energetic effects during phase transition under freezing-thawing load in porous media–a continuum multiphase description and fe-simulation. ZAMM-J. Appl. Math. Mech./Zeitschrift für Angewandte Mathematik und Mechanik **94**(7–8), 586–608 (2014)

N. Boettcher, N. Watanabe, J.-U. Goerke, O. Kolditz, *Geoenergy Modeling I: Geothermal Processes in Fractured Porous* (Springer, Heidelberg, 2016), http://www.springer.com/de/book/9783319313337

A.E. Bond, I. Bruský, T. Cao, N. Chittenden, R. Fedors, X.-T. Feng, J.-P. Gwo, O. Kolditz, P. Lang, C. McDermott, I. Neretnieks, P.-Z. Pan, J. Šembera, H. Shao, N. Watanabe, H. Yasuhara, H. Zheng, A synthesis of approaches for modelling coupled thermal-hydraulic-mechanical-chemical processes in a single novaculite fracture experiment. Environ. Earth Sci. **76**(1) (2017) https://www.scopus.com/inward/record.uri?eid=2-s2.0-85006741754&doi=10.1007%2fs12665-016-6326-6&partnerID=40&md5=7df0bdedf689bd6ea30ad6d869ddf223

A.E. Bond, I. Bruský, N. Chittenden, X.-T. Feng, O. Kolditz, P. Lang, R. Lu, C. McDermott, I. Neretnieks, P.-Z. Pan, J. Šembera, H. Shao, H. Yasuhara, H. Zheng, Development of approaches for modelling coupled thermal-hydraulic-mechanical-chemical processes in single granite fracture experiments. Environ. Earth Sci. **75**(19) (2016), https://www.scopus.com/inward/record.uri?eid=2-s2.0-84989332469&doi=10.1007%2fs12665-016-6117-0&partnerID=40&md5=8e7ee7e180c9492c3a0ea744de7448a1

P. Bossart, F. Bernier, J. Birkholzer, C. Bruggeman, P. Connolly, S. Dewonck, M. Fukaya, M. Herfort, M. Jensen, J.-M. Matray, J.C. Mayor, A. Moeri, T. Oyama, K. Schuster, N. Shigeta, T. Vietor, K. Wieczorek. Mont terri rock laboratory, 20years of research: introduction, site characteristics and overview of experiments. Swiss J. Geosci. 110(1), 3–22 (2017), https://www.scopus.com/inward/record.uri?eid=2-s2.0-85013382569&doi=10.1007%2fs00015-016-0236-1&partnerID=40&md5=281ca38992d0e138c13ef18bfdc14d69

D. Browne, A. Deletic, G.M. Mudd, T.D. Fletcher, A new saturated/unsaturated model for stormwater infiltration systems. Hydrol. Process. **22**(25), 4838–4849 (2008). ISSN 08856087. https://doi.org/10.1002/hyp.7100

P. Brunner, P.G. Cook, C.T. Simmons, Hydrogeologic controls on disconnection between surface water and groundwater. Water Res. Res. **45**(1) (2009). ISSN 00431397. https://doi.org/10.1029/2008WR006953

M. Cacace, A.B. Jacquey, Flexible parallel implicit modelling of coupled thermal-hydraulic-mechanical processes in fractured rocks. Solid Earth Discuss. (March), 1–33 (2017), http://www.solid-earth-discuss.net/se-2017-33/. ISSN 1869-9537. https://doi.org/10.5194/se-2017-33

D.E. Carlson, Linear thermoelasticity, in *Encyclodedia of Physics, volume VIa/2, Mechanics of solids II*, ed. by S. Flügge, C. Truesdell (Springer, Berlin, Heidelberg, New York, 1972), pp. 297–345

R.F. Carsel, R.S. Parrish, Developing joint probability distributions of soil water retention characteristics. Water Res. Res. **24**(5), 755–769 (1988). ISSN 19447973. https://doi.org/10.1029/WR024i005p00755

Ch. Lehmann, O. Kolditz, Th. Nagel, Models of thermochemical heat storage, in *Computational Modeling of Energy Systems*, vol. 1, 1st edn ed. by T. Nagel, H. Shao, . (Springer, New York, 2017), p. 100

T. Chen. *Upscaling permeability for fractured porous rocks and modeling anisotropic flow and heat transport*. Ph.D. thesis, Faculty of Georesources and Materials Engineering, RWTH Aachen University, 7 (2017)

A.H.D. Cheng, *Poroelasticity* (Springer International Publishing, Switzerland, 2016)

R.V. Churchill, *Operational Mathematics*, 2nd edn. (McGraw-Hill Book Company, New York, 1958)

K.S. Crump, Numerical inversion of Laplace transforms using a Fourier series approximation. J. Asso. Comp. Mach **23**(1), 89–96 (1976)

T. Davis, Algorithm 832: umfpack, an unsymmetric-pattern multifrontal method. ACM Trans. Math. Softw. **30**, 196–199 (2004)

B. de Saint-Venant. Théorie du mouvement non-permanent des eaux, avec application aux crues des rivières et à l'introduction de marées dans leurs lits [eng.: Theory of non-permanent water movement with application to river flooding and the introduction of the tides in their tidal beds.]. C.R. de l'Acad. Sci. **73**, 147–154 (1871L)

B. Derode, Y.G. Guglielmi, L. De Barros, F. Cappa. Seismic responses to fluid pressure perturbations in a slipping fault: Fault reactivation by fluid pressures. Geophys. Res. Lett. (2015). https://doi.org/10.1002/2015GL063671

W. Dershowitz, G. Lee, J. Geier, T. Foxford, P. LaPointe, A. Thomas. FracMan, interactive discrete feature data analysis, geometric modeling, and exploration simulation. (Golder Associates Inc., Seattle, Washington, 1998). User Documentation

H.-J. Diersch, *FEFLOW: Finite Element Modeling of Flow, Mass and Heat Transport in Porous and Fractured Media* (Springer Science & Business Media, Berlin, 2013)

H.J. Diersch, *FEFLOW Finite Element Modeling of Flow, Mass and Heat Transport in Porous and Fractured Media* (Springer, Berlin, Heidelberg, 2014)

Docker - build, ship, and run any app, anywhere (2017), http://docker.io. Accessed 05 Sept 2017

Y. Dong, J.S. McCartney, N. Lu, Critical review of thermal conductivity models for unsaturated soils. Geotech. Geol. Eng. **33**(2), 207–221 (2015). ISSN 0960-3182. https://doi.org/10.1007/s10706-015-9843-2

W. Ehlers, A single-surface yield function for geomaterials. Arch. Appl. Mech. **65**(4), 246–259 (1995)

W. Ehlers, O. Avci, Stress-dependent hardening and failure surfaces of dry sand. Int. J. Numer. Anal. Methods Geomech. **37**(8), 787–809 (2013)

W. Ehlers, C. Luo, A phase-field approach embedded in the theory of porous media for the description of dynamic hydraulic fracturing. Comput. Methods Appl. Mech. Eng. **315**, 348–368 (2017)

J.W. Elder. Transient convection in a porous medium. J. Fluid Mech. **27**, 609–623 (1967). https://doi.org/10.1017/S0022112067000576

S. Ergun, Fluid flow through packed columns. Chem. Eng. Prog. **48**, 89–94 (1952)

ESRI, Esri shapefile technical description. Esri white paper, U.S. Geological Survey (1998)

I. Faoro, D. Elsworth, T. Candela. Evolution of the transport properties of fractures subject to thermally and mechanically activated mineral alteration and redistribution. Geofluids **16**(3), 96–407 (2016), https://www.scopus.com/inward/record.uri?eid=2-s2.0-84946203942&doi=10.1111%2fgfl.12157&partnerID=40&md5=8379fa6441e34782259a558455acb196

FEBio software suite (2017), https://febio.org. Accessed 5 Sept 2017

FEFLOW (2017), https://www.mikepoweredbydhi.com/products/feflow. Accessed 5 Sept 2017

T. Fischer, D. Naumov, S. Sattler, O. Kolditz, M. Walther, GO2OGS 1.0: A versatile workflow to integrate complex geological information with fault data into numerical simulation models. Geosci. Model Develop. **8**(11), 3681–3694 (2015). ISSN 19919603. https://doi.org/10.5194/gmd-8-3681-2015

B. Flemisch, M. Darcis, K. Erbertseder, B. Faigle, A. Lauser, K. Mosthaf, S. Müthing, P. Nuske, A. Tatomir, M. Wolff et al., Dumu^x: Dune for multi–{phase, component, scale, physics} flow and transport in porous media. Adv. Water Res. **34**(9), 1102–1112 (2011)

R.A. Freeze, J.A. Cherry, *Groundwater* (Prentice-Hall, Englewood Cliffs, New Jersey, 1979)

A. Ghirian, M. Fall, Long-term coupled behaviour of cemented paste backfill in load cell experiments. Geomech. Geoeng **11**(4) 237–251 (2016), https://www.scopus.com/inward/record.uri?eid=2-s2.0-84958529020&doi=10.1080%2f17486025.2016.1145256&partnerID=40&md5=1e4cfac5e1b1a56b1bef50996c557843

J. Gironás, L.A. Roesner, L.A. Rossman, J. Davis, Software, data and modelling news a new applications manual for the storm water management model (SWMM). Environ. Model. Softw. **25**, 813–814 (2009), http://www.epa.gov/ednnrmrl/models/. ISSN 13648152. https://doi.org/10.1016/j.envsoft.2009.11.009

G. Goertzel, An algorithm for the evaluation of finite trigonometric series. Am. Math. Month. **65**(1), 34–35 (1958)

D. Gross, S. Kolling, R. Mueller, I. Schmidt, Configurational forces and their application in solid mechanics. Eur. J. Mech. - A/Solids **22**(5), 669–692 (2003), http://www.sciencedirect.com/science/article/pii/S0997753803000767. ISSN 0997-7538. https://doi.org/10.1016/S0997-7538(03)00076-7

G. Guennebaud, B. Jacob et al., Eigen v3 (2010), http://eigen.tuxfamily.org

Y.G. Guglielmi, F. Cappa, J.P. Avouac, P. Henry, D. Elsworth. Seismicity triggered by fluid-injection-induced aseismic slip. Science, 348 (2015a)

Y.G. Guglielmi, P. Henry, C. Nussbaum, P. Dick, C. Gout, F. Amann. Underground research laboratories for conducting fault activation experiments in shales. ARMA (2015b)

D.J. Gunn. Transfer of heat or mass to particles in fixed and fluidised beds. Int. J. Heat Mass Transf. **21**(4), 467–476 (1978), http://www.sciencedirect.com/science/article/pii/0017931078900807. ISSN 0017-9310. https://doi.org/10.1016/0017-9310(78)90080-7

M.E. Gurtin, The linear theory of elasticity, in *Encyclodedia of physics, volume VIa/2, Mechanics of solids II*, ed. by S. Flügge, C. Truesdell (Springer, Berlin, Heidelberg, New York, 1972), pp. 1–295

M. Hanif Chaudhry. *Open Channel Flow* (2008). ISBN 9780387301747. https://doi.org/10.1007/978-0-387-68648-6

A.W. Harbaugh, E.R. Banta, M.C. Hill, M.G. McDonald, MODFLOW-2000, the U.S. geological survey modular ground-water model –user guide to modularization concepts and the ground-water flow process. open-File Report 00-92, U.S. Geological Survey (2000)

W.K. Heidug, Intergranular solid-fluid phase transformations under stress: the effect of surface forces. J. Geophys. Res.: Solid Earth (1978–2012) **100**(B4), 5931–5940 (1995)

C. Helbig, L. Bilke, H.-S. Bauer, M. Böttinger, O. Kolditz, Meva - an interactive visualization application for validation of multifaceted meteorological data with multiple 3d devices. PLoS ONE **10**(4) (2015), https://www.scopus.com/inward/record.uri?eid=2-s2.0-84928634274&doi=10.1371%2fjournal.pone.0123811&partnerID=40&md5=19f5267e144890cfd07a1eadda39c2c0. https://doi.org/10.1371/journal.pone.0123811

C. Helbig, D. Dransch, M. Böttinger, C. Devey, A. Haas, M. Hlawitschka, C. Kuenzer, K. Rink, C. Schäfer-Neth, G. Scheuermann, T. Kwasnitschka, A. Unger. Challenges and strategies for the visual exploration of complex environmental data. Int. J. Digit. Earth **10**(10), 1070–1076 (2017), https://www.scopus.com/inward/record.uri?eid=2-s2.0-85019202520&doi=10.1080%2f17538947.2017.1327618&partnerID=40&md5=f26f74aed48605f3896f3ca202248458. https://doi.org/10.1080/17538947.2017.1327618

R. Helmig, Multiphase flow and transport processes in the subsurface: a contribution to the modeling of hydrosystems (1997)

R.B. Hetnarski, M.R. Eslami, *Thermal Stresses - Advanced Theory and Applications* (Springer Science, Berlin, 2009)

IAPWS (The International Association for the Properties of Water and Steam). Revised Release on the IAPWS Industrial Formulation 1997 for the Thermodynamic Properties of Water and Steam (1997), http://www.iapws.org/IF97-Rev.pdf

E. Jang, J. Boog, W. He, Th. Kalbacher, *OpenGeoSys-Tutorial: Computational Hydrology III: OGS#IPhreeqc Coupled Reactive Transport Modeling* (Springer, Heidelberg, 2018), http://www.springer.com/us/book/9783319671529

W. Jefferson, W.G. Worley, B.A. Robinso, C.O. Grigsby, J.L. Feerer, Correlating quartz dissolution kinetics in pure water from 25 to 625 c. Geochimica et Cosmochimica Acta **58**(11), 2407–2420 (1994)

Jenkins pipeline syntax (2017a), https://jenkins.io/doc/book/pipeline/syntax. Accessed 05 Sept 2017

Jenkins pipeline: Extending with shared libraries (2017b), https://jenkins.io/doc/book/pipeline/shared-libraries/. Accessed 05 Sept 2017

M. Jing, F. Heße, W. Wang, T. Fischer, M. Walther, M. Zink, A. Zech, R. Kumar, L. Samaniego, O. Kolditz, S. Attinger, Improved representation of groundwater at a regional scale – coupling of mesocale hydrologic model (mhm) with opengeosys (ogs). Geosci. Model Dev. Discuss. **2017**, 1–28 (2017), https://www.geosci-model-dev-discuss.net/gmd-2017-231/. https://doi.org/10.5194/gmd-2017-231

Kanalnetzberechnung - Hydrodynamische Abfluss-Transport- und Schmutzfrachtberechnung. HYSTEM-EXTRAN 7 Modellbeschreibung. itwh, Institut für technisch-wissenschaftliche Hydrologie GmbH Hannover (2010)

M.F. Kanfar, Z. Chen, S.S. Rahman, Analyzing wellbore stability in chemically-active anisotropic formations under thermal, hydraulic, mechanical and chemical loadings. J. Nat. Gas Sci. Eng. **41**, 93–111 (2017), https://www.scopus.com/inward/record.uri?eid=2-s2.0-85014878007&doi=10.1016%2fj.jngse.2017.02.006&partnerID=40&md5=efafc6a595d2c7845773c05959beeda0

O. Kolditz, S. Bauer, L. Bilke, N. Böttcher, J.O. Delfs, T. Fischer, U.J. Görke, T. Kalbacher, G. Kosakowski, C.I. McDermott et al., OpenGeoSys: an open-source initiative for numerical simulation of thermo-hydro-mechanical/chemical (THM/C) processes in porous media. Environ. Earth Sci. **67**(2), 589–599 (2012a)

O. Kolditz, S. Bauer, L. Bilke, N.B.öttcher, J.-O. Delfs, T. Fischer, U.J. Görke, T. Kalbacher, C.I. Georg Kosakowski, McDermott, et al., OpenGeoSys: an open-source initiative for numerical simulation of thermo-hydro-mechanical/chemical (THM/C) processes in porous media. Environ. Earth Sci. **67**(2), 589–599 (2012b)

O. Kolditz, U.-J. Görke, H. Shao, W. Wang, *Thermo-Hydro-Mechanical-Chemical Processes in Porous Media: Benchmarks and Examples*, vol. 86 (Springer Science & Business Media, 2012c)

O. Kolditz, H. Shao, W. Wang, *Thermo-Hydro-Mechanical-Chemical Processes in Fractured Porous Media* (Springer Publishing Company, Incorporated, 2012d). ISBN 3319118951, 9783319118956

O. Kolditz, U.-J. Görke, H. Shao, W. Wang, S. Bauer eds. by *Thermo-Hydro-Mechanical-Chemical Processes in Fractured Porous Media: Modelling and Benchmarking. Benchmarking Initiatives.* Terrestrial Environmental Sciences. (Springer, GmbH, 2016a), http://www.ebook.de/de/product/26168086/thermo_hydro_mechanical_chemical_processes_in_fractured_porous_media_modelling_and_benchmarking.html

O. Kolditz, U.-J. Görke, H. Shao, W. Wang, S. Bauer eds. by *Thermo-Hydro-Mechanical-Chemical Processes in Fractured Porous Media: Modelling and Benchmarking: Benchmarking Initiatives* (Springer, Heidelberg, 2016b), http://www.springer.com/de/book/9783319292236

O. Kolditz, H. Shao, W. Wang, S. Bauer, *Thermo-Hydro-Mechanical-Chemical Processes in Fractured Porous Media: Modelling and Benchmarking* (Springer, Berlin, 2016c)

O. Kolditz, H. Shao, W. Wang, S. Bauer, *Thermo-Hydro-Mechanical-Chemical Processes in Fractured Porous Media: Modelling and Benchmarking: Closed-Form Solutions* (Springer, Heidelberg, 2015), http://www.springer.com/de/book/9783319118932

C. Kuhn, *Numerical and analytical investigation of a phase field model for fracture*. Ph.D. thesis, Technische Universität Kaiserslautern (2013)

E.R. Lapwood, Convection of a fluid in a porous medium. Math. Proc. Camb. Philos. Soc. **44**, 508–552 (1948). https://doi.org/10.1017/S030500410002452X

T. Laszewski, P. Nauduri, Migrating to the cloud: client/server migrations to the oracle cloud, in *Migrating to the Cloud*, pp. 1–19 (2012). ISBN 978-1-59749-647-6. https://doi.org/10.1016/B978-1-59749-647-6.00001-6

E. Laurien, H. Oertel Jr., *Numerische Strömungsmechanik: Grundgleichungen und Modelle-Lösungsmethoden-Qualität und Genauigkeit* (Springer, Berlin, 2013)

S.G. Lekhnitskii, *Theory of Elasticity of an Anisotropic Body* (Mir Publishers, Moscow, 1981)

R.W. Lewis, B.A. Schrefler, The finite element method in the deformation and consolidation of porous media (1987)

X. Li, Q. Li, B. Bai, N. Wei, W. Yuan, The geomechanics of shenhua carbon dioxide capture and storage (ccs) demonstration project in ordos basin, china. J. Rock Mech. Geotech. Eng. **8**(6), 948–966 (2016a), https://www.scopus.com/inward/record.uri?eid=2-s2.0-85003691410&doi=10.1016%2fj.jrmge.2016.07.002&partnerID=40&md5=7478d27508192f670a99308f170bafff

X.-C. Li, W. Yuan, B. Bai, A review of numerical simulation methods for geomechanical problems induced by co2 geological storage. Yantu Lixue/Rock Soil Mech. **37**(6)1762–1772, 2016b. https://www.scopus.com/inward/record.uri?eid=2-s2.0-84974691572&doi=10.16285%2fj.rsm.2016.06.029&partnerID=40&md5=f63c16aaebe97e983de224990dff4cdd

L. Li, C.A. Peter, M.A. Celia, Upscaling geochemical reaction rates using pore-scale network modeling. Adv. water Res. **29**(9), 1351–1370 (2006)

Lis: library of iterative solvers for linear systems (2017), http://www.ssisc.org/lis. Accessed 05 Sept 2017

Y.M. Liu, S.F. Cao, J. Wang, J.L. Xie, L.K. Ma, X.G. Zhao, L. Chen, Stress evolution in thmc gmz-bentonite china-mock-up for geological disposal of high level radioactive waste in china. pages 721–726 (2016), https://www.scopus.com/inward/record.uri?eid=2-s2.0-85012214857&partnerID=40&md5=35caaab20e28dc2d9b1b4f78f53c8b85

Lorenzo Adolph Richards, Capillary conduction of liquids through porous mediums. Physics **1**(5), 318–333 (1931b)

G. Lu, M. Fall, Modelling blast wave propagation in a subsurfacegeotechnical structure made of an evolutive porous material. Mech. Mater. **108**, 21–39 (2017), https://www.scopus.com/inward/record.uri?eid=2-s2.0-85015096120&doi=10.1016%2fj.mechmat.2017.03.003&partnerID=40&md5=825696553051765c37b6f9d367a47bb6

G. Lu, M. Fall, L. Cui, A multiphysics-viscoplastic cap model for simulating blast response of cemented tailings backfill. J. Rock Mech. Geotech. Eng. **9**(3), 551–564 (2017a), https://www.scopus.com/inward/record.uri?eid=2-s2.0-85019043699&doi=10.1016%2fj.jrmge.2017.03.005&partnerID=40&md5=42c9d65a348bbbc65ca48df150a28798

R. Lu, N. Watanabe, W. He, E. Jang, H. Shao, O. Kolditz, H. Shao, Calibration of water-granite interaction with pressure solution in a flow-through fracture under confining pressure. Environ. Earth Sci. **76**, 417–430 (2017b). https://doi.org/10.1007/s12665-017-6727-1

V.I. Malkovsky, F. Magri, Thermal convection of temperature-dependent viscous fluids within three-dimensional faulted geothermal systems: Estimation from linear and numerical analyses. Water Res. Res. **52**, 2855–2867 (2016). https://doi.org/10.1002/2015WR018001

B. Massart, M. Paillet, V. Henrion, J. Sausse, C. Dezayes, A. Genter, A. Bisset, Fracture characterization and stochastic modeling of the granitic basement in the HDR Soultz Project (France), in *Proceedings World Geothermal Congress 2010* (Bali, Indonesia, 25–29 April 2010). IGA

J. Maßmann, J. Wolfer, M. Huber, K. Schelkes, V. Hennings, A. Droubi, M. Al-Sibai, WEAP-MODFLOW as a decision support system (DSS) for integrated water resources management: Design of the coupled model and results from a pilot study in Syria, in *Groundwater Quality Sustainability*, ed. by P. Maloszewski, S. Witczak, G. Malina (International Association of Hydrogeologists selected papers (CRC Press, Boca Raton, 2012)

G.A. Maugin, Material forces: concepts and applications. Appl. Mech. Rev. **48**(5), 213–245 (1995), https://doi.org/10.1115/1.3005101. ISSN 0003-6900. https://doi.org/10.1115/1.3005101

C. McDermott, A. Bond, A.F. Harris, N. Chittenden, K. Thatcher, Application of hybrid numerical and analytical solutions for the simulation of coupled thermal, hydraulic, mechanical and chemical processes during fluid flow through a fractured rock. Environ. Earth Sci. **74**(12), 7837–7854 (2015), https://www.scopus.com/inward/record.uri?eid=2-s2.0-84948738702&doi=10.1007%2fs12665-015-4769-9&partnerID=40&md5=6547b0739fab70dea16d0ca1dc275c89

D.B. McWhorter, D.K. Sunada, Exact integral solutions for two-phase flow. Water Res. Res. **26**(3), 399–413 (1990)

X.-Y. Miao, O. Kolditz, T. Nagel, Phase-field modeling of fracture in poroelastic solids for thermal energy storage, in *Poromechanics VI: Proceedings of the Sixth Biot Conference on Poromechanics* (American Society of Civil Engineers, 2017a), pp. 1976–1983

X.-Y. Miao, D. Naumov, T. Zheng, O. Kolditz, T. Nagel, Phase-field modeling of cracking processes in geomaterials for subsurface geotechnical engineering and energy storage (2017b)

C. Miehe, F. Welschinger, M. Hofacker, Thermodynamically consistent phase-field models of fracture: variational principles and multi-field fe implementations. Int. J. Numer. Methods Eng. **83**(10), 1273–1311 (2010a)

C. Miehe, M. Hofacker, F. Welschinger, A phase field model for rate-independent crack propagation: robust algorithmic implementation based on operator splits. Comput. Methods Appl. Mech. Eng. **199**(45), 2765–2778 (2010b)

R. Mueller, S. Kolling, D. Gross, On configurational forces in the context of the finite element method. Int. J. Numer. Methods Eng. **53**(7), 1557–1574 (2002)

T. Nagel, H. Shao, C. Roßkopf, M. Linder, A. Wörner, O. Kolditz, The influence of gas-solid reaction kinetics in models of thermochemical heat storage under monotonic and cyclic loading. Appl. Energy **136**(0), 289–302 (2014), http://www.sciencedirect.com/science/article/pii/S0306261914009271. ISSN 0306-2619. https://doi.org/10.1016/j.apenergy.2014.08.104

T. Nagel, H. Shao, A.K. Singh, N. Watanabe, C. Roßkopf, M. Linder, A. Wörner, O. Kolditz, Non-equilibrium thermochemical heat storage in porous media: Part 1 – Conceptual model. Energy **60**, 254–270 (2013), http://www.sciencedirect.com/science/article/pii/S0360544213005239. ISSN 0360-5442. https://doi.org/10.1016/j.energy.2013.06.025

T. Nagel, W. Minkley, N. Böttcher, D. Naumov, U.-J. Görke, O. Kolditz, Implicit numerical integration and consistent linearization of inelastic constitutive models of rock salt. Comput. Struct. **182**, 87–103 (2017b)

A. Nishida, Experience in developing an open source scalable software infrastructure in Japan, in *Computational Science and Its Applications–ICCSA 2010* (Springer, 2010), pp. 448–462

T. Nishimura. Investigation on creep behavior of geo-materials with suction control technique, vol. 9 (2016), https://www.scopus.com/inward/record.uri?eid=2-s2.0-85013658141&doi=10.1051%2fe3sconf%2f20160918002&partnerID=40&md5=45071110e12b223ac7480f9d7b7ccc08

W. Nowacki, *Thermoelasticity*, 2nd edn. (Pergamon Press, Polish Scientific Publisher, Warszawa, 1986)

Ogs-6 ci-platform jenkins (2017b), https://jenkins.opengeosys.org. Accessed 05 Sept 2017

Ogs-6 source code (2017c), https://github.com/ufz/ogs. Accessed 05 Sept 2017

Olgierd Cecil Zienkiewicz, R.L. Taylor, J.Z. Zhu, *The Finite Element Method Set*, 6th edn. (Elsevier Butterworth-Heinemann, Oxford, 2005–2006), http://www.worldcat.org/oclc/636432933. ISBN 978-0-7506-6321-2

P. Ostermeier, A. Vandersickel, S. Gleis, H. Spliethoff. Numerical simulation of gas-solid fixed bed reactors: approaches, sensitivity and verification, in *14th Multiphase Flow Conference, Helmholtz Zentrum Dresen Rossendorf (HZDR)* (2016)

J.L. Palandri, Y.K. Kharaka, A compilation of rate parameters of water-mineral interaction kinetics for application to geochemical modeling. Technical report, DTIC Document (2004)

P.-Z. Pan, X.-T. Feng, H. Zheng, A. Bond, An approach for simulating the thmc process in single novaculite fracture using epca. Environ. Earth Sci. **75**(15) (2016), https://www.scopus.com/inward/record.uri?eid=2-s2.0-84981747337&doi=10.1007%2fs12665-016-5967-9&partnerID=40&md5=334f663659fa24299f4d0e8b98ff8f47

D.L. Parkhurst, C.A.J. Appelo, et al., User's guide to phreeqc (version 2): A computer program for speciation, batch-reaction, one-dimensional transport, and inverse geochemical calculations (1999)

W.T. Pfeiffer, B. Graupner, S. Bauer, The coupled non-isothermal, multiphase-multicomponent flow and reactive transport simulator opengeosys-eclipse for porous media gas storage. Environ. Earth Sci. **75**(20) (2016), https://www.scopus.com/inward/record.uri?eid=2-s2.0-84991358614&doi=10.1007%2fs12665-016-6168-2&partnerID=40&md5=25919a972bf82c1675e0ec9bf0dab280

B.E. Poling, J.M. Prausnitz, O.J. Paul, R.C. Reid, *The Properties of Gases and Liquids*, vol. 5 (McGraw-Hill New York, 2001)

P.Y. Polubarinova-Kochina, *Theory of Groundwater Movement* (Princeton University Press, New Jersey, 1962)

A. Preissmann, Propagation des intumescences dans les canaux et rivières, in *Proceedings of 1st Congres de l'Assoc. Francaise de Calcul* (1961), pp. 433–442

L.A. Richards, Capillary conduction of liquids through porous mediums. J. Appl. Phys. **1**(5), 318–333 (1931a). ISSN 01486349. https://doi.org/10.1063/1.1745010

T. Ricken, J. Bluhm, Modeling fluid saturated porous media under frost attack. GAMM-Mitteilungen. **33**(1), 40–56 (2010). ISSN 1522-2608. https://doi.org/10.1002/gamm.201010004

K. Rink, C. Chen, L. Bilke, Z. Liao, K. Rinke, M. Frassl, T. Yue, O. Kolditz, Virtual geographic environments for water pollution control. Int. J. Digital Earth (2016), pp.1–11. https://doi.org/10.1080/17538947.2016.1265016, https://www.scopus.com/inward/record.uri?eid=2-s2.0-85006868895&doi=10.1080%2f17538947.2016.1265016&partnerID=40&md5=a03ad7829daacbee33acff275e34c708

L.A. Rossman, Storm Water Management Model Quality Assurance Report: Dynamic Wave Flow Routing. Technical Report EPA/600/R-06/097 (2006), http://www.epa.gov/water-research/storm-water-management-model-swmm

K. Roth, Soil physics. *Lecture Notes, Institute of Environmental physics*, University of Heidelberg (2006), http://www.iup.uniheidelberg.de/institut/forschung/groups/ts/students

J. Rutqvist, An overview of tough-based geomechanics models. Comput. Geosci. (2016), https://www.scopus.com/inward/record.uri?eid=2-s2.0-85008704695&doi=10.1016%2fj.cageo.2016.09.007&partnerID=40&md5=313018542c3b569d4f7f5e10bb5f2ee6

J. Rutqvist, A.P. Rinaldi, F. Cappa, G.J. Moridis, Modeling of fault activation and seismicity by injection directly into a fault zone associated with hydraulic fracturing of shale-gas reservoirs. J. Pet. Sci. Eng. **127**, 377–386 (2015). https://doi.org/10.1016/j.petrol.2015.01.019

A. Sachse, E. Nixdorf, E. Jang, K. Rink, Th. Fischer, B. Xi, C. Beyer, S. Bauer, M. Walther, Y. Sun, Y. Song, *OpenGeoSys-Tutorial: Computational Hydrology II: Groundwater Quality Modeling* (Springer, Heidelberg, 2017), http://www.springer.com/us/book/9783319528083

A. Sachse, K. Rink, W. He, O. Kolditz, *OpenGeoSys-Tutorial: Computational Hydrology I: Groundwater Flow Modeling* (Springer, Heidelberg, 2015) http://www.springer.com/de/book/9783319133348

L. Samaniego, R. Kumar, S. Attinger, Multiscale parameter regionalization of a grid-based hydrologic model at the mesoscale. Water Res. Res. **46**(5) (2010)

J. Sausse, C. Dezayes, L. Dorbath, A. Genter, J. Place, 3D model of fracture zones at Soultz-sous-Forêts based on geological data, image logs, induced microseismicity and vertical seismic profiles. Comptes Rendus Geosci. **342**, 531–545 (2010)

F. Schaube, L. Koch, A. Wörner, H. Müller-Steinhagen, A thermodynamic and kinetic study of the de- and rehydration of $Ca(OH)_2$ at high H_2O partial pressures for thermo-chemical heat storage. Thermochimica Acta, **538**, 9–20 (2012), http://www.sciencedirect.com/science/article/pii/S0040603112001153. ISSN 0040-6031. https://doi.org/10.1016/j.tca.2012.03.003

A.E. Scheidegger, General theory of dispersion in porous media. J. Geophys. Res. **66**(10), 3273–3278 (1961)

M.J. Setzer, R. Auberg, S. Kasparek, S. Palecki, P. Heine, Cif-test-capillary suction, internal damage and freeze thaw test. Mater. Struct. **34**(9), 515–525 (2001)

H. Shao, P. Hein, A. Sachse, O. Kolditz, Geoenergy modeling II - shallow geothermal processes, in *Computational Modeling of Energy Systems*, vol. 2, 1st edn., ed. By Th. Nagel, H. Shao (Springer, Heidelberg, 2017), p. 120

H. Shao, T. Nagel, C. Roßkopf, M. Linder, A. Wörner, O. Kolditz, Non-equilibrium thermo-chemical heat storage in porous media: Part 2 – A 1D computational model for a calcium hydroxide reaction system. Energy **60**, 271–282 (2013), http://www.sciencedirect.com/science/article/pii/S0360544213006695. ISSN 0360-5442. https://doi.org/10.1016/j.energy.2013.07.063

N.R. Siriwardene, A. Deletic, T.D. Fletcher, Clogging of stormwater gravel infiltration systems and filters: Insights from a laboratory study. Water Res. **41**(7), 1433–1440 (2007), https://doi.org/10.1016/j.watres.2006.12.040. ISSN 00431354

R. Skaggs, E.J. Monke, L.F. Huggins, An approximate method for determining the hydraulic conductivity of unsaturated soils. Technical Report 2, Water Resources Research Center (Purdue University, Lafayette, Indiana, 1970)

I.S. Sokolnikoff, *Mathematical Theory of Elasticity*, 2nd edn. (McGraw-Hill Book Company, New York, 1956)

K. Stephan, R. Krauss, A. Laesecke, Viscosity and thermal conductivity of nitrogen for a wide range of fluid states. J. Phys. Chem. Ref. Data **16**(4), 993–1023 (1987)

J. Taron, D. Elsworth, Constraints on compaction rate and equilibrium in the pressure solution creep of quartz aggregates and fractures: controls of aqueous concentration. J. Geophys. Res. Solid Earth **115**(B7) (2010)

Th. Nagel, S. Beckert, C. Lehmann, R. Gläser, O. Kolditz, Multi-physical continuum models of thermochemical heat storage and transformation in porous media and powder beds—a review. Appl. Energy **178**, 323–345 (2016a). ISSN 03062619. https://doi.org/10.1016/j.apenergy.2016.06.051

Th. Nagel, N. Böttcher, U.-J. Görke, O. Kolditz, Computational geotechnics I - storage of energy carriers, in *Computational Modeling of Energy Systems*, vol. 1, 1st edn., ed. by Th. Nagel, H. Shao (Springer, New York, 2017a), p. 120

Th. Nagel, U.-J. Görke, K.M. Moerman, O. Kolditz, On advantages of the kelvin mapping in finite element implementations of deformation processes. Environ. Earth Sci. **75**(11), 1–11 (2016b), http://dx.doi.org/10.1007/s12665-016-5429-4. ISSN 1866-6299. https://doi.org/10.1007/s12665-016-5429-4

M.Th. van Genuchten. A Closed-form Equation for Predicting the Hydraulic Conductivity of Unsaturated Soils1. Soil Sci. Soc. Am. J. **44**(5), 892 (1980), https://www.soils.org/publications/sssaj/abstracts/44/5/SS0440050892. ISSN 0361-5995. https://doi.org/10.2136/sssaj1980.03615995004400050002x

D. Tian, D. Liu, A new integrated surface and subsurface flows model and its verification. Appl. Math. Model. **35**(7), 3574–3586 (2011). ISSN 0307904X. https://doi.org/10.1016/j.apm.2011.01.035

S. Timoshenko, J.N. Goodier, *Theory of Elasticity* (McGraw-Hill Book Company, New York, 1951)

T.C.T. Ting, *Anisotropic Elasticity* (Oxford University Press, New York, Oxford, 1996)

J. Toth, A theory of groundwater motion in small drainage basins in central Alberta. Canada. J. Geophys. Res. **67**(11), 4375–4387 (1962)

ufz/dockerfiles source code (2017a), https://github.com/ufz/dockerfiles. Accessed 05 Sept 2017

E. Veveakis, S. Alevizos, T. Poulet, Episodic tremor and slip (ets) as a chaotic multiphysics spring. Phys. Earth Planet. Inter. **264**, 20–34 (2017), https://www.scopus.com/inward/record.uri?eid=2-s2.0-85004073276&doi=10.1016%2fj.pepi.2016.10.002&partnerID=40&md5=a221cd11b35d71db25229213ef232458

T. Vietor, H.J. Alheid, P. Blümling, P. Bossart, D. Gibert, S. Heusermann, F. Nicolin, C. Nussbaum, I. Plischke, K. Schuster, K. Shin, T. Spies. Rock mechanics experiment. *Mont Terri Rock Labo-*

ratory - Project, Programme 1996 to 2007 and Results, ed. by P. Bossart, M. Thury, Report Swiss Geol. Surv. 3 (Swisstopo, Wabern, 2008), pp. 117–131. ISBN 978-3-302-40016-7

T. Vogel. Swmii-numerical model of two-dimensional flow in a variably saturated porous medium. Technical Report 87, Department of Hydraulics and Catchment Hydrology, Agricultural University, Wageningen, Netherlands (1987)

vtk.js (2017), https://kitware.github.io/vtk-js. Accessed 05 Sept 2017

W Wagner, JR Cooper, A Dittmann, J Kijima, H-J Kretzschmar, A Kruse, R Mares, K Oguchi, H Sato, I Stocker, et al., The iapws industrial formulation 1997 for the thermodynamic properties of water and steam. J. Eng. Gas Turbines Power $122(1)$, 150–184 (2000), http://manufacturingscience.asmedigitalcollection.asme.org/article.aspx?articleid=1421003

R. Wallace, H. Morris, Characteristics of faults and shear zones in deep mines. PAGEOPH, **124**, 107–125 (1986). https://doi.org/10.1007/BF00875721

J.B. Walsh, Effect of pore pressure and confining pressure on fracture permeability. Int. J. Rock Mech. Mining Sci. Geomech. Abstracts **18**, 429–435 (1981). Elsevier

M. Wang, D.R. Kassoy, P.D. Weidman, Onset of convection in a vertical slab of saturated porous media between two impermeable conducting blocks. Int. J. Heat Mass Transfer **30**, 1331–1341 (1987). https://doi.org/10.1016/0017-9310(87)90165-7

L. Wang, S. Wang, R. Zhang, C. Wang, Y. Xiong, X. Zheng, S. Li, K. Jin, Z. Rui, Review of multi-scale and multi-physical simulation technologies for shale and tight gas reservoirs. J. Nat. Gas Sci. Eng. **37**, 560–578 (2017), https://www.scopus.com/inward/record.uri?eid=2-s2.0-85007048874&doi=10.1016%2fj.jngse.2016.11.051&partnerID=40&md5=51fb571fd01b124f34a23403c540a0a6

Y. Wang. On subsurface fracture opening and closure. J. Petr. Sci. Eng. **155**, 46–53 (2017), https://www.scopus.com/inward/record.uri?eid=2-s2.0-85021324039&doi=10.1016%2fj.petrol.2016.10.051&partnerID=40&md5=06b2dad1ee3c4f4ad37d07dc6477f326

A.W. Warrick, J.W. Biggar, D.R. Nielsen, Simultaneous solute and water transfer for an unsaturated soil [J]. Water Resour. Res. $7(5)$, 1216–1225 (1971)

N. Watanabe, G. Blöcher, M. Cacace, S. Held, Th. Kohl, Geoenergy modeling III - enhanced geothermal processes, *Computational Modeling of Energy Systems*, vol. 3, 1st edn., ed. by Th. Nagel, H. Shao (Springer, Heidelberg, 2017), p. 120

N. Watanabe, W. Wang, J. Taron, U.J. Görke, O. Kolditz, Lower dimensional interface elements with local enrichment: application to coupled hydro-mechanical problems in discretely fractured porous media. Int. J. Numer. Meth. Eng. **90**, 1010–1034 (2012). https://doi.org/10.1002/nme.3353. URL wileyonlinelibrary.com

F. Wechsung, *Auswirkungen des globalen Wandels auf Wasser, Umwelt und Gesellschaft im Elbegebiet*, vol. 6 (Weißensee Verlag, 2005)

P.A. Witherspoon, J.S.Y. Wang, K. Iwai, J.E. Gale, Validity of cubic law for fluid flow in a deformable rock fracture. Water Res. Res. $16(6)$, 1016–1024 (1980)

S. Woinowsky-Krieger, Der Spannungszustand in dicken elastischen Platten. Ing.-Arch. $4(203–226)$, 305–331 (1933)

W. Woodside, J.H. Messmer, Thermal conductivity of porous media. i. unconsolidated sands. J. Appl. Phys. $32(9)$, 1688–1699 (1961). ISSN 0021-8979. https://doi.org/10.1063/1.1728419

H. Yasuhara, N. Kinoshita, K. Kishida, Predictions of rock permeability by thmc model considering pressure solution, vol. 2 (2016a), pp. 1242–1245, https://www.scopus.com/inward/record.uri?eid=2-s2.0-85010311045&partnerID=40&md5=b65cc6f530940eda316a82c1bdea8242

H. Yasuhara, N. Kinoshita, S. Ogata, D.-S. Cheon, K. Kishida, Coupled thermo-hydro-mechanical-chemical modeling by incorporating pressure solution for estimating the evolution of rock permeability. Int. J. Rock Mech. Mining Sci. **86**, 104–114 (2016b) , https://www.scopus.com/inward/record.uri?eid=2-s2.0-84962815648&doi=10.1016%2fj.ijrmms.2016.03.015&partnerID=40&md5=e9016f3b7fc491c108017971e765779a

H. Yasuhara, N. Kinoshita, H. Ohfuji, D. Sung Lee, S. Nakashima, K. Kishida, Temporal alteration of fracture permeability in granite under hydrothermal conditions and its interpretation by coupled chemo-mechanical model. Appl. Geochem. $26(12)$, 2074–2088 (2011)

D. Yates, J. Sieber, D. Purkey, A. Huber-Lee, WEAP21 - a demand-, priority-, and preference-driven water planning model. part 1: model characteristics. IWRA. Water Int. **30**(4), 487–500 (2005)

R. Zhang, Y.-S. Wu, Sequentially coupled model for multiphase flow, mean stress, and reactive solute transport with kinetic chemical reactions: applications in co2 geological sequestration. J. Porous Media **19**(11), 1001–1021 (2016), https://www.scopus.com/inward/record.uri?eid=2-s2.0-84994850601&partnerID=40&md5=538d19c0da49e0246f2031d4b1683230

R. Zhang, Y. Xiong, P.H. Winterfeld, X. Yin, Y.-S. Wu, A novel computational framework for thermal-hydrological-mechanical-chemical processes of co2 geological sequestration into a layered saline aquifer and a naturally fractured enhanced geothermal system. Greenhouse Gases: Sci. Technol. **6**(3), 370–400 (2016a), https://www.scopus.com/inward/record.uri?eid=2-s2.0-84952359825&doi=10.1002%2fghg.1571&partnerID=40&md5=bdf355839c009625fd85e29b4f81448d

R. Zhang, X. Yin, P.H. Winterfeld, Y.-S. Wu, A fully coupled thermal-hydrological-mechanical-chemical model for co2 geological sequestration. J. Nat. Gas Sci. Eng. **28**, 280–304 (2016b), https://www.scopus.com/inward/record.uri?eid=2-s2.0-84949654989&doi=10.1016%2fj.jngse.2015.11.037&partnerID=40&md5=3f4135d6dc5f4ef2f3628294124f9dcc

T. Zheng, H. Shao, S. Schelenz, P. Hein, T. Vienken, Z. Pang, O. Kolditz, T. Nagel, Efficiency and economic analysis of utilizing latent heat from groundwater freezing in the context of borehole heat exchanger coupled ground source heat pump systems. Appl. Thermal Eng. **105**, 314–326 (2016)

T. Zheng, X.-Y. Miao, D. Naumov, H. Shao, O. Kolditz, T. Nagel, A thermo-hydro-mechanical finite element model of freezing in porous media-thermo-mechanically consistent formulation and application to ground source heat pumps. Coupled Prob. **1008–1019**, 2017 (2017)

Printed in the United States
By Bookmasters